Laser Optoacoustics

Laser Optoacoustics

V. E. Gusev
A. A. Karabutov

Translated by
Kevin Hendzel

American Institute of Physics New York

Printed in the United States of America

American Institute of Physics
335 East 45th Street
New York, NY 10017-3483

Library of Congress Cataloging-in-Publication Data

Gusev, V. E. (Vitalii Eduardovich)
[Lazernaia optoakustika. English]
Laser optoacoustics / V. E. Gusev, A. A. Karabutov; translated by Kevin Hendzel.
p. cm.
Translation of: Lazernaia optoakustika.
Includes bibliographical references and index.
ISBN 1-56396-036-2
1. Laser beams. 2. Solids--Optical properties. 3. Solids--Acoustical properties. 4. Liquids--Optical properties. 5. Liquids--Acoustical properties. I. Karabutov, A. A. (Aleksandr Alekseevich) II. Title.
QC689.5.L37G8713 1992 92-43951
543'.0858--dc20 CIP

Contents

Foreword

Optoacoustics, which has existed for over a century, is currently experiencing a renaissance. The total number of publications on this topic is approaching 2000 today. The majority of this research, which concerns optoacoustic spectroscopy, has been presented in monographs and in a variety of volumes.

Optoacoustics has an extraordinarily broad range of applications, including optical and IR spectroscopy, nondestructive testing and microtechnology, chemical analysis and ecology, etc. Research at the same time rarely extends beyond the scope of laboratory or experimental systems. This is due to both the expensive and high-precision nature of optoacoustic technology, which requires highly trained personnel, as well as possible difficulties in interpreting derived data.

The wide variety of designs, setups, and solutions used in optoacoustics highlights the importance of selecting an optimum configuration of a setup for application to each specific problem. This is why a precise understanding of the mechanisms of optical excitation of sound in various media and their relative effectiveness is important.

Optoacoustic spectroscopy makes it possible to obtain two types of spectra: optical and acoustic spectra. This capability is most fully implemented in pulsed excitation of sound. Pulsed optoacoustics has been the subject of a significant number of studies (of the order of 20% of the total number of studies). However, such research lacks a common approach to analyzing pulsed optical excitation of sound, as well as a rigorous discussion of the capabilities, advantages, and drawbacks of this excitation technique. The authors therefore consider methodological issues to be worthy of special attention.

The material in this book is essentially presented on three "levels." The first level (the Introduction and Chap. 1) carries out a qualitative analysis of the principal mechanisms and properties of optoacoustic excitation of sound. Chaps. 2 and 3 (the second level) are largely methodological in nature and provide a basis for analyzing optoacoustic phenomena in more complex cases. Optoacoustic interactions under conditions of various nonlinearities of the medium are described on the next level (Chaps. 4–6). The last chapter, Chap. 7, contains a survey of experimental pulsed optoacoustic methods; this analysis is based on concepts developed in Chaps. 2 and 3. The authors have attempted to provide a presentation from a unified methodological viewpoint. Naturally, we were not able to provide a review of (or even cite) all studies on pulsed optoacoustics, although we hope that the fundamental and more important results have been discussed in the present volume.

We wish to express our most sincere gratitude to S. A. Akhmanov, who initiated this study, as well as Yu. V. Gulyaev, O. V. Rudenko, and S. I. Soluyan for their support and attention.

Introduction

Optoacoustics, as suggested by its name, overlaps two fields of physics—optics and acoustics—and is one of the fields that describes the interaction of electromagnetic and acoustic waves. Optoacoustics and acoustooptics are differentiated based on which of the wave fields has the strongest effect on the other. In this regard, acoustooptics covers the effects of acoustic oscillations of the medium on a lightwave traveling through the medium. Optoacoustics is the reverse, and deals with the excitation of sound from the action of electromagnetic radiation on the medium.

Acoustic waves represent the natural response of the medium to a varying external effect. One of the manifestations of this general mechanism is the optoacoustic effect: the excitation of sound in a medium that absorbs a variable luminous flux. This effect was discovered in 1880 by Bell,[1,2] who heard a "pure musical tone" in a closed gas volume that had absorbed a modulated light beam. This discovery was subsequently confirmed by Tyndall[3] and Roentgen.[4] However, interest in the optoacoustic effect dropped off rapidly, since the lack of microphones and the poor level of development of technology made precision measurement impossible.

After the development of acoustic instruments the optoacoustic (OA) effect was used in our century to measure gas absorption in the IR portion of the spectrum[5] and to investigate the population dynamics of molecular excited states.[6] The application of lasers raised OA spectroscopy to a qualitatively new level. The results achieved in this field have been discussed in sufficient detail in Refs. 7–10. Photoacoustic spectroscopy[11] represented the further development of this field; this application makes it possible to measure light absorption by condensed matter based on the excitation of sound in an adjacent gas. This field of optoacoustics is the most advanced area today and has been discussed in detail in the literature (see, for example, Refs. 8–12 and 13).

The acoustic excitation of sound waves by modulated radiation from a light source is used in what has become the classical version of photoacoustics. The optoacoustic signal level is proportional to the variable part of the luminous flux. Since pulsed lasers can produce substantially higher intensities than cw lasers, excitation of a broad acoustic spectrum—acoustic wide-band pulses—is typical of pulsed laser optoacoustics.

The pioneering theoretical and experimental studies in this field were carried out in 1963 by Askar'yan, Prokhorov, Chanturiya, and Shipulo[16] in the U.S.S.R. and White[16,18] in the United States. A significant research program on pulsed optoacoustics was initiated in the U.S.S.R. on the basis of a review written by Bunkin and Komissarov and published in 1973. Although a large number of reviews[17–24] and original studies have been devoted to this problem, there has been no unified viewpoint regarding the specific nature of the pulsed mode, its advantages, or drawbacks. The present study carries out an analysis of laser optoacoustics based on a unified approach: the transfer function method.[25,26]

When the medium is exposed to electromagnetic radiation, sound excitation can arise via a variety of mechanisms. Such mechanisms can be divided into two broad classes: field-linear and field-quadratic mechanisms. The field-linear mechanisms—the piezoelectric and piezomagnetic mechanisms—will excite sound of the same frequency as the electromagnetic wave. Since the acoustic phonon spectrum has an upper limit imposed by the Debye frequency ($\omega_D \sim 10^{13}\ s^{-1}$), sound will be excited by these mechanisms in a quasistatic field.* Therefore, under laser irradiation, sound excitation results from nonlinear quadratic effects: electrostriction and magnetostriction, the thermal effect, and light pressure. In this case, acoustic oscillations are not excited at the lightwave frequency (which is much higher than the Debye frequency) but rather at the intensity modulation frequency, which already falls within the acoustic range.

We consider the simplest case of electromagnetic irradiation of a homogeneous isotropic dielectric, neglecting the dispersion of the refractive index. For this purpose we assume a transparent medium (no expression has been derived for the force exerted by the electromagnetic field on the medium in the case of an absorbing medium). The spatial density of forces acting on the dielectric in a static external electromagnetic field can be given as[29]

$$\mathbf{f} = -\nabla p - \nabla\varepsilon \frac{\langle \mathbf{E}^2 \rangle}{8\pi} - \nabla\mu \frac{\langle \mathbf{H}^2 \rangle}{8\pi} + \nabla\left[\left(\rho \frac{\partial\varepsilon}{\partial\rho}\right)_T \frac{\langle \mathbf{E}^2 \rangle}{8\pi} + \left(\rho \frac{\partial\mu}{\partial\rho}\right)_T \frac{\langle \mathbf{H}^2 \rangle}{8\pi}\right] + \frac{\varepsilon\mu - 1}{4\pi c} \frac{\partial}{\partial t} \langle [\mathbf{E} \times \mathbf{H}] \rangle. \tag{I.1}$$

Here p is the pressure in the medium (for a given density ρ and temperature T) in zero field, ε and μ are the permittivity and magnetic permeability, and c is the speed of light. The angular brackets denote averaging over a time period far greater than the characteristic alternation period of light. The terms containing $\nabla\varepsilon$ and $\nabla\mu$ can be dropped given the homogeneity of the medium. Moreover, magnetostriction effects as a rule are not significant at optical frequencies (magnetic permeability $\mu \approx 1$). The last term in (I.1), called the Abraham force, cannot be given as a gradient. This force has the direction $\mathbf{n}$ of the Umov-Poynting vector of the electromagnetic wave:

$$\mathbf{f}_A = \frac{\varepsilon\mu - 1}{4\pi c} \frac{\partial}{\partial t} \langle [\mathbf{E} \times \mathbf{H}] \rangle = \frac{\varepsilon\mu - 1}{c^2} \frac{\partial I}{\partial t} \mathbf{n}, \tag{I.2}$$

where I is light intensity.

Then, expressing $\langle E^2 \rangle$ through I as well and introducing the refractive index $n = \sqrt{\varepsilon}$, we transform the striction force equation to

$$\mathbf{f}_{str} = \nabla\left[\left(\rho \frac{\partial\varepsilon}{\partial\rho}\right)_T \frac{\langle \mathbf{E}^2 \rangle}{8\pi}\right] = \nabla\left[\left(\rho \frac{\partial n}{\partial\rho}\right)_T \frac{I}{c}\right] = \frac{Y}{c} \nabla I, \tag{I.3}$$

* The Debye frequency is therefore nearly always less than the laser optical frequency.

where $Y=\rho(\partial n/\partial\rho)_T$ is the optoacoustic coupling parameter. We estimate the ratio of striction forces and the Abraham force, assuming $Y \simeq 1$, $n^2-1\sim 1$ (if $n^2-1\ll 1$, then $Y\ll 1$ as well):

$$|f_A|/|f_{\text{str}}|\sim\omega a/c,$$

where a is the spatial scale of the variation in light intensity (i.e., beam diameter, light path length, etc.), while ω is the characteristic light intensity modulation frequency (the frequency of the excited sound). These relations clearly reveal that in the typical case $a\lesssim 1$ cm the Abraham force is far less than the striction force at frequencies through the single gigahertz range: $\omega\lesssim 10^{10}$ s^{-1}. Consequently, it is sufficient to account for only striction forces in Eq. (I.1). Now the expression for the density of forces acting on the dielectric will take the form

$$\mathbf{f}=-\nabla p+(Y/c)\nabla I. \tag{I.4}$$

Therefore in a homogeneous transparent medium, the fundamental forces in the optoacoustic effect are striction forces.

The quantity $p=p(\rho,T)$ represents the pressure in the medium in (I.4) in zero electromagnetic field. However, we are in fact not interested in the pressure itself but rather its deviation from equilibrium, since p is under the gradient. The temperature dependence of p is responsible for sound excitation by the thermal effect of light. Consistent with Ref. 28 the expression for the pressure increment p' can be given as

$$p'=p-p(\rho_0,T_0)=c_0^2\rho'+c_0^2\rho_0\beta T'. \tag{I.5}$$

Here ρ_0 is the equilibrium density of the medium, $\rho'=\rho-\rho_0$ is the density increment, c_0 is the adiabatic speed of sound, $T'=T-T_0$ is the temperature increment, and $\beta=-\rho^{-1}(\partial\rho/\partial T)_p$ is the thermal coefficient of volume expansion of the medium. The second term in (I.5) describing the change in pressure with temperature is responsible for the thermo-optical (thermal) mechanism of laser excitation of sound.

The change in temperature of the medium is given by the heat conduction equation

$$\rho_0 C_p\left(\frac{\partial T'}{\partial t}\right)=\kappa\Delta T'-\left(\frac{c}{4\pi}\right)\mathrm{div}\langle[\mathbf{E}\times\mathbf{H}]\rangle=\kappa\Delta T'-(\mathbf{n}\nabla)I. \tag{I.6}$$

The thermal diffusion length at frequencies $\omega>\alpha^2\kappa/\rho_0 c_p$ is shorter than the heating zone (κ is the thermal conductivity, α is the light absorptivity), and thermal conduction has no significant effect on the sound excitation process.

We compare the efficiency of the thermo-optical and striction sound excitation mechanisms. The amplitude of pressure oscillations attributable to the first excitation mechanism can be estimated as

$$p'_T\sim\rho_0 c_0^2\beta\frac{\alpha I}{\rho_0 c_p\omega}=\frac{\beta c_0^2}{c_p}\frac{\alpha I}{\omega}$$

[since $|(n\nabla)I|\sim\alpha I$]. The striction pressure in accordance with Eq. (I.4) is equal to

$$p'_{\text{str}} \sim YI/c.$$

Hence

$$\frac{p'_{\text{str}}}{p'_T} \sim \frac{Y}{\beta c_0^2/c_p} \frac{\omega}{\alpha c}.$$

As discussed above, the optoacoustic coupling parameter Y for condensed matter is of the order of unity: $Y \sim 1$ ($Y \approx n - 1 \ll 1$ for gases). The quantity that defines the efficiency of the thermal sound excitation mechanism $c_0^2\beta/c_p$ is related to the Greunhausen parameter and characterizes the fraction of heat converted into mechanical work. Given below are the values of this parameter for several different media:

TABLE I.

Parameter	Air	Water	Acetone	Quartz	Aluminum	Mercury
$\beta c_0^2/c_p$	0.42	0.11	0.93	0.03	2.1	2.7

As a rule, this quantity is of the order of unity and $c_0^2\beta/c_p = \gamma - 1$ for gases, where γ is the adiabatic index. The thermal sound generation mechanism is therefore dominant over the striction mechanism at relatively low frequencies $\omega \ll \alpha c$.

The frequency αc has a simple physical meaning: it is defined as the transit time of light over the absorption length $(\alpha c)^{-1}$. We can therefore assume that if the light intensity does not change appreciably during its transmission through the medium, the thermal mechanism will be the predominant sound excitation mechanism. In the opposite case, where light is rather weakly absorbed and its intensity changes rather substantially during its transmission through the medium, the striction mechanism will be the fundamental sound excitation mechanism. In this case, the light intensity gradients related to its temporal variation will be greater than those related to wave absorption. In fact, electrostriction will only be significant in transparent media ($\alpha < 1$ cm^{-1}) and at high ultrasonic frequencies.

Exceptions to this rule are possible when the absorbed light energy is not immediately or entirely thermalized.

It is possible to achieve a significant delay of up to tens of microseconds between the instant of light absorption and the instant when the absorbed energy is completely converted into thermal motion, if the light quantum energy is sufficient to cause the valence electrons to break away from the atoms. This is due to the fact that a free electron created in such a process may not return to an equilibrium state for an extended period. In this case, a significant fraction of absorbed light energy is stored as electron–ion interaction energy. Clearly, electron stripping from the atoms will alter the interaction forces between atoms. In the case of solids, obviously, this will alter the density of the matter in a manner entirely independent of heating.

Electron stripping in covalent crystals will break the covalent bonds between neighboring atoms. It is therefore clear from a qualitative viewpoint that a change in density ρ' of matter will in a first approximation be proportional to the concentration n of photoexcited electrons:

$$|\rho'|/\rho_0 \sim n/n_0. \tag{I.7}$$

A significant change in density ($|\rho'| \sim \rho_0$) can be expected when each of the atoms is additionally ionized by light. In this case, not only is interatomic interaction important, but atomic interaction with the free electron system also becomes important. The atomic bonds in the crystal become increasingly similar to metallic bonds. Therefore n_0 in estimate (I.7) will be of the order of the number of atoms per unit of volume.

Equations (I.5) and (I.7) can be used to derive the following representation for pressure p'_n in the medium resulting from photoexcitation of free charge carriers:

$$|p'_n| \sim \rho_0 c_0^2 n/n_0. \tag{I.8}$$

Consistent with Eq. (I.8), sound excitation may result from the change in carrier concentration. This optical sound generation mechanism is called the electronic mechanism[29] or the concentration-deformation or, most commonly, the deformation[30] mechanism. The so-called deformation potential constant d is introduced for a phenomenological description of this mechanism:

$$p'_n = -dn. \tag{I.9}$$

The physical meaning of the parameter d will be refined in Sec. 3.4. Photoionization of the lattice atoms may both compress the crystals ($d>0$) and expand the crystals ($d<0$). A comparison of (I.8) and (I.9) makes it possible to estimate d within an order of magnitude: $|d| \sim \rho_0 c_0^2/n_0$. Assuming $\rho_0 \sim 5$ g/cm^3, $c \sim 5\times10^5$ cm/s, $n_0 \sim 10^{23}$ g/cm^3, we obtain a typical value of $|d| \sim 10$ eV.

The electronic mechanism of the optoacoustic effect may play an important role in semiconductor material when visible and IR lasers are used. In this case, the optical quantum energy $\hbar\omega_L$ must exceed the semiconductor bandgap E_g in order to achieve ionization. Interband light absorption will create an electron in the conduction band and a hole in the valence band. The resulting electron-hole pair stores the energy ($\sim E_g$) through recombination. For typical semiconductors $E_g \lesssim 2$ eV.

We compare the efficiencies of the thermal and deformation sound excitation mechanisms in semiconductors. For this purpose we assume that the growth of concentration of photoexcited charge carriers is not limited under laser irradiation by either carrier diffusion or recombination, while we assume heat conduction has little effect on the temperature distribution. Then

$$\frac{\partial n}{\partial t} \sim \frac{\alpha I}{\hbar\omega_L}, \quad \frac{\partial T}{\partial t} \sim \frac{(\hbar\omega_L - E_g)}{\hbar\omega_L} \frac{\alpha I}{\rho_0 c_p}.$$

Here we assume that part $\hbar\omega_L - E_g$ of the optical radiation energy $\hbar\omega_L$ is rapidly thermalized, while the energy E_g is stored by the electron-hole (EH) pair. Further assuming for purposes of this estimate that these energies are equal ($\hbar\omega_L \sim 2E_g$), we have

$$-p'_T/p'_n \sim \beta E_g c_0^2/dc_p = B. \tag{I.10}$$

Numerical estimates have demonstrated[31] that $|B|$ is of the order of 0.1 in such semiconductors as Ge, Si, and GaAs and hence the deformation mechanism may be dominant in the optoacoustic effect. However, in the general case, saturation of growth of the photoexcited carrier concentration (due, for example, to their recombination) may lead to a strong predominance of the thermal mechanism.

The mechanisms examined above (thermal, striction, and deformation) give rise to additional mechanical stresses in matter under laser irradiation. It is these stresses that are the sources of acoustic waves.[32] Their sound excitation process will be described in Chap. 1.

REFERENCES

1. A. G. Bell, Am. J. Sci. **20**, 305 (1880).
2. A. G. Bell, Phil. Magn. **11** (68), 510 (1881).
3. J. Tyndall, Proc. R. Soc. **31** (208), 307 (1881).
4. W. C. Roentgen, Phil. Magn. **11** (68), 308 (1881).
5. M. L. Veyngerov, Dokl. Akad. Nauk SSSR **9**, 9 (1938) [Sov. Phys. Dokl. (1938)].
6. G. S. Gorelik, Dokl. Akad. Nauk SSSR **54**, 779 (1946) [Sov. Phys. Dokl. (1946)].
7. D. O. Gorelik and B. B. Sakharov, *Optiko-akusticheskiy effekt v fiziko-khimicheskikh izmereniyakh* [*Optoacoustic Effect in Physiochemical Measurements*] (Izd-vo standartov, Moscow, 1969).
8. V. P. Zharov and V. S. Letokhov, *Lazernaya optiko-akusticheskaya spektroskopiya* [*Laser Optoacoustic Spectroscopy*] (Nauka, Moscow, 1984).
9. *Optiko-akusticheskiy metod v lazernoy spektroskopii molekulyarnykh gazov* [*Optoacoustic Method in Laser Spectroscopy of Molecular Gases*], edited by Yu. S. Makushkin (Nauka, Novosibirsk, 1984).
10. B. G. Ageev, Yu. N. Ponomarev, and B. A. Tikhomirov, *Nelineynaya optiko-akusticheskaya spektroskopiya molekulyarnykh gazov* [*Nonlinear Optoacoustic Spectroscopy of Molecular Gases*] (Nauka, Novosibirsk, 1987).
11. A. Rosencwaig and A. Gersho, J. Appl. Phys. **47**, 64 (1976).
12. A. Rosencwaig, *Photoacoustics and Photoacoustic Spectroscopy* (Wiley, New York, 1980).
13. *Fotoakustika i rodstvennye metody: Bibliograficheskiy ukazatel'. Vyp. I. Fotoakustika v kondensirovannykh sredakh. 1973–1984* [*Photoacoustics and Associated Methods: Bibliographic Index. Vol. 1. Photoacoustics of Condensed Matter. 1973–1984*], compiled by Ya. Ya. Gedrovits (Riga, 1987).
14. G. A. Askar'yan, A. M. Prokhorov, G. F. Chanturiya, and G. P. Shipulo, Zh. Eksp. Teor. Fiz. **44**, 2180 (1963) [Sov. Phys. JETP **17**, 1463 (1963)].
15. R. M. White, J. Appl. Phys. **34**, 2123 (1963).
16. R. M. White, J. Appl. Phys. **34**, 3559 (1963).
17. F. V. Bunkin and V. M. Komissarov, Akust. Zh. **19**, 305 (1973) [Sov. Phys. Acoust. **19**, 203 (1973)].
18. L. M. Lyamshev, Usp. Fiz. Nauk **135**, 637 (1981) [Sov. Phys. Usp. **24**, 977 (1981)].
19. C. K. N. Patel and A. C. Tam, Rev. Mod. Phys. **53**, 517 (1981).
20. L. M. Lyamshev and L. V. Sedov, Akust. Zh. **27**, 5 (1981) [Sov. Phys. Acoust. **27**, 4 (1981)].
21. L. M. Lyamshev and K. A. Naugol'nykh, Akust. Zh. **27**, 641 (1981) [Sov. Phys. Acoust. **27**, 641 (1981)].
22. A. A. Karabutov, Usp. Fiz. Nauk **147**, 604 (1985) [Sov. Phys. Usp. **28**, 1042 (1985)].
23. A. I. Bozhkov, F. V. Bunkin *et al.*, *Tr. FIAN* [*Proceedings of the Lebedev Physics Institute*] (Nauka, Moscow, 1985), Vol. 156, pp. 123–176.
24. M. W. Sigrist, J. Appl. Phys. **60**, R83 (1986).
25. L. V. Burmistrova, A. A. Karabutov *et al.*, Akust. Zh. **24**, 655 (1978) [Sov. Phys. Acoust. **24**, 369 (1978)].
26. V. K. Novikov, O. V. Rudenko, and V. I. Timoshenko, *Nelineynaya gidroakustika* [*Nonlinear Hydroacoustics*] (Sudostroenie, Leningrad, 1981).
27. L. D. Landau and E. M. Lifshitz, *Elektrodinamika sploshnykh sred* [*The Electrodynamics of Continuous Media*] (Nauka, Moscow, 1982).

28. L. D. Landau and E. M. Lifshitz, *Gidrodinamika* [*Hydrodynamics*] (Nauka, Moscow, 1986).
29. W. B. Gauster and D. H. Habing, Phys. Rev. Lett. **18**, 1058 (1967).
30. Yu. V. Pogorelyskiy, Fiz. Tverd. Tela **24**, 2361 (1982) [Sov. Phys. Solid State **24**, 1340 (1982)].
31. S. M. Avanesyan, V. E. Gusev, and N. I. Zheludev, Appl. Phys. A **40**, 163 (1986).
32. S. A. Akhmanov and U. E. Gusev, Usp. Fiz. Nauk **162(3)**, 3 (1992) [Sov. Phys. Usp. **35**, March (1902)].

Chapter 1

Qualitative Theory of Thermooptical Sound Excitation in a Liquid

We consider a typical case of optoacoustic sound excitation (Fig. 1.1). Assume a light beam of intensity $I = I_S f(t) H(x,y)$ from transparent medium I is incident on the boundary with absorbing medium II: the xy plane. The dimensionless function $f(t)$ describes its time dependence, $H(x, y)$ is the intensity distribution in the beam cross section, I_S is the characteristic radiation intensity. A portion of the energy is specularly reflected at the boundary $z = 0$, while a portion is transmitted through the absorbing medium and the remaining fraction is scattered. Neglecting scattering, we assume $R = 1 - A$ (R is the reflection coefficient, A is the transmission coefficient). The light intensity in absorbing medium II ($z > 0$) is therefore given by

$$I = I_0 f(t) H(x,y) e^{-\alpha z} \tag{1.1}$$

(light is incident normal to the interface surface), $I_0 = A\, I_S$ is the light intensity at the boundary of the absorbing medium. The light absorption coefficient α is independent of the coordinates. The spatial dimensions of the thermal source distribution are determined by the light beam radius (the spot radius) a and the light path length α^{-1}. If $\alpha a \gg 1$ (strong absorption) the heating zone takes the form of a thin disk adjacent to the boundary. In the opposite case $\alpha a \ll 1$ (weak absorption) heat will be delivered in an extended cylinder lying along the z axis.

1.1 Thermooptical Sound Excitation in a Weakly Absorbing Medium

The weak absorption case $\alpha a \ll 1$ is characteristic of photoacoustic spectroscopy of gases and other transparent media. Let the luminous flux obey a harmonic law over time with an angular frequency ω:

$$f(t) = 1 + m \cos \omega t \tag{1.2}$$

(m is the depth of intensity modulation).

We select a beam segment of length l at a distance z from the boundary such that $\alpha l \ll 1$ (Fig. 1.1). Then, the luminous flux will remain virtually unchanged within this segment. Hence, the quantity of heat delivered in a volume of gas of length l and area πa^2 per unit of time is given by

$$\dot{Q} = \alpha l \pi a^2 I_0 (1 + m \cos \omega t) e^{-\alpha z}.$$

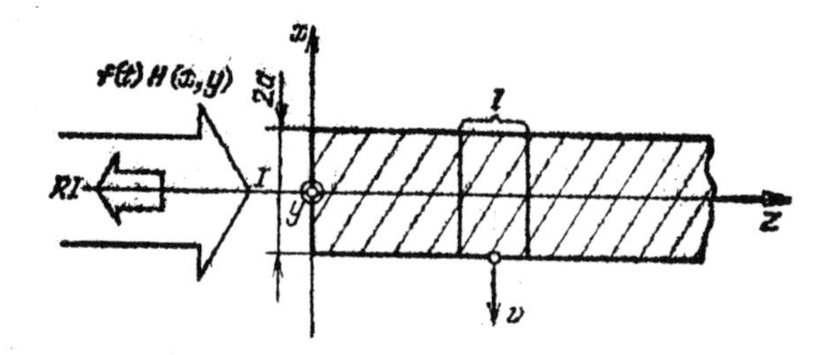

FIG. 1.1. Geometry of the heating zone for the case of weak light absorption, $\alpha a \ll 1$.

The variable portion of the thermal flux is the only component responsible for sound excitation, and the change in temperature of the medium corresponding to this flux takes the form

$$T' = \int \frac{\dot{Q}}{\rho_0 c_p (\pi a^2 l)}\, dt = \frac{\alpha I_0 m}{\rho_0 c_p \omega} e^{-\alpha z} \sin \omega t$$

(ρ_0 is the uniform density of the medium, c_p is its specific heat at constant pressure).

The medium will undergo thermal expansion only transverse to the beam due to the weak absorption of light. The relative change in volume $V = \pi a^2 l$ is given by $\beta T'$ where β is the coefficient of thermal expansion. Therefore, if the modulation frequency of the light is not too great, $\omega a / c_0 \ll 1$ (the transverse dimensions of the heated region are far less than the length of the acoustic wave), the oscillations will be in phase within the spot itself and the vibrational velocity of the particles of the medium v at the beam boundary (Fig. 1.1) can be expressed as

$$v = \frac{1}{2\pi a l} \frac{dV}{dt} = \frac{\pi a^2 l}{2\pi a l} \beta \dot{T}' = \frac{\alpha a m I_0 \beta}{2 \rho_0 c_p} e^{-\alpha z} \cos \omega t. \qquad (1.3)$$

Therefore, the column of the medium heated by the variable luminous flux takes the form of a thin cylindrical acoustic radiator producing sound with a vibrational velocity amplitude proportional to the density of the absorbed radiation power $\alpha m I_0$ and independent of modulation frequency ω. Sound generation will result from the absorption of not only light, but any other type of penetrating radiation including electrons, neutrinos, protons, etc. The fundamental principles of the sound excitation process are conserved in this case.

A divergent cylindrical wave appears at distances r large compared to the beam radius (where the acoustic field is ordinarily recorded). The acoustic power radiated by the section of the heated surface of length l, $W_{ac} = 2\pi a l I_{ac}$, can be found from the wave intensity I_{ac}. For thin cylindrical radiators, I_{ac} is given by[1]

$$I_{ac} = \rho_0 \omega a \langle v^2 \rangle \qquad (I.4)$$

(the angular brackets denote time averaging). Substituting Eq. (1.3) into Eq. (1.4), we obtain

$$W_{ac} = \frac{\omega l \beta^2}{4\pi \rho_0 c_p^2} (\alpha m W_0 e^{-\alpha z})^2, \qquad (1.5)$$

where $W_0 = \pi a^2 I_0$ is the average light radiation power. The conversion efficiency, i.e., the ratio of acoustic power radiated by a source of length l to the variable portion of luminous radiation power absorbed by this section is

$$\eta = \frac{W_{ac}}{\alpha m W_0 l e^{-\alpha z}} = \frac{\omega \beta^2}{4\pi \rho_0 c_p^2} \alpha m W_0 e^{-\alpha z}. \tag{1.6}$$

The thermooptical sound excitation efficiency is therefore proportional to the light radiation power. Hence, the efficiency grows with radiation power. This is one of the reasons why the optoacoustic effect was not widely used prior to the development of lasers. The radiated acoustic power grows with increasing modulation frequency [we recall that Eqs. (1.5) and (1.6) were obtained in a low-frequency approximation, when the wavelength of sound far exceeds the beam diameter]. The vibrational velocity in the wave will be independent of frequency. The increase in acoustic power with frequency is simply due to the increasing size of the acoustic antenna relative to the wavelength.

The energy conversion efficiency is given by a characteristic combination of thermophysical parameters of the medium

$$\Gamma = c_0 \beta^2 / \rho_0 c_p^2 = (\gamma - 1)^2 / \rho_0 c_0^3. \tag{1.7}$$

The quantity Γ is essentially dependent on the type of medium. Thus, it will be of the order of 10^{-7} cm^2/W for gases, 10^{-9}–10^{-10} cm^2/W for liquids, and 10^{-12}–10^{-15} cm^2/W for solids. It is clear from these figures that the energy conversion efficiency is quite low. Since the average power level of cw lasers is far less than the peak power of pulsed lasers, the optoacoustic conversion scheme examined here is largely used for spectroscopy of gases[2-4] which have the highest excitation efficiency.

Equation (1.3) defines the vibrational velocity at the light-beam boundary, although ordinarily the wave will be received at distances exceeding the spot dimensions. It is easiest to determine the amplitude of the vibrational velocity v_a at large distances (in the far field) based on a radiated power expression. Thus, outside the beam $r > a$ the acoustic wave is a purely diverging wave, while in the far field $r \gg c_0/\omega$ the radiated power can be given as[1]

$$W_{ac} = \rho_0 c_0 \cdot 2\pi r l \langle v^2 \rangle. \tag{1.8}$$

We find from Eqs. (1.8) and (1.5)

$$v_a = \frac{\alpha \beta m W_0 e^{-\alpha z}}{2\pi \rho_0 c_p} \sqrt{\frac{\omega}{2\pi r c_0}}. \tag{1.9}$$

As would be expected, the amplitude of vibrational velocity decays with distance as $r^{-1/2}$ (cylindrical wave). However, the amplitude of the vibrational velocity in the far field is determined by the absorbed power rather than light intensity as in Eq. (1.3). This means that the specific form of the intensity distribution at the spot is not significant in the far field, at least in the low frequency range $\omega < c_0/a$.

Here we have considered the thermal sound excitation mechanism in a weakly absorbing medium $\alpha a \ll 1$. However, it is in the case of weak absorption of light that the striction mechanism may also be significant. This mechanism will be manifested

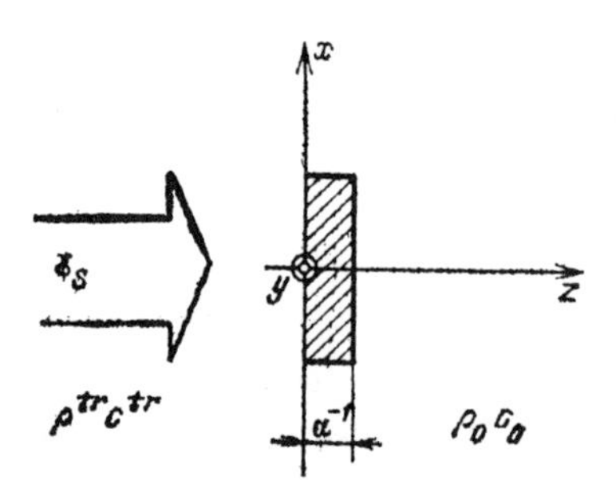

FIG. 1.2. Geometry of the heating zone for the case of strong light absorption, $\alpha a \gg 1$.

at frequencies $\omega \gtrsim \alpha c \beta c_0^2/(c_p Y)$ determined by the light transit time along the absorption length $(\alpha c)^{-1}$. Therefore, the effect of striction sound excitation cannot be neglected in the case of very small absorption coefficients ($\alpha < 10^{-5}$ cm^{-1}), since the radiation modulation frequency ω will ordinarily lie in the acoustic range. An analysis of the effect of striction excitation of sound can be found in Refs. 2–4.

1.2 Thermooptical Sound Excitation in a Strongly Absorbing Medium

The thermooptical energy conversion efficiency is proportional to the light radiation power αW_0 absorbed by the medium [see Eq. (1.6)]. We can therefore anticipate that under strong absorption, i.e., for $\alpha a \gg 1$, sound will be excited much more efficiently than in the case of weak absorption. Moreover, since Q-switched lasers have a much higher power level than cw lasers, pulsed operation makes it possible to achieve significantly higher pressure amplitudes.

Assume, as in the preceding case, a broad laser beam $\alpha a \gg 1$ is incident along the normal to the interface (Fig. 1.2) of a transparent medium and an absorbing medium. We denote the surface density of incident energy by $\mathscr{E}_S$. Then the volumetric heat flow density delivered to the medium is given by $\alpha \mathscr{E}_0 \exp(-\alpha z)$ (where $\mathscr{E}_0 = A \mathscr{E}_S$ is the absorbed energy flow density), while the change in temperature of the medium

$$T' = (\alpha \mathscr{E}_0/\rho_0 c_p) e^{-\alpha z}. \tag{1.10}$$

Let the laser pulse duration τ_L be sufficiently small compared to the transit time of sound through the depth of penetration of light $(\alpha c_0)^{-1}$: $\alpha c_0 \tau_L \ll 1$. Then, near-instantaneous heating of the medium can be assumed: the density of the medium cannot change substantially during laser pulse action, and a stress field is created as a result of the inhomogeneous temperature field. The pressure distribution in the medium $(z > 0)$ is therefore described by Eq. (I.5) with $\rho' = 0$:

$$p' = \rho_0 c_0^2 \beta T' = (\alpha c_0^2 \beta \mathscr{E}_0/c_p) e^{-\alpha z} = p_a e^{-\alpha z} \tag{1.11}$$

(p_a is the pressure amplitude). These stresses [Fig. 1.3(a), solid line] are distributed evenly between the wave exiting the boundary $z = 0$ and the wave traveling toward the boundary [the dashed and dot-dashed lines in Fig. 1.3(a), while the arrows indicate the direction of wave propagation]. The wave profiles in an infinite medium are shown in Fig. 1.3(b) at a certain time following laser irradiation. A complex state of standing and traveling acoustic waves [Fig. 1.3(c)] arises in a bounded

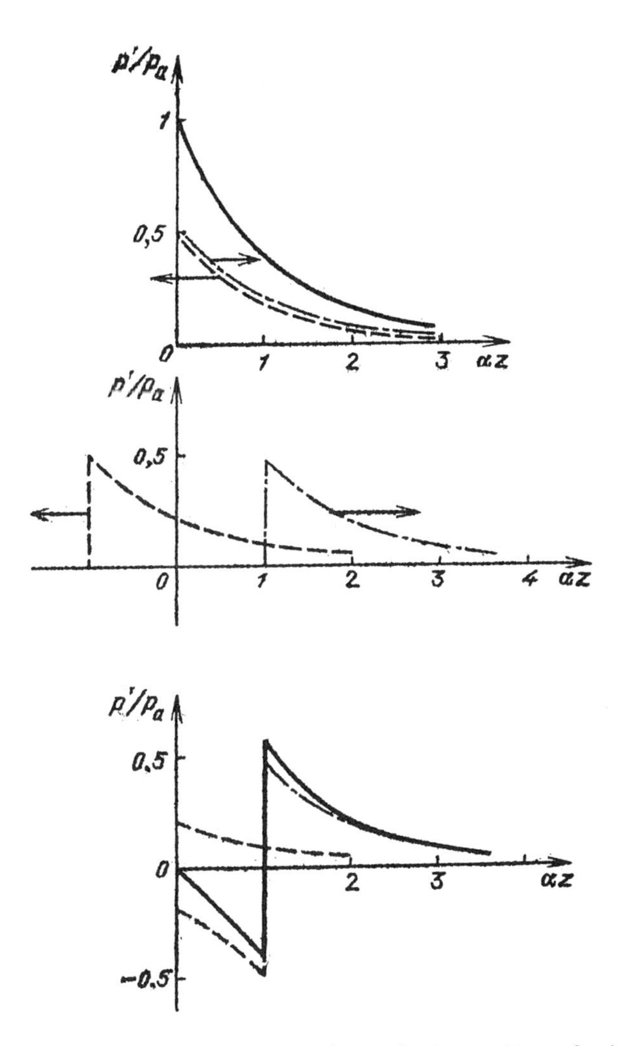

FIG. 1.3. Traveling wave formation from a short laser pulse ($\alpha c_0 \tau_L \ll 1$) at a free boundary.

medium within a region of the order of $3\alpha^{-1}$ in size where significant heating occurs. The pressure (solid line) is generated by the forward wave as well as the boundary-reflected wave.

The only wave outside the heating zone $z > 3\alpha^{-1}$ is the wave traveling from the boundary into the bulk of the absorbing medium. Its temporal profile is produced by both the forward loading wave and the wave traveling toward the boundary and reflected by the boundary.

The reflection coefficient of an acoustic wave reflected by the interface of the absorbing and transparent media will depend on the ratio of their acoustic impedances:

$$N = \rho_0 c_0 / \rho^{\text{tr}} c^{\text{tr}}. \tag{1.12}$$

TABLE 1.1 Relative acoustic impedance of an absorbing medium.

Absorbing medium	Transparent medium Air	Water	Quartz
Water	3.5×10^3	1	0.12
Aluminum	3.9×10^4	11	1.4
Mercury	4.6×10^4	13	1.6

Therefore, the wave form of the acoustic pulse excited by absorption of a light pulse is essentially dependent on N.

Ordinarily, two limiting cases are considered: when the acoustic impedance of a transparent medium $\rho^{tr}c^{tr}$ is much less than the impedance of the absorbing medium $\rho_0 c_0$ [$N \gg 1$, free (or soft) boundary of the absorbing medium] and the reverse case where the acoustic impedance of the transparent medium far exceeds the impedance of the absorbing medium [$N \ll 1$, rigid (or constrained) boundary of the absorbing medium]. The relative impedance values N are given in Table 1.1 for several pairs of substances.

An air-mercury boundary is an example of an acoustic free boundary of an absorbing medium, while a water-quartz boundary is an example of an acoustically rigid boundary. In the general case the boundary is an impedance boundary.

The reflection coefficient for pressure wave reflection by the boundary R_{ac} is related to the impedance ratio by the following equation:

$$p'_{ref}/p'_{inc} = R_{ac} = (1-N)/(1+N). \qquad (1.13)$$

Hence, with a short laser pulse $\alpha c_0 \tau_L \ll 1$ a profile of the excited acoustic wave is described by the following equation:

$$p' = \frac{\alpha c_0^2 \beta \mathscr{E}_0}{2c_p} \begin{cases} [(1-N)/(1+N)]\exp[\alpha(z-c_0 t)], & z < c_0 t, \\ \exp[-\alpha(z-c_0 t)], & z > c_0 t \end{cases} \qquad (1.14)$$

(the time $t=0$ represents the time of laser pulse arrival). The wave form (profile) of pulses excited for different impedances of the boundary of the absorbing medium is shown in Fig. 1.4. Since the leading edge of the pulse ($t < z/c_0$, $\tau = t - z/c_0 < 0$)

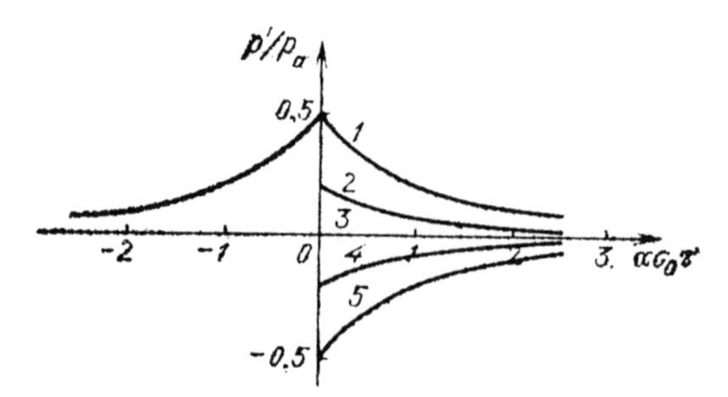

FIG. 1.4. OA-signal profiles for different acoustic impedances N: 0 (1), 0.4 (2), 1 (3), 2.5 (4), ∞ (5).

is generated by the forward wave directed off the boundary, its profile is a replica of the spatial distribution of the stress sources and is independent of N.

The relation $p'(\tau<0)$ has a universal exponential form for uniformly absorbing media ($\alpha=\text{const}$): $p'(\tau<0)\sim\exp(-\alpha c_0\tau)$. The wave reflected by boundary $z=0$ and denoted $p'(\tau>0)$ also has an exponential form: $p'(\tau>0)\sim\exp(-\alpha c_0\tau)$, while its relative value is determined by the acoustic reflection coefficient R_{ac} (1.13). For a rigid boundary ($N=0$) $R_{ac}=1$ and for a free boundary ($N\to\infty$) $R_{ac}=-1$. If the absorbing medium is acoustically softer than a transparent medium ($N<1$) the wave will be reflected with the same phase as the incident wave: $R_{ac}>0$. In the opposite case ($N>1$) the wave will be phase shifted by π : $R_{ac}<0$. The pulse is therefore a compressional wave and a trailing rarefactional (for $N>1$) or compressional (for $N<1$) wave. The crossover time between phases is within an order of magnitude of the laser pulse duration τ_L. The total acoustic pulse duration is determined by the sound transit time through the light absorption region and is equal to (4–6)$(\alpha c_0)^{-1}$ [laser pulse duration $\tau_L \ll (\alpha c_0)^{-1}$].

It follows from Eq. (1.14) that the shape of a pulse excited at an impedance boundary will be the weighted sum of the pulses at the rigid $p_r(N=0)$ and free $p'_f(N\to\infty)$ boundaries:

$$p'(N)=(p'_r+Np'_f)/(1+N).$$

This equation remains valid with an arbitrary laser pulse duration.

The light absorptivity α may vary over an extraordinarily broad range. The thermooptical mechanism can therefore be used to excite acoustic pulses of the broadest range of durations through the shortest pulses of the order of 10^{-11} s. No other means can be used to excite such short acoustic pulses.

The acoustic pulse amplitude p_a (pressure amplitude) is equal to one-half of the maximum thermoelastic stress of the medium:

$$p_a=(\alpha c_0^2\beta\mathscr{E}_0/2c_p)=\alpha\mathscr{E}_0(\gamma-1)/2,$$

and is proportional to the delivered volumetric heat density $\alpha\mathscr{E}_0$. The coefficient $c_0^2\beta/2c_p$ is determined by the thermophysical properties of the medium which reduce to the Greunhausen parameter (see the derivation given in the Introduction). Pressure amplitudes up to 10^8–10^9 Pa can be achieved by pulsed thermooptical excitation of sound. This level is achieved at a delivered energy density approaching the heat of evaporation. Higher pressure levels involve utilizing media with higher light absorptivities. However, this requires even shorter pulses: $\tau_L\ll(\alpha c_0)^{-1}$. Combining high amplitude and short (through the subnanosecond range) duration optoacoustic signals has yielded record deformation rates of materials that are inaccessible by other means.

The efficiency of a thermooptical acoustic pulse generator can be defined as the ratio of the surface acoustic energy density to the absorbed energy density:

$$\eta'=\frac{\mathscr{E}_{ac}}{\mathscr{E}_0}=\frac{1}{\rho_0c_0\mathscr{E}_0}\int_{-\infty}^{\infty}p'^2\,dt=\frac{c_0^2\beta^2}{8\rho_0c_p^2}\alpha\mathscr{E}_0(1+R_{ac}^2). \qquad (1.15)$$

Therefore, as in the case of weak absorption, thermooptical conversion efficiency (1.15) is proportional to the heating density $\alpha\mathscr{E}_0$. Recasting Eq. (1.15) as

$$\eta' = \left(\frac{\gamma - 1}{2}\right)^2 \frac{\alpha \mathscr{E}_0}{\rho_0 c_0^2} \frac{1 + R_{ac}^2}{2},$$

it is clearly evident that the thermooptical conversion efficiency is determined by the ratio of the delivered heat density $\alpha \mathscr{E}_0$ to the "internal energy density" $\rho_0 c_0^2$. The thermooptical conversion efficiency will therefore not exceed a few percent through heating densities comparable to the heat of evaporation $\alpha \mathscr{E}_0 \approx \rho_0 c_p T'_{ev}$ (see below). Any increase in laser pulse energy will ordinarily result in overheating of the medium, vapor, or plasma formation, or optical breakdown. In this case, the luminous-to-mechanical energy conversion efficiency rises to 10%–20%. However, these mechanisms will damage the medium.

TABLE 1.2

Parameter	Water	Ethanol	Aluminum	Iron	Mercury
$c_p T'_{ev}/c_0^2$	0.15	0.1	0.015	0.019	0.9

The optoacoustic effect makes it possible to obtain high-amplitude acoustic video signals with an easily manipulated wave form. These signals are used to measure the linear and nonlinear properties of media such as the sound speed, the absorption coefficients of light and sound, the thermal coefficients of volumetric expansion, and the parameters of an acoustic nonlinearity.[5–9] Video signals are used for remote flaw detection of materials and for measuring the distribution of parameters of inhomogeneous media.[10,11]

1.3 Thermooptical Excitation of Sound by a Moving Beam

In the cases examined above the thermal acoustic sources were fixed relative to the absorbing medium. A differentiating feature of optoacoustic sources, however, is a capacity to travel through the medium with an arbitrary, including a sonic, velocity. This permits a synchronous acoustic wave excitation regime.

It is intuitively clear that efficient excitation of a wave requires the sources and the perturbations to be in the same region of space for an extended period. This is possible when the light beam is traveling at near sound speed V_0: $|V_0/c_0 - 1| \ll 1$. Then, elementary perturbations generated at different instants in time (and, consequently, at different points in space) will be clustered and summed in phase. Therefore, as the traveled distance increases, the perturbation grows and may become significant even with a comparatively weak beam. The thermooptical effect permits realization of such a sound generation regime; sound generation by continuous radiation is made possible by this mechanism.

We consider sound excitation by a moving laser beam. Let the beam cross section take the form of a prolate rectangle with sides $a \times b$ ($a \ll b$), while the beam moves at a velocity V_0 perpendicular to the beam plane (Fig. 1.5). If the light absorptivity is small ($\alpha a \ll 1$), homogeneous heating can be assumed over a sufficient distance. In this case the thermal sources have a thin layer geometry.

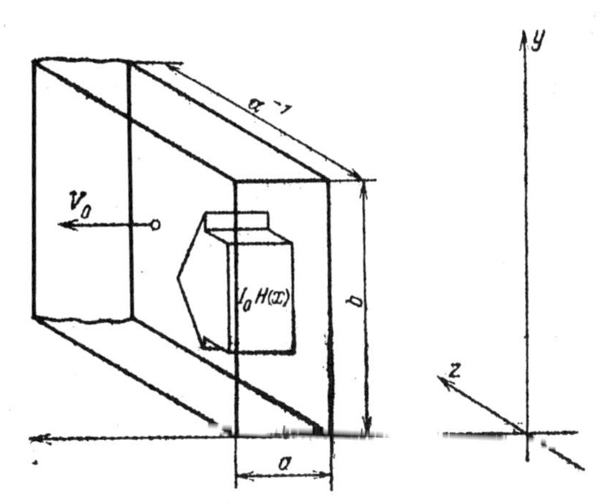

FIG. 1.5. OA-source geometry for the case of sound excitation by a moving laser beam.

Let the laser radiation be applied at the instant $t = 0$ with the beam center at the point $x = 0$ at this instant. We place the x axis along beam velocity V_0. We denote the function describing the light intensity distribution in the beam cross section by $H(x)$. Then the radiation energy absorbed by the medium over time t', $t' + dt'$ will increase the temperature of the medium by

$$dT = (\alpha W_0/\rho_0 c_p b) H(x - V_0 t') dt'.$$

This change in temperature will excite two acoustic waves of equal amplitude that travel in opposite directions with velocities $\pm c_0$,

$$dp' = (\alpha W_0/2\rho_0 c_p b)[H(x - c_0(t - t') - V_0 t') + H(x + c_0(t - t') - V_0 t')] dt' \rho_0 c_0^2 \beta. \tag{1.16}$$

In order to find the wave form of the wave excited by the traveling laser beam, Eq. (1.16) must be integrated with respect to time:

$$p'(x,t) = \int_0^t dp' = \left(\frac{\alpha W_0 c_0^2 \beta}{2c_p b}\right) \times \int_0^t dt' [H(x - c_0 t - (V_0 - c_0) t') + H(x + c_0 t - (V_0 + c_0) t')]. \tag{1.17}$$

We introduce the function

$$\mathscr{H}(\xi) = \int_{-\infty}^{\xi} H(y) dy. \tag{1.18}$$

Since the light intensity distribution H is a bell distribution, the function $\mathscr{H}$ takes the form of a flattened step (Fig. 1.6). Then subject to Eq. (1.18), Eq. (1.17) can be written as

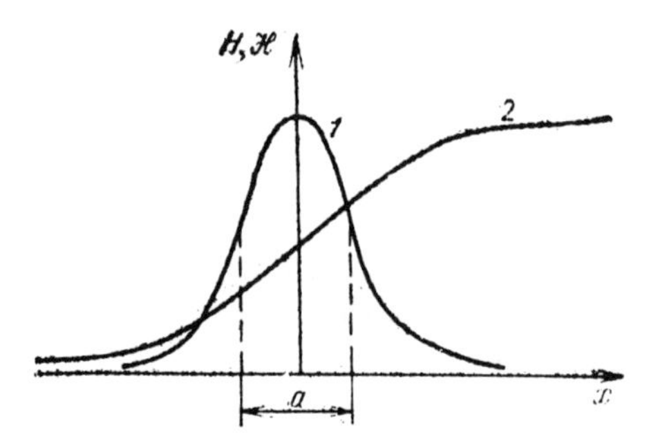

FIG. 1.6. Intensity distribution $H(x)$ (1) and heat distribution $\mathscr{H}(x)$ (2) in the beam cross section.

$$p' = \frac{\alpha W_0 \beta c_0^2}{2bc_p}\left[\frac{\mathscr{H}(x-c_0t)-\mathscr{H}(x-V_0t)}{V_0-c_0}+\frac{\mathscr{H}(x+c_0t)-\mathscr{H}(x-V_0t)}{V_0+c_0}\right]. \tag{1.19}$$

This solution is quite remarkable. It consists of two waves $\mathscr{H}(x-c_0t)$ and $\mathscr{H}(x+c_0t)$ traveling at a speed of $\pm c_0$ as well as stimulated waves $\mathscr{H}(x-V_0t)$ comoving with the sources. The amplitude ratio of waves running to the right and left on the x axis (in the direction of beam motion and opposite this motion) is given by $|(V_0+c_0)/(V_0-c_0)|$. Therefore, if the beam velocity is small compared to the sound velocity c_0 ($V_0 \ll c_0$), both waves will be excited at nearly an identical efficiency. The situation changes radically if the beam velocity is near the sound speed: $|V_0-c_0| \ll c_0$. In this case, the co-moving wave will have a greater amplitude than the counterpropagating wave. Thus, for example, if $V_0=0.9c_0$, their amplitude ratio will be 19. Obviously, in this regime, which is called the synchronous regime, excitation of a comoving wave will be more efficient than excitation by immobile sources.

We consider the case $|V_0-c_0| \ll c_0$ in greater detail. Under this regime the first term in brackets in Eq. (1.19) will have a predominant amplitude value over the second term. The wave form of the acoustic pulse is given in Fig. 1.7 ($V_0<c_0$ is assumed for the example). The pulse profile consists of a short powerful pulse of duration $(V_0-c_0)t$ and amplitude

$$p_a = \alpha c_0^2 \beta W_0/(2bc_p|V_0-c_0|) \tag{1.20}$$

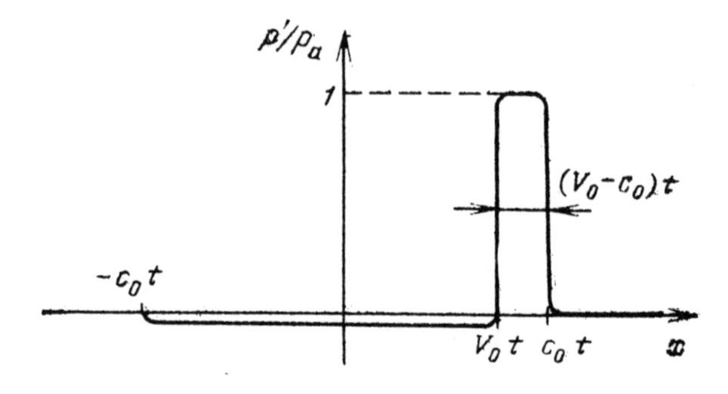

FIG. 1.7. Profile of a moving beam-excited OA signal.

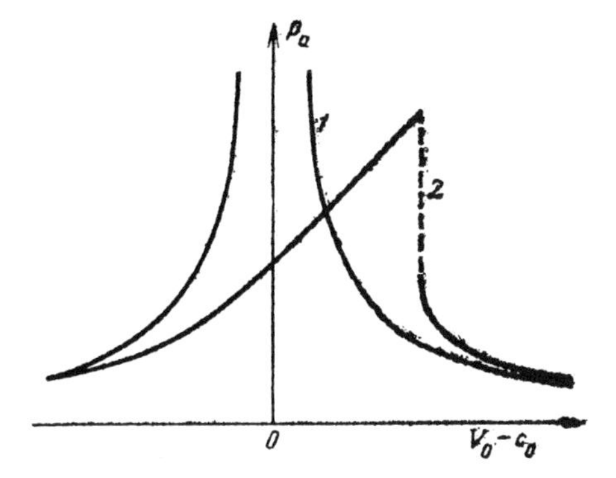

FIG. 1.8. Acoustic wave amplitude plotted as a function of beam velocity in accordance with the linear (1) and nonlinear (2) theories.

and a long loop of duration $(c_0 + V_0)t$ and an amplitude that is

$$|V_0 + c_0|/|V_0 - c_0|$$

smaller than Eq. (1.20).

Figure 1.8 shows the dependence of wave amplitude on beam velocity. As the beam velocity V_0 approaches the sound speed c_0, the wave amplitude grows (while the pulse duration narrows correspondingly). At precise phase matching $V_0 = c_0$ the perturbation grows linearly with time (without limit within the framework of this model):

$$p' = (\alpha c_0^2 \beta W_0/2bc_p)tH(x - c_0 t).$$

Acoustic nonlinear effects will limit the wave amplitude. The dependence of wave amplitude on the beam velocity accounting for these effects is also given in Fig. 1.8 (curve 2). Nonlinear amplitude limiting is examined in greater detail in Chap. 6.

The total area under curve (1.19) is equal to zero consistent with the law of conservation of momentum. However, the mass conveyed by each of the terms in (1.19) grows with time. Since $\rho' = p'/c^2$ in a freely propagating wave,

$$m_+ = \frac{\alpha\beta W_0}{2bc_p} \int_{-\infty}^{\infty} \frac{\mathcal{H}(x - c_0 t) - \mathcal{H}(x - V_0 t)}{V_0 - c_0} dx\Sigma,$$

where Σ is the area of the heating zone in the beam cross section: $\Sigma = \alpha^{-1} b$. Consequently,

$$m_+ = \beta W_0 t/2c_p,$$

i.e., the mass conveyed by the wave will be determined by the absorbed beam energy $W_0 t$. Under ordinary conditions of a gaseous medium, each kilojoule of absorbed energy causes a 2 g increase in the "mass" of the perturbation. Since the perturbation travels at a velocity $(V_0 + c_0)/2 \approx c_0$ the fraction of absorbed energy converted into kinetic perturbation energy $m_+ c_0^2/2$ will be

$$\eta'' = m_+ c_0^2/2W_0 t = c_0^2 \beta/4c_p.$$

Using thermodynamic identities and state equations of an ideal gas, we easily find that

$$\eta'' = (\gamma - 1)/4.$$

This quantity is of the order of 10^{-1} for gases.

Therefore, the use of synchronous sources makes it possible to achieve maximum laser radiation-to-mechanical energy conversion efficiencies. In this case if the number of degrees of freedom of the gas molecule is equal to j, then

$$\eta'' = (2j)^{-1}.$$

In other words, the thermal energy is uniformly distributed over j degrees of freedom, while the synchronicity of the sources permits utilization of one-half of the energy of one translational degree of freedom (molecular motion with the beam). It should be noted that such a high conversion efficiency (characteristic of instantaneous point heating) is achieved at comparatively low radiation power densities: far from breakdown density levels.

We have considered the simplest versions of thermo-optic sound excitation: excitation by a harmonically modulated beam in a weakly absorbing medium, by a short pulse in a strongly absorbing medium, and by a continuous traveling beam. This analysis was carried out on a qualitative level, although the results obtained here are also valid in the corresponding limiting cases. Subsequent chapters are devoted to a quantitative description of the optoacoustic effect.

REFERENCES

1. M. A. Isakovich, *Obshchaya akustika* [*General Acoustics*] (Nauka, Moscow, 1973).
2. V. P. Zharov and V. S. Letokhov, *Lazernaya optiko-akusticheskaya spektroskopiya* [*Laser Optoacoustic Spectroscopy*] (Nauka, Moscow, 1984).
3. *Optiko-akusticheskiy metod v lazernoy spektroskopii molekulyarnykh gazov* [*Optoacoustic Method in the Laser Spectroscopy of Molecular Gases*], edited by Yu. S. Makushkin (Nauka, Novosibirsk, 1984).
4. B. G. Ageev, Yu. N. Ponomarev, and B. A. Tikhomirov, *Nelineynaya optiko-akusticheskaya spektroskopiya molekulyarnykh gazov* [*Nonlinear Optoacoustic Spectroscopy of Molecular Gases*] (Nauka, Novosibirsk, 1984).
5. L. M. Lyamshev, Usp. Fiz. Nauk **135**, 637 (1981) [Sov. Phys. Usp. **24**, 977 (1981)].
6. C. K. N. Patel and A. C. Tam, Rev. Mod. Phys. **53**, 517 (1981).
7. L. M. Lyamshev and L. V. Sedov, Akust. Zh. **27**, 5 (1981). [Sov. Phys. Acoust. **27**, 4 (1981)].
8. A. I. Bozhkov and F. V. Bunkin *et al.*, *Tr. FIAN* [*Proceedings of the Lebedev Physics Institute of the USSR Academy of Sciences*] (Nauka, Moscow, 1985), Vol. 156, pp. 123–176.
9. M. W. Sigrist, J. Appl. Phys. **60**, R83 (1986).
10. J. Phys. C **6** (1983).
11. IEEE Trans. Ultrason. Ferroelectr. Freq. Control **UFFC-33** (1986).

Chapter 2

Spectral Characteristics of Optoacoustic Signals in a Liquid

Chapter 1 described a qualitative theory of the thermooptical effect in its simplest form that is partially based on intuitive concepts of the processes related to this effect. A rigorous analysis of this phenomenon requires solving a nonstationary thermoelasticity problem[1] that in the general case is a complicated problem. This theory is examined as it applies to liquids in the present chapter when only longitudinal acoustic waves are excited in both an absorbing and a transparent medium.

2.1. Nonstationary Thermoelasticity Equation in the Case of Liquids

It is convenient to use equations in Euler variables (Lagrangian variables are suitable for one-dimensional motion) to describe thermooptical excitation of sound in liquids. The continuity equation of the medium takes the form

$$\frac{\partial \rho}{\partial t} + \operatorname{div}(\rho \boldsymbol{v}) = 0, \tag{2.1}$$

where ρ is density and $\boldsymbol{v}$ is the vibrational velocity of the particles in the medium. The equation of motion of a viscous liquid (the Navier-Stokes equation) is given as follows:

$$\rho \frac{\partial \boldsymbol{v}}{\partial t} + \rho(\boldsymbol{v}\nabla)\boldsymbol{v} = -\nabla p + \eta \Delta \boldsymbol{v} + \left(\xi + \frac{\eta}{3}\right) \operatorname{grad} \operatorname{div} \boldsymbol{v}. \tag{2.2}$$

Here p is the pressure of the medium, while ξ and η are the bulk and shear viscosities of the medium.

Equations (2.1) and (2.2) are derived from the laws of conservation of mass and momentum. One additional equation that defines the motion of a viscous liquid is the heat conduction equation

$$\rho T\left(\frac{\partial s}{\partial t} + (\boldsymbol{v}\nabla)s\right) = \operatorname{div}(\kappa \nabla T) + (\eta/2)\left[\frac{\partial v_i}{\partial x_k} + \frac{\partial v_k}{\partial x_i} - \left(\frac{2}{3}\right)\delta_{ik}\frac{\partial v_l}{\partial x_l}\right]^2 + \xi(\operatorname{div} \boldsymbol{v})^2 - \operatorname{div}\langle \mathbf{S} \rangle, \tag{2.3}$$

where s is entropy, v_i are the components of the velocity vector; repeating indices infer summation, while $\langle \mathbf{S} \rangle$ is the Umov-Poynting vector of the incident wave (the radiation intensity in the medium) averaged over the electromagnetic oscillation period. The latter term describes the heating density due to absorption of the light wave.

In order to obtain a closed system of equations it is necessary to add to equation set (2.1)–(2.3) the equations of state of the medium

$$p = p(\rho, s), \quad T = T(\rho, s).$$

However, since there is no analytic expression of these relations and relative density and entropy perturbations are small,

$$|\rho - \rho_0| \ll \rho_0, \quad |s - s_0| \ll c_V,$$

it is sufficient to use an expansion of the equations of state near the equilibrium values of parameters ρ_0, s_0. Thermodynamic relations[2] can be used to represent these quantities as

$$p = p_0 + c_0^2 \rho' + T_0 c_0^2 \rho_0 \beta s'/c_p, \tag{2.4}$$

$$T = T_0 + T_0 c_0^2 \rho' \beta/\rho_0 c_p + T_0 s'/c_V. \tag{2.5}$$

Here $\rho' = \rho - \rho_0$; $s' = s - s_0$; $c_0^2 = (\partial p/\partial \rho)_S$ is the squared adiabatic sound velocity; $\beta = V^{-1}(\partial V/\partial T)_p$ is the thermal coefficient of volume expansion of the medium, c_p, c_V are the specific heats at a constant pressure and volume, while the "0" subscript denotes equilibrium parameter values. The last term in Eq. (2.4) is responsible for thermoelastic effects, while the second term in Eq. (2.5) is responsible for the change in temperature in the acoustic wave.

We simplify fundamental equations (2.1)–(2.3), assuming minor deviations from equilibrium values:

$$\frac{|\rho - \rho_0|}{\rho_0} \sim \frac{|v|}{c_0} \sim \frac{|p - p_0|}{p_0} \sim \frac{|s - s_0|}{c_V} \sim \mu \ll 1.$$

As in the case of linear acoustics we retain only perturbation-linear terms of Eqs. (2.1)–(2.3):

$$\frac{\partial \rho'}{\partial t} + \rho_0 \operatorname{div} \boldsymbol{v} = 0, \tag{2.6}$$

$$\rho_0 \frac{\partial \boldsymbol{v}}{\partial t} = -c_0^2 \operatorname{grad} \rho' - (T_0 c_0^2 \rho_0 \beta / c_p) \operatorname{grad} s' + \eta \Delta \boldsymbol{v} + (\xi + \eta/3) \operatorname{grad} \operatorname{div} \boldsymbol{v}, \tag{2.7}$$

$$\rho_0 T_0 \left(\frac{\partial s}{\partial t} \right) = T_0 \kappa \, \Delta s'/c_V + (\kappa T_0 c_0^2 \beta / \rho_0 c_p) \Delta \rho' - \operatorname{div} \langle \mathbf{S} \rangle. \tag{2.8}$$

Since the thermal balance equations at the boundaries of the media are expressed through the temperature rather than entropy, heat conduction equation (2.8) is more appropriate to reduce to the form

$$\rho_0 c_V \left(\frac{\partial T'}{\partial t} \right) = \kappa \Delta T' - \operatorname{div} \langle \mathbf{S} \rangle, \tag{2.9}$$

thereby eliminating entropy from Eq. (2.7) by means of Eq. (2.5). The external sources only enter into heat conduction equation (2.9). Since the problem of determining the temperature field $T'(t,\mathbf{r})$ can be solved independent of the acoustic part of the problem, the expression found for s' can in turn be treated as an external source of acoustic waves. It follows that the acoustic waves propagate freely in the range where temperature changes can be ignored. Since the wave sources have a potential nature, liquid motion will be potential motion. We therefore set

$$v = \text{grad}\ \varphi, \tag{2.10}$$

where φ is the scalar potential of the velocity field. Then system (2.6)–(2.9) reduces to the equations

$$\frac{\partial^2\varphi}{\partial t^2} - c_0^2\,\Delta\varphi - \left[\left(\xi + \frac{4\eta}{3}\right)\rho_0^{-1}\right]\frac{\partial\Delta\varphi}{\partial t} = -\left(\frac{c_0^2\beta}{c_p}\right)\frac{\partial(T_0 s')}{\partial t}, \tag{2.11}$$

$$\frac{\partial(T_0 s')}{\partial t} = \gamma\chi\Delta(T_0 s') + \frac{\chi T_0 c_0^2\beta\,\Delta\rho'}{\rho_0} - \text{div}\,\langle\mathbf{S}\rangle/\rho_0. \tag{2.12}$$

Here $\chi = \kappa/\rho_0 c_p$ is the thermal diffusivity.

In order to identify the role of the second term on the right-hand side of Eq. (2.12), we consider free-wave propagation:

$$\text{div}\langle\mathbf{S}\rangle = 0.$$

Then equation set (2.11), (2.12) will be solved via harmonic wave

$$\varphi = \varphi_0' \exp(-i\omega t + i\mathbf{kr}),\ s' = s_0' \exp(-i\omega t + i\mathbf{kr}).$$

Substituting this form of the solution into Eqs. (2.11), (2.12) subject to Eqs. (2.6) and (2.10), we obtain

$$[-\omega^2 + c_0^2k^2 - i\omega k^2(\xi + 4\eta/3)/\rho_0]\varphi_0' - i\omega c_0^2\beta T_0 s_0'/c_p = 0,$$

$$ik^4\chi T_0 c_0^2\beta\varphi_0'/\omega + (-i\omega + \gamma k^2\chi)T_0 s_0' = 0.$$

Based on the nontrivial nature of the solution of this equation system, we obtain a dispersion equation that, subject to the known relation

$$\gamma - 1 = T_0 c_0^2\beta^2/c_p,$$

reduces to the form

$$[-\omega^2 + c_0^2k^2 - i\omega k^2(\xi + 4\eta/3)/\rho_0](-i\omega + k^2\chi\gamma)$$
$$-\chi c_0^2 k^4(\gamma - 1) = 0.$$

The square brackets on the left-hand side contain the dispersion relation for acoustic waves for the case of no wave coupling to thermal waves.* We assume the following relation holds:

$$\omega k^2(\xi + 4\eta/3)/\rho_0 \ll (\omega^2, c_0^2k^2),$$

*The second set of round parentheses contain the dispersion relation for thermal waves.

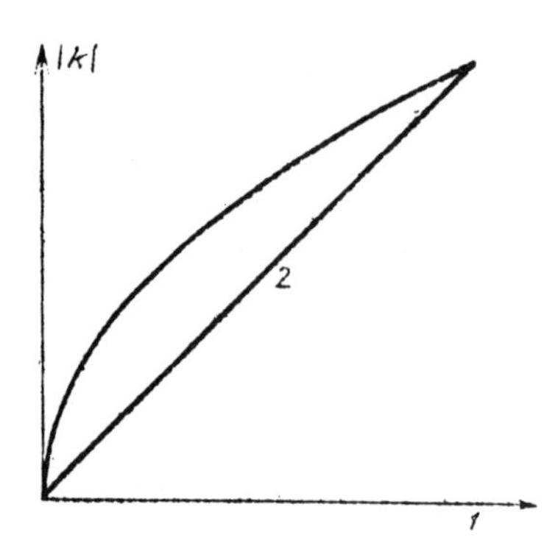

FIG. 2.1. Dispersion curves for thermal (1) and acoustic (2) waves.

since otherwise the acoustic wave would decay over distances of the order of the wavelength. This inequality holds through the gigahertz frequency band.

Since the wave vector of the acoustic wave is much less than the wave vector of the thermal wave across the entire frequency range $\omega \ll \omega_\chi = c_0^2/\chi$ (Fig. 2.1)

$$\omega/c_0 \ll (\omega/\gamma\chi)^{1/2},$$

the acoustic and thermal modes are weakly coupled. Hence, their interaction will only yield small corrections to the dispersion equations. They will appear, as follows, for the acoustic and thermal waves, respectively:

$$-\omega^2 + c_0^2 k^2 - i\omega k^2 \frac{\xi + 4\eta/3}{\rho_0} - \kappa i\omega k^2 \frac{c_V^{-1} - c_p^{-1}}{\rho_0} = 0,$$

$$-i\omega + k^2\chi = 0.$$

Accounting for interaction of the acoustic and thermal modes led to secondary acoustic wave absorption and modified to some degree (a factor of $\sqrt{\gamma}$) the wave vector of the thermal wave. Hence, the equations of thermooptical sound excitation in a viscous heat-conducting medium can be given as

$$\frac{\partial^2\varphi}{\partial t^2} - c_0^2\Delta\varphi - \frac{b}{\rho_0}\frac{\partial}{\partial t}\Delta\varphi = -\frac{c_0^2\beta}{c_p}\frac{\partial}{\partial t}T_0 s',$$

$$\frac{\partial}{\partial t}T_0 s' = \chi\Delta(T_0 s') - \frac{\operatorname{div}\langle \mathbf{S}\rangle}{\rho_0}.$$

As is the convention, we neglect the difference of χ and $\gamma\chi$ in the heat conduction equation, since $\gamma - 1 \ll 1$ ordinarily holds for condensed matter. The thermal sources due to external heating are on the right-hand side of the wave equation.

These equations can be further simplified. First, it is possible to neglect the dissipative term in the sound excitation zone, since sound absorption is low at distances of the order of the wavelength. Moreover, at acoustic frequency $\omega > (\alpha^2\chi, \chi/a^2)$ heat conduction is not significant (α^{-1}, a represent the dimensions of the heating zone). This will hold across the entire ultrasonic frequency band for the

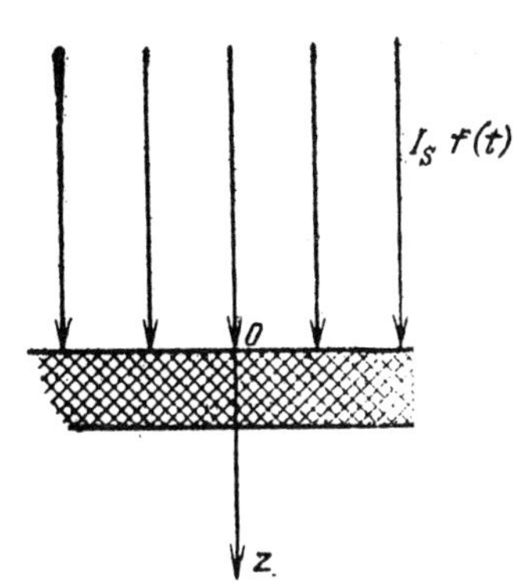

FIG. 2.2. OA-source geometry for the case of strong light absorption.

majority of liquids (with the possible exception of mercury). The following equation can therefore be used to describe pulsed thermooptical sound excitation in liquids

$$\frac{\partial^2 \varphi}{\partial t^2} - c_0^2 \, \Delta\varphi = \left(\frac{c_0^2 \beta}{\rho_0 c_p}\right) \operatorname{div}\langle \mathbf{S} \rangle. \tag{2.13}$$

Its solutions will be analyzed here.

2.2. The Transfer Function Method

Under thermooptical excitation an acoustic pulse wave form will be determined by both the characteristics of the medium—the light absorptivity and sound speed—as well as parameters of the laser radiation: pulse duration and beam diameter. The primary problem lies in differentiating the effect of the properties of the medium and the laser pulse wave form on the acoustic signal profile. The transfer function method can be used for this purpose.

We first consider the one-dimensional problem. Assume a plane light wave (Fig. 2.2) from a transparent medium is incident along the normal to the boundary with an absorbing medium, and the change in wave intensity with time is given by Burgher's law

$$I = I_S f(t). \tag{2.14}$$

Let the medium be homogeneous and the absorption coefficient of light α be constant. Then the Umov-Poynting vector of the medium will be determined by

$$\langle \mathbf{S} \rangle = I_0 e^{-\alpha z} f(t) \mathbf{n}_z, \; I_0 = A I_S,$$

where $\mathbf{n}_z$ is the unit vector on the z axis. thermooptical sound excitation is then described by the equation

$$\frac{\partial^2 \varphi}{\partial t^2} - c_0^2 \frac{\partial^2 \varphi}{\partial z^2} = -\left(\frac{\alpha c_0^2 \beta}{\rho_0 c_p}\right) I_0 e^{-\alpha z} f(t) \tag{2.15}$$

[only the z derivative need be retained in the Laplace operator, since $\varphi = \varphi(t, z)$].

Equation (2.15) is easily solved by the spectral method.[3] We find φ as

$$\varphi = (2\pi)^{-1} \int_{-\infty}^{\infty} e^{-i\omega t} \tilde{\varphi}(\omega,z) d\omega. \tag{2.16}$$

Substituting Eq. (2.16) into Eq. (2.15) we find

$$\frac{d^2}{dz^2} \tilde{\varphi} + \frac{\omega^2}{c_0^2} \tilde{\varphi} = \frac{\alpha\beta}{\rho_0 c_p} I_0 e^{-\alpha z} \tilde{f}(\omega), \tag{2.17}$$

where

$$\tilde{f}(\omega) = \int_{-\infty}^{\infty} e^{i\omega t} f(t) dt \tag{2.18}$$

is the spectrum of laser radiation intensity (2.14) (the spectrum of the laser pulse envelope). The solution of ordinary differential equation (2.17) can be represented as

$$\tilde{\varphi} = C_+ e^{i(\omega/c_0)z} + C^- e^{-i(\omega/c_0)z} + \frac{\beta I_0}{\rho_0 c_p} \frac{\alpha}{\alpha^2 + \omega^2/c_0^2} \tilde{f}(\omega) e^{-\alpha z}. \tag{2.19}$$

The last term in Eq. (2.19) describes the perturbations localized near the surface in the absorption range of laser radiation. This term diminishes exponentially with increasing depth, and may be taken as zero for $\alpha z > 3$–5. The first two terms correspond to [see Eq. (2.16)] acoustic waves traveling into the medium (the amplitude C_+) and exiting the medium toward the boundary (amplitude C_-).

The coefficients C_+ and C_- in Eq. (2.19) are determined by the auxiliary boundary conditions. One such condition is the emission of radiation condition. Since an absorbing medium is semibounded by the causality principle, there will be no wave traveling towards the boundary:

$$C_- = 0. \tag{2.20}$$

Solution (2.19) therefore demonstrates that an optoacoustic signal consists of both a localized perturbation caused by heating of the medium and the acoustic wave traveling into the medium. Consequently, only an acoustic traveling wave will be present outside the heating zone $\alpha z \gtrsim 3$–5. It is this wave that we will consider.

The wave spectrum $C_+(\omega)$ is determined by auxiliary boundary conditions on the surface $z = 0$. Equality of the velocities and pressure increments in the transparent and absorbing media along their interface must be used as these conditions. Ordinarily the two limiting cases are considered: a rigid (hard) and a free (soft) boundary of the absorbing medium.

In the case of a rigid boundary, the points on the surface $z = 0$ are fixed [see Eq. (2.10)] and the velocity

$$v\Big|_{z=0} = \frac{\partial \varphi}{\partial z}\Big|_{z=0} = 0. \tag{2.21}$$

Substituting solution (2.19) into Eq. (2.21), subject to (2.20) we find

$$C_+(\omega) = -i \frac{c_0}{\omega} \frac{\beta I_0}{\rho_0 c_p} \frac{\alpha^2}{\alpha^2 + \omega^2/c_0^2} \tilde{f}(\omega). \tag{2.22}$$

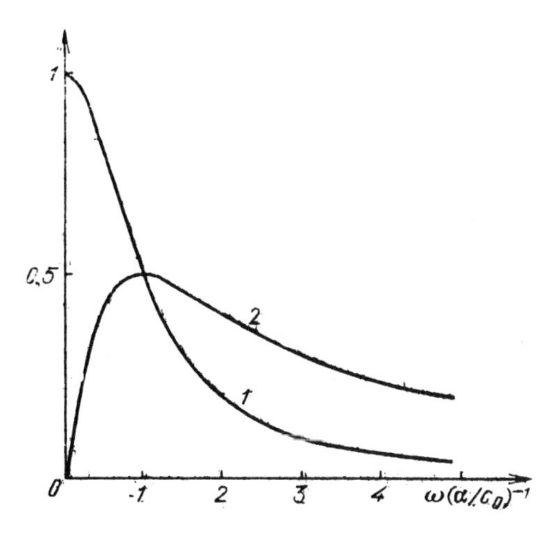

FIG. 2.3. Profiles of the transfer function moduli for rigid (1) and free (2) boundaries.

However, the experimentally measured quantity is either the vibrational velocity or the pressure. Since in a planar acoustic traveling wave they are related by[4]

$$v/c_0 = p'/\rho_0 c_0^2 = \rho'/\rho_0,$$

the spectrum of either wave is all that is needed. Using Eqs. (2.21), (2.22), and (2.10), we find ($\tau = t - z/c_0$)

$$v = (2\pi)^{-1} \frac{\beta I_0}{\rho_0 c_p} \int_{-\infty}^{\infty} \frac{\alpha^2}{\alpha^2 + \omega^2/c_0^2} \tilde{f}(\omega) e^{-i\omega\tau} \, d\omega. \tag{2.23}$$

The OA-signal spectrum therefore is the product of the spectrum of the laser radiation intensity $I_0 \tilde{f}(\omega)$ and the transfer function $K(\omega)$, which is determined only by the properties of the medium (primarily the absorption coefficient of light). In the case of a rigid boundary

$$K_r(\omega) = \frac{\beta}{\rho_0 c_p} \frac{\alpha^2}{\alpha^2 + \omega^2/c_0^2}. \tag{2.24}$$

The pressure increment $p'|_{z=0} = 0$ vanishes on the surface in the case of a free boundary (for example, a transparent medium in the form of a gas and an absorbing medium in the form of a liquid). Subject to Eqs. (2.4), (2.6), and (2.15) it reduces to

$$\varphi|_{z=0} = 0.$$

Intermediate calculations analogous to those given above yield the following expression for the transfer function in the case of a free boundary of an absorbing medium:

$$K_f(\omega) = \frac{\beta}{\rho_0 c_p} \frac{-i\alpha\omega/c_0}{\alpha^2 + \omega^2/c_0^2}. \tag{2.25}$$

The profiles of the moduli of the transfer functions (2.24) and (2.25) are shown in Fig. 2.3. These clearly indicate that for $\omega < \alpha c_0$ sound excitation will be more efficient at a rigid boundary, while in the range $\omega > \alpha c_0$, at a free boundary. The characteristic frequency

$$\omega_a = \alpha c_0$$

has a simple meaning: the wave vector of the acoustic wave at this frequency is equal to the absorption coefficient of light. This frequency defines the upper boundary of efficient sound excitation with a rigid boundary. The frequency $\omega = \omega_a$ corresponds to the peak efficiency of sound excitation with a free boundary. The efficient excitation frequency band in the case of a free boundary is more than two times broader than in the case of a rigid boundary, and lies within the range

$$(2 - \sqrt{3};\ 2 + \sqrt{3})\omega_a \approx (0.27;\ 2.73)\omega_a.$$

We now analyze the temporal wave form of the OA signals. We consider the case of a rigid boundary.

For the case of sound excitation by a short laser pulse $\alpha c_0 \tau_L \ll 1$ the spectral range of light intensity is significantly broader than the range of transfer function (2.24). The light intensity spectrum can be treated as constant across the entire frequency range of efficient sound excitation:

$$I_0 \tilde{f}(\omega) = I_0 \tilde{f}(\omega = 0) = \mathscr{E}_0,$$

where $\mathscr{E}_0 = I_0 \int_{-\infty}^{\infty} f(t)dt$ is the radiation energy flow density. Taking account of this relation

$$v_r(\tau) = (2\pi)^{-1} \frac{\beta \mathscr{E}_0}{\rho_0 c_p} \int_{-\infty}^{\infty} \exp(-i\omega\tau) \frac{\alpha^2}{\alpha^2 + \omega^2/c_0^2} d\omega$$

$$= \frac{\alpha c_0 \beta \mathscr{E}_0}{2\rho_0 c_p} \begin{cases} \exp(\alpha c_0 \tau), & \tau \leqslant 0, \\ \exp(-\alpha c_0 \tau), & \tau > 0, \end{cases} \qquad (2.26)$$

which corresponds to Eq. (1.14). The signal is a symmetrical compressional pulse with exponential leading and trailing edges and an amplitude proportional to the absorbed energy density $\alpha \mathscr{E}_0$.

For the case of laser excitation by long laser pulses $\alpha c_0 \tau_L \gg 1$ there is an inverse relation between the spectral ranges of the transfer function and the laser radiation intensity. Therefore we can assume $K_r(\omega) \approx \beta / \rho_0 c_p$ across the complete excited frequency range $\omega \lesssim (\pi \tau_L)^{-1}$. In this case

$$v_r(\tau) = (\beta I_0 / \rho_0 c_p) f(\tau)$$

and the acoustic pulse follows the wave form of the laser pulse. This is understandable, since in accordance with $\alpha c_0 \tau_L \gg 1$ there takes place a quasistatic expansion of a layer (α^{-1}) thin compared to the characteristic length $(c_0 \tau_L)$ of the acoustic wave. The pulse amplitude is determined by the heating intensity I_0.

In the general case of an arbitrary laser pulse duration, the OA-signal wave form is given by integral (2.23), which, subject to Eq. (2.26), can be converted to

$$v_r(\tau) = \frac{\alpha c_0 \beta I_0}{2\rho_0 c_p} \int_{-\infty}^{\infty} f(t) \exp[-\alpha c_0 |\tau - t|] dt. \qquad (2.27)$$

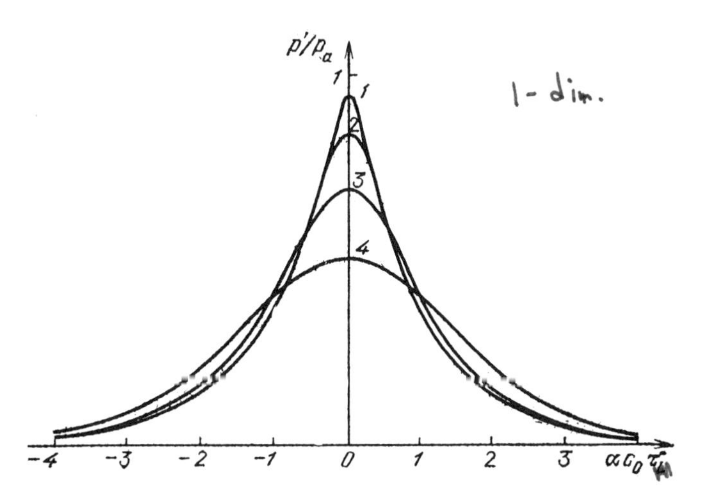

FIG. 2.4. OA-pulse profiles for a rigid boundary with different parameters $\alpha c_0 \tau_L$ equal to 0.1 (1), 0.3 (2), 0.7 (3), and 1.5 (4).

For small, yet finite values of $\alpha c_0 \tau_L$, the finite nature of the laser pulse duration will be manifested as a smooth peak of the OA signal. Figure 2.4 shows the profiles of the OA pulses for different values of the parameter $\alpha c_0 \tau_L$ for a Gaussian laser pulse

$$f(t) = (\pi)^{-1/2} \exp[-(t/\tau_L)^2].$$

Here integral (2.27) is expressed analytically through the error function, Φ,

$$v_r = \frac{c_0 \alpha \beta \mathscr{E}_0}{2\rho_0 c_p} \cdot \frac{1}{2} \exp\left(\frac{\alpha^2 c_0^2 \tau_L^2}{4}\right) \left\{ e^{\alpha c_0 \tau} \left[1 - \Phi\left(\frac{\tau}{\tau_L} + \frac{\alpha c_0 \tau_L}{2}\right)\right] \right.$$

$$\left. + e^{-\alpha c_0 \tau} \left[1 + \Phi\left(\frac{\tau}{\tau_L} - \frac{\alpha c_0 \tau_L}{2}\right)\right]\right\}.$$

The dependence of signal amplitude on $\alpha c_0 \tau_L$ is shown in Fig. 2.5 for a constant absorbed energy density $\alpha \mathscr{E}_0 = \text{const}$. For small $\alpha c_0 \tau_L$ it is constant, while for $\alpha c_0 \tau_L > 1$ it diminishes due to the diminishing radiation intensity.

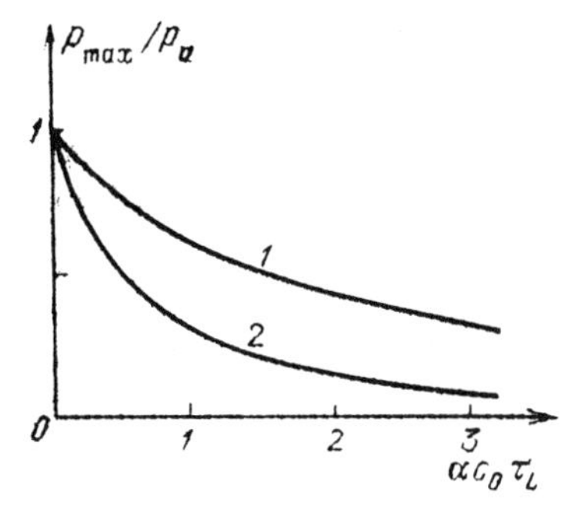

FIG. 2.5. OA-signal amplitude plotted as a function of laser pulse duration for rigid (1) and free (2) boundaries.

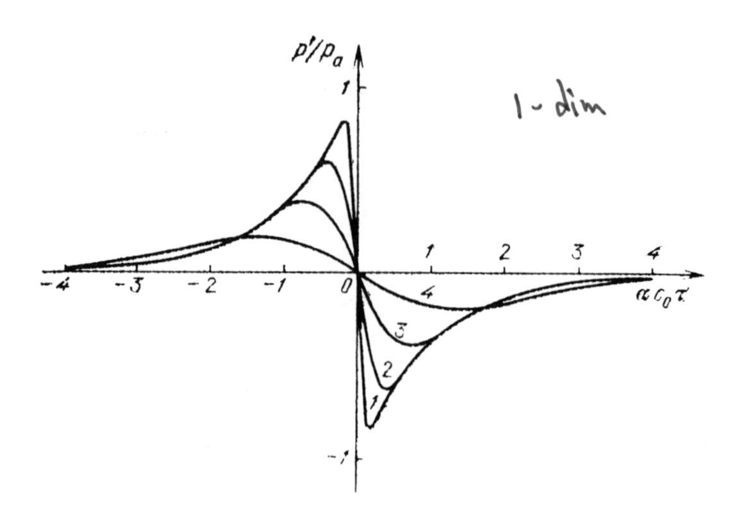

FIG. 2.6. OA-pulse profiles for a free boundary for different parameters $\alpha c_0 \tau_L$ equal to 0.1 (1), 0.3 (2), 0.7 (3), and 1.5 (4).

We now consider thermooptical sound excitation for the case of a free boundary of the absorbing medium. It is easily determined from a comparison of Eqs. (2.25) and (2.24) that the transfer functions of the OA signals for free and rigid boundaries are related as follows:

$$K_f(\omega) = -(i\omega/\alpha c_0)K_r(\omega).$$

Consequently, the temporal wave forms of the OA signals for the case of a free and a rigid boundary are related by

$$v_f(\tau) = \frac{1}{\alpha c_0}\frac{d}{d\tau} v_r(\tau). \tag{2.28}$$

Hence

$$v_f(\tau) = -\frac{\alpha c_0 \beta I_0}{2\rho_0 c_p}\int_{-\infty}^{\infty} f(t)\exp[-\alpha c_0|\tau - t|]\mathrm{sgn}(\tau - t)dt.$$

The corresponding OA-signal profiles in the case of a free boundary for a Gaussian laser pulse profile are given in Fig. 2.6, while their amplitude is shown in Fig. 2.5.

With a free boundary, the OA signal consists of the compressional phase followed by the rarefactional phase which is related to the sign reversal of the reflection wave at the free surface. For short laser pulses ($\alpha c_0 \tau_L \ll 1$) the profile will consist of two exponents with a pulse rise time $(\alpha c_0)^{-1}$ and a transition zone with a duration of the order of τ_L. For long laser pulses ($\alpha c_0 \tau_L \gg 1$) the OA-signal profile in accordance with Eqs. (2.27) and (2.28) is given by the derivative of the laser pulse wave form:

$$v_f(\tau) = (\beta I_0/\rho_0 c_p \alpha c_0)\frac{df}{d\tau}.$$

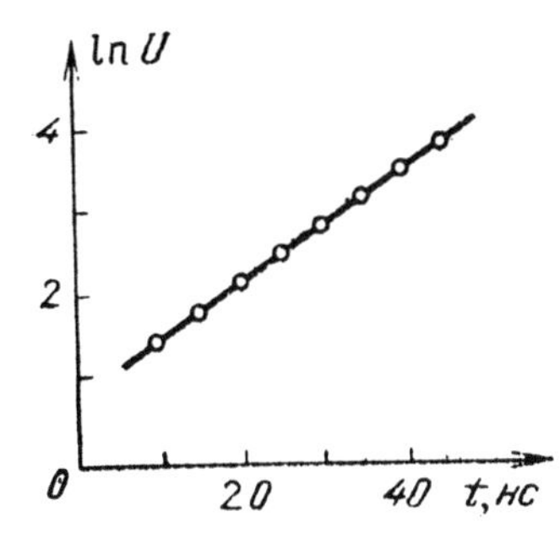

FIG. 2.7. OA-signal wave front profile plotted on a semilogarithmic scale for the case of a short laser pulse.

Hence the OA-signal amplitude is constant when the absorbed energy density $I_0\tau_L$ is constant and the pulses are short in duration, while the amplitude diminishes with increasing pulse duration (due to the diminishing radiation intensity).

The mechanisms behind changes in the OA-signal wave form were analyzed as early as the initial experimental studies on laser optoacoustics.[5] The experimentally observed OA-signal profiles for different radiation absorption coefficients are in good agreement with the theoretical relations.

The exponential time dependence of the vibrational velocity at the leading edge of the wave was analyzed in Refs. 6 and 7. Figure 2.7 plots the experimental points of the wave front of an acoustic pulse excited thermooptically in an aqueous copper chloride solution (to increase the light absorptivity). These points are accurately approximated by a straight line on a semilogarithmic scale corresponding to the exponent. The light absorptivity can be determined by the slope of the line.

Reference 7 carried out a quantitative analysis of the spectral transfer functions for the case of thermooptical sound excitation. Sound excitation by CO_2 laser pulses in transformer oil was analyzed. A surface of oil was covered with a barium fluoride window to simulate the rigid boundary (the impedance ratio of the absorbing and transparent media $N \approx 0.08$). The correlation between the transfer functions and the theoretical form (2.24), (2.25) was tested by plotting these functions on special coordinates (Fig. 2.8) on which they should appear as a linear relation (the experimental points are accurately represented by a straight line, confirming the anticipated form of the transfer functions).

As demonstrated above—Eqs. (2.24) and (2.25)—the transfer functions and, consequently, the spectral characteristics of the thermooptical signals are determined by the light absorptivity. These equations are valid for a homogeneous medium. If the absorptivity is dependent on the coordinate, Eqs. (2.24) and (2.25) must be generalized. Here the discussion concerns a solution of the wave equation for the particle velocity v in an acoustic wave of the type

$$\frac{\partial^2 v}{\partial t^2} - c_0^2 \frac{\partial^2 v}{\partial z^2} = -\rho_0^{-1} \frac{\partial^2 G(t,z)}{\partial t\, \partial z} \tag{2.29}$$

or the derived equation for the scalar potential

$$\frac{\partial^2 \varphi}{\partial t^2} - c_0^2 \frac{\partial^2 \varphi}{\partial z^2} = -\rho_0^{-1} \frac{\partial G(t,z)}{\partial t}. \tag{2.30}$$

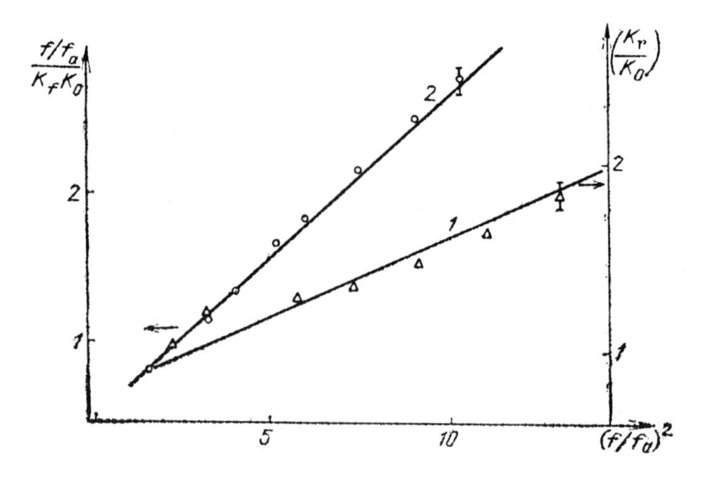

FIG. 2.8. Calculated (solid curves) and experimental values of the transfer functions for rigid (1) and free (2) boundaries.

In terms of physical meaning function $G(t,z)$ is the potential of the force field **f** initiating motion of the medium:

$$f_z(t,z) = -\frac{\partial G(t,z)}{\partial z}.$$

Equation (2.30) can be solved by the spectral method. Applying a Fourier time transform we obtain

$$\frac{d^2\tilde{\varphi}}{dz^2} + \frac{\omega^2}{c_0^2}\tilde{\varphi} = -\frac{i\omega}{\rho_0 c_0^2}\tilde{G}(\omega,z). \tag{2.31}$$

It is most advantageous to use a Laplace z-coordinate transform

$$\hat{\varphi}(\omega,p) = \int_0^\infty \exp(-pz)\tilde{\varphi}(\omega,z)dz$$

to solve this equation. Taking the corresponding integral of (2.31), we obtain the following algebraic relation:

$$\left(p^2 + \frac{\omega^2}{c_0^2}\right)\hat{\varphi} = \left.\frac{d\tilde{\varphi}}{dz}\right|_0 + p\tilde{\varphi}|_0 + \frac{-i\omega}{\rho_0 c_0^2}\hat{G}(\omega,p). \tag{2.32}$$

Here $\hat{G}(\omega,p)$ are the Laplace and Fourier transforms of the space-time distribution of the sources:

$$\hat{G}(\omega,p) = \int_{-\infty}^{\infty} dt \int_0^\infty G(t,z)\exp(i\omega t - pz)dz, \tag{2.33}$$

while $\varphi|_0$ and $(d\varphi/dz)|_0$ are the Fourier spectra of the potential φ and its derivative at the boundary of the medium. These Fourier transforms are determined by the boundary conditions [analogous to the amplitude of traveling waves C_+, C_- in Eq. (2.19)].

The radiation condition corresponds to the absence of poles for function $\hat{\varphi}$ in the lower semiplane of complex parameter p. In our case we may have two poles $p = \pm i\omega/c_0$. Hence, $\tilde{\varphi}|_0$ and $(d\tilde{\varphi}/dz)|_0$ will correspond to vanishing of the right-hand side of Eq. (2.32) for $p = -i\omega/c_0$:

$$\frac{d\tilde{\varphi}}{dz}\bigg|_0 + \left(-i\frac{\omega}{c_0}\right)\tilde{\varphi}\bigg|_0 + \frac{1}{\rho_0 c_0}\left(-i\frac{\omega}{c_0}\right)\hat{G}\left(\omega, -i\frac{\omega}{c_0}\right) = 0. \qquad (2.34)$$

The second condition will be vanishing of either $(d\tilde{\varphi}/dz)|_0$ [for a rigid boundary, compare to Eq. (2.21)] or $\tilde{\varphi}|_0$ (in the case of a free boundary). Therefore for a rigid boundary we obtain

$$\hat{\varphi}_r = -\left[\rho_0 c_0\left(p^2 + \frac{\omega^2}{c_0^2}\right)\right]^{-1}\left[p\hat{G}\left(\omega, -i\frac{\omega}{c_0}\right) + i\frac{\omega}{c_0}\hat{G}(\omega,p)\right],$$

while for a free boundary

$$\hat{\varphi}_f = -\left[\rho_0 c_0\left(p^2 + \frac{\omega^2}{c_0^2}\right)\right]^{-1}\left[\hat{G}\left(\omega, -i\frac{\omega}{c_0}\right) - \hat{G}(\omega,p)\right].$$

Carrying out an inverse Laplace transform

$$\tilde{\varphi} = (2\pi i)^{-1}\int_{\delta - i\infty}^{\delta + i\infty} e^{pz}\hat{\varphi}(\omega,p)\,dp,$$

we find for the traveling wave

$$\tilde{\varphi}_{r,f} = -(2\rho_0 c_0)^{-1}\left[\hat{G}\left(\omega, i\frac{\omega}{c_0}\right) \pm \hat{G}\left(\omega, -i\frac{\omega}{c_0}\right)\right]\exp\left(i\frac{\omega}{c_0}z\right).$$

We therefore obtain for the spectrum of the vibrational velocity of a traveling acoustic wave

$$\tilde{v}_{r,f} = (2\rho_0 c_0)^{-1}\left(-i\frac{\omega}{c_0}\right)\left[\hat{G}\left(\omega, i\frac{\omega}{c_0}\right) \pm \hat{G}\left(\omega, -i\frac{\omega}{c_0}\right)\right]. \qquad (2.35)$$

An inverse Fourier transform makes it possible to employ Eq. (2.35) to obtain the following relation between the acoustic wave profile and the space-time distribution of sound sources:

$$v_{r,f}(\tau) = (2\rho_0 c_0^2)^{-1}\int_0^\infty \left[\frac{\partial G}{\partial \tau}\left(\tau + \frac{z'}{c_0}, z'\right) \pm \frac{\partial G}{\partial \tau}\left(\tau - \frac{z'}{c_0}, z'\right)\right]dz'. \quad (2.36)$$

Expression (2.36) is to be used when description of the sources in the spectral language is not convenient. For example, such a situation occurs when the space-time evolution of the sources is determined by nonlinear differential equations (Chaps. 5 and 6). If the sources are conveniently characterized in the spectral language, Eq. (2.35) is preferred.

We use Eq. (2.35) to analyze thermoelastic sound generation for the case where sound absorption is dependent on the coordinate: $\alpha=\alpha(z)$. The acoustic wave sources can be represented as

$$\frac{\partial G(t,z)}{\partial t}=-\frac{c_0^2\beta}{c_p}I_0 f(t)\frac{d}{dz}\left[\exp\left(-\int_0^z \alpha(\xi)d\xi\right)\right].$$

We then obtain for traveling acoustic wave (2.36) in the rigid boundary case

$$v_r(\tau)=\frac{\beta I_0}{\rho_0 c_p}\int_{-\infty}^{\infty}\tilde{f}(\omega)e^{-i\omega\tau}d\omega\int_0^{\infty}\cos\left(\frac{\omega}{c_0}\xi\right)g(\xi)d\xi, \tag{2.37}$$

where $g(\xi)$ describes the spatial distribution of the sources G.

$$v_f(\tau)=\frac{\beta I_0}{\rho_0 c_p}\int_{-\infty}^{\infty}\tilde{f}(\omega)e^{-i\omega\tau}d\omega\int_0^{\infty}\sin\left(\frac{\omega}{c_0}\xi\right)g(\xi)d\xi. \tag{2.38}$$

Equations (2.37) and (2.38) can be generalized by continuing the source function $g(z)$ (this function is defined only for $z>0$)into the domain $z<0$ either oddly [for a free boundary $\bar{g}_f(-z)=-\bar{g}_f(z)$], or evenly (for a rigid boundary) $\bar{g}_r(-z)=\bar{g}_r(z)$. Taking this into account

$$v=\frac{\beta I_0}{2\rho_0 c_p}\int_{-\infty}^{\infty}\tilde{f}(\omega)e^{-i\omega\tau}d\omega\int_{-\infty}^{\infty}e^{-i\omega\tau}\bar{g}(\xi)d\xi, \tag{2.39}$$

where $\bar{g}(z)$ is the corresponding continuation. Therefore the thermooptical transfer function is the Fourier transform of the spatial distribution of the thermal sources. In other words, the OA-signal spectrum is the product of the laser radiation density spectrum and the Fourier spectrum of the spatial distribution of absorption (taking the continuation into account).

This is the fundamental result from using the transfer function method. The transfer function and the spatial distribution of optoacoustic sources can be recovered from the known light intensity spectrum (the laser pulse wave form) and the measured spectrum of the acoustic signal. This technique was implemented in Ref. 8. When laser pulses of sufficiently short duration are used and when the spectrum $\tilde{f}(\omega)$ is broader than the range of the transfer function, the leading edge of the acoustic pulse mimics the source distribution:[3]

$$v\sim\alpha(z-c_0t)\exp\left(-\int_0^{z-c_0t}\alpha(\xi)d\xi\right),\quad z>c_0t.$$

Here $\alpha(z)$ can be determined from this equation. Figure 2.9 shows the results of experiments to recover the absorption distribution in a model medium.[8] Such experiments have demonstrated that recovery is possible to a depth z of the order of the optical thickness of the medium

$$\int_0^z \alpha(\xi)d\xi\approx 2-3.$$

The other consequence of Eq. (2.39) lies in the possibility for employing inhomogeneously absorbing media to generate an acoustic signal with a prescribed wave form. For example, a quasiperiodic acoustic signal can be generated by means of a

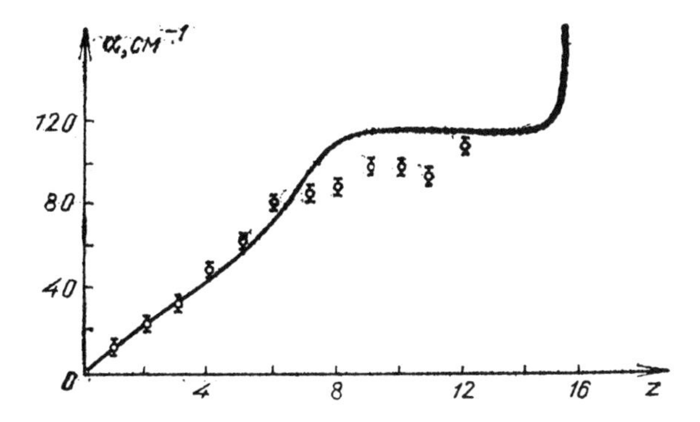

FIG. 2.9. OA diagnostics of an inhomogeneously absorbing medium. Solid curve—absorption coefficient vs depth; circles—experiment; z—number of absorbing layer.

single laser pulse.[3] In some sense this is the inverse of the previous case. This technique has also been implemented experimentally in Ref. 8.

It should be pointed out that in many versions of OA spectroscopy a traveling acoustic wave is not measured but rather different characteristics of the field in the excitation region are measured. We will nonetheless represent their temporal spectrum as the product of the radiation intensity spectrum and a certain frequency function independent of the type of radiation modulation. The transfer function method is therefore also applicable in these modifications of optoacoustic spectroscopy. Some of the more commonly encountered versions will be examined below.

2.3. Stage-by-stage Analysis of Thermooptical Excitation of Sound

The transfer function method can be employed to find the spectrum of an optoacoustic signal to a depth of a few (3–5) absorption lengths of light α^{-1}. However, sound is commonly recorded at far greater distances experimentally. The effects neglected in the generation zone can no longer be ignored here. These include sound dissipation, finite sound amplitude, and diffraction.

As a rule, a disk-type acoustic source geometry is typical of laser optoacoustics and a stage-by-stage approach can be used to account for the cited effects. This approach is as follows. If the lengths over which dissipation L_{DS}, nonlinearity L_{NL}, and diffraction L_{DF} are manifested for efficiently thermooptically excited frequencies are much greater than the dimensions of the thermal sources

$$(\alpha L_{DS}, \alpha L_{NL}, \alpha L_{DF}) \gg 1,$$

the problem is divided into two stages. In the first stage the transfer function method, for example, is employed to determine the traveling wave spectrum and profile (see Sec. 2.2; all cited effects are not incorporated). The resulting solution is represented as a boundary regime for the case of evolution of an acoustic beam of finite amplitude in a dissipative medium.[9]

The equation describing such evolution—the Khokhlov-Zabolotskaya equation—can be written as[10]

$$\frac{\partial}{\partial\tau}\left(\frac{\partial v}{\partial z}-\frac{\varepsilon}{c_0^2}v\frac{\partial v}{\partial\tau}-\frac{b}{2\rho_0 c_0^3}\frac{\partial^2 v}{\partial\tau^2}\right)=\frac{c_0}{2}\Delta_\perp v. \tag{2.40}$$

This equation is more conveniently expressed via the vibrational velocity or pressure. Here, v represents the z component of the vibrational particle velocity, while ε is the nonlinear acoustic parameter of the medium, and b is the dissipation coefficient.

In the general case Eq. (2.40) can only be solved numerically.[10] Analytic results can be obtained more easily when the scales of the individual effects are separated. Thus, if the dissipation length $L_{DS} \ll L_{NL}, L_{DF}$, then Eq. (2.40) reduces to the parabolic equation

$$\frac{\partial v}{\partial z}-\frac{b}{2\rho_0 c_0^3}\frac{\partial^2 v}{\partial\tau^2}=0.$$

The solution of this equation is more easily represented in spectral form:

$$\tilde{v}(\omega,z)=\tilde{v}(\omega,z=0)\exp(-b\omega^2 z/2\rho_0 c_0^3), \tag{2.41}$$

i.e., each spectral component diminishes exponentially with distance, with the exponent being proportional to the square of the frequency: $L_{DS}=2\rho_0 c_0^3/b\omega^2$. The spectral evolution of an optoacoustic signal in a dissipative medium is therefore given by

$$\tilde{v}(\omega,z)=K(\omega)\tilde{f}(\omega)\exp(-b\omega^2 z/2\rho_0 c_0^3),$$

where $K(\omega)$ is the corresponding transfer function, or

$$v(z,\tau)=(2\pi)^{-1}\int_{-\infty}^{\infty}K(\omega)\tilde{f}(\omega)\exp\left(-i\omega\tau-\frac{b\omega^2 z}{2\rho_0 c_0^3}\right)d\omega. \tag{2.42}$$

Solution (2.42) of the parabolic equation reveals that dissipation serves to smooth out the sharp gradients of the profile. At large distances $z>2\rho_0 c_0^3 b^{-1}[\min(\tau_L^{-1},\alpha c_0)]^{-2}$ the acoustic signal spectrum is in fact only determined by the dissipative factor, and the optoacoustic signal profile is universal

$$v_r(z,\tau)=(\beta\mathscr{E}_0/\rho_0 c_p)(\rho_0 c_0^3/2\pi bz)^{1/2}\exp(-\rho_0 c_0^3\tau^2/2bz)$$

which is a Gaussian profile for a rigid boundary, while its derivative for a free boundary [in accordance with Eq. (2.28)]

$$v_f(z,\tau)=(\beta\mathscr{E}_0/\rho_0 c_p)(\rho_0 c_0^3/2\pi bz)^{1/2}(-\tau\rho_0 c_0^2/b\alpha z)$$
$$\times\exp(-\rho_0 c_0^3\tau^2/2bz).$$

The signal excited with a rigid boundary diminishes with distance as $z^{-1/2}$ and expands in proportion to $z^{1/2}$. An OA signal excited with a free boundary will diminish faster: as z^{-1}.

The wave profile becomes universal at distances $z>2\rho_0 c_0^3/(b\alpha c_0)^2$, where no detailed features of light absorption are yet manifested in the optoacoustic pulse. In practice sound dissipation begins to dominate in optoacoustic experiments only in the hypersonic frequency range.

The ratio of the dissipation length L_{DS} to the nonlinearity length L_{NL} is characterized by the Reynold's number Re of the excited waves:

$$\mathrm{Re} \sim \frac{L_{DS}}{L_{NL}} \sim \frac{2\rho_0 c_0^3/b\omega^2}{c_0^2/\omega\varepsilon v_a} = \frac{2\rho_0 c_0 \varepsilon}{b\omega} v_a,$$

where v_a is the amplitude of the particle (vibrational) velocity. The wave amplitude may be quite significant for optoacoustic signals (acoustic Mach numbers $v_a/c_0 \sim 10^{-2}$). As a rule their Reynold's number will also be large. Consequently, we can expect acoustic nonlinear effects to be manifested and if $L_{NL} \ll L_{DS}, L_{DF}$ then Eq. (2.40) becomes an equation of simple waves:

$$\frac{\partial v}{\partial z} - \left(\frac{\varepsilon v}{c_0^2}\right)\frac{\partial v}{\partial \tau} = 0. \tag{2.43}$$

This equation and its solution have been analyzed in detail in nonlinear acoustics.[11–13] Only those results that are important for optoacoustics will be provided here.

The solution of Eq. (2.43) can be given in implicit form:

$$v = v_0(\tau + \varepsilon z v/c_0^2), \tag{2.44}$$

where $v_0(\tau)$ is the particle velocity profile at the boundary $z=0$. In the case of a stage-by-stage analysis $v_0(\tau)$ is the OA-signal profile obtained from a solution of the sound excitation problem. Figure 2.10 shows the nonlinear transformation of profiles typical of thermooptical excitation. Solution (2.44) becomes ambiguous beginning at certain distances $z \sim (c_0^2/\varepsilon)(dv/d\tau)^{-1} = L_{NL}$. Physically this corresponds to the appearance of a shock front whose position is determined from the "equal area" rule.[11] An asymptotic OA-signal wave form in a nonlinear medium will therefore consist of the rectilinear segments of the profile and their interconnecting discontinuities.

The variation in the slope of the rectilinear section of the profile with distance is universal. Equation (2.43) has the solution

$$v = \tau(C - \varepsilon z/c_0^2)^{-1},$$

where C is an auxiliary constant. This equation is conveniently represented as

$$\frac{\partial \tau}{\partial v} = \left(\frac{\partial \tau}{\partial v}\right)\bigg|_{z=0} - \frac{\varepsilon z}{c_0^2}. \tag{2.45}$$

The slope of the profile varies linearly with distance. If the initial value $(\partial\tau/\partial v)|_0 > 0$ (pulse wave front), then $\partial\tau/\partial v$ diminishes and a shock front is formed. In the opposite case $(\partial\tau/\partial v)|_0 < 0$ (trailing edge of the pulse) the pulse will be distended and the rectilinear segment will be elongated (Fig. 2.10).

In the vicinity of the shock front we can no longer neglect the effect of dissipation, which stabilizes the width of the shock region. Burgers' equation

$$\frac{\partial v}{\partial z} - \frac{\varepsilon}{c_0^2} v \frac{\partial v}{\partial \tau} - \frac{b}{2\rho_0 c_0^3}\frac{\partial^2 v}{\partial \tau^2} = 0,$$

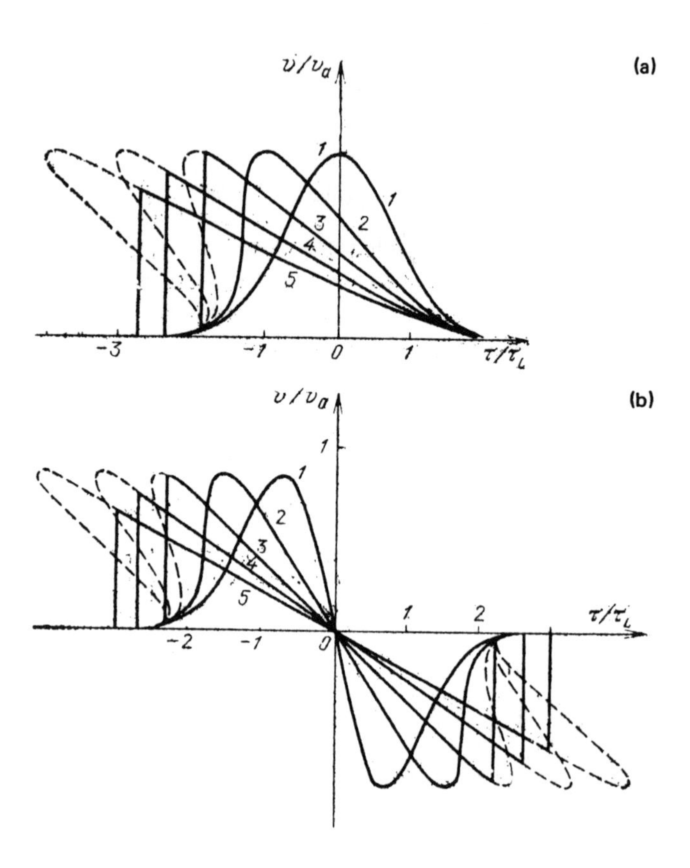

FIG. 2.10. Nonlinear transformation of OA-signal profiles for the case of propagation in an ideal medium for rigid (a) and free (b) boundaries. The parameter z/L_{NL} is equal to 0 (1), 1 (2), 2 (3), 3 (4), and 4 (5).

describing sound propagation in a nonlinear dissipative medium can be obtained from Eq. (2.40) for $L_{NL}, L_{DS} \ll L_{DF}$. This equation has an analytical solution.[11] The wave behavior is given by a single dimensionless parameter: the Reynold's number. If $Re \ll 1$, the dissipation length is much less than the nonlinearity length and no shock front forms and nonlinear effects are weakly manifested. In the opposite case, $Re \gg 1$, dissipation will have a significant effect only in the neighborhood of the shock wave front. Figure 2.11 shows the evolution of OA-signal profiles in a nonlinear dissipative medium; this evolution confirms these qualitative conclusions. A more detailed analysis can be found in Ref. 13.

Two additional dimensionless scaling parameters arise when diffraction is accounted for by analogy to the Reynold's number: dissipation/diffraction L_{DF}/L_{DS} and nonlinearity/diffraction L_{DF}/L_{NL} (these parameters have no special designations). We consider these parameters. The quantity

$$\frac{L_{DF}}{L_{DS}} = \frac{\omega a^2}{2c_0} \frac{b\omega^2}{2\rho_0 c_0^3} = \frac{ba^2\omega^3}{4\rho_0 c_0^4}$$

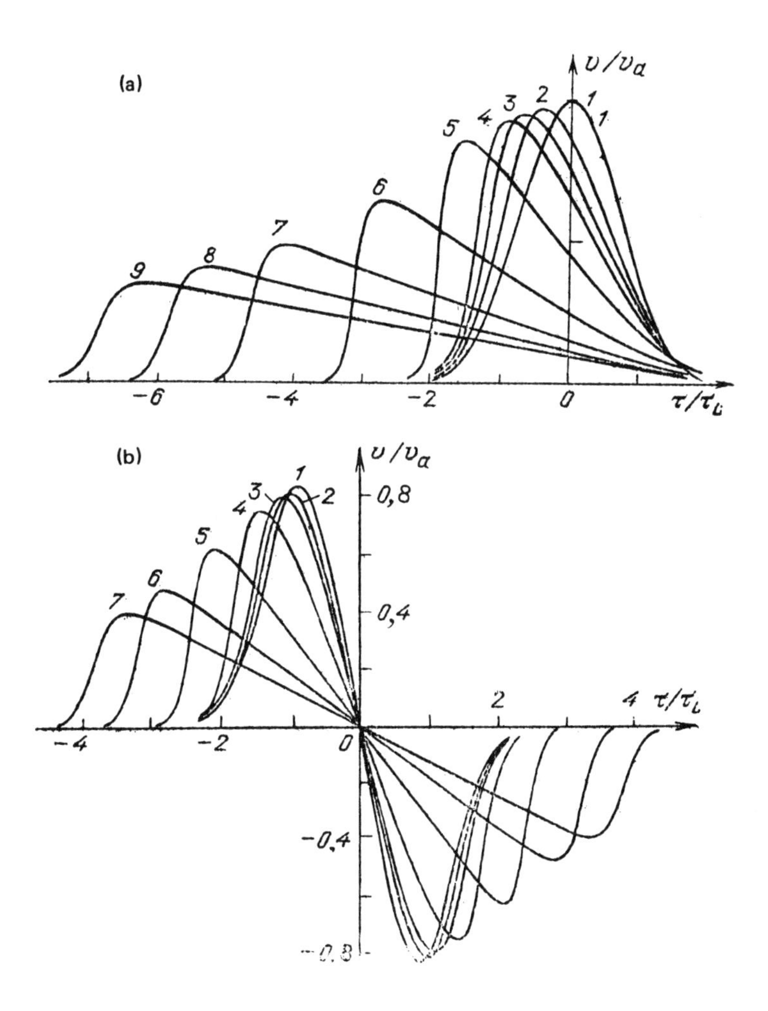

FIG. 2.11. Nonlinear transformation of OA-signal profiles for the case of propagation in a dissipative medium for rigid (a) and free (b) boundaries. The parameter z/L_{NL} is equal to 0 (1), 0.4 (2), 0.7 (3), 1 (4), 2 (5), 5 (6), 10 (7), 15 (8), 20 (9); the parameter Re = 10.

is proportional to the cube of the frequency, and in most cases diffraction will appear much earlier in optoacoustic experiments for $a \sim 0.1$ cm compared to dissipation across the entire ultrasonic frequency band. Therefore, the effect of dissipation will only be manifested at the shock wave front with an OA signal (if one is formed), while the primary competition is between nonlinear and diffraction effects.

The nonlinearity/diffraction parameter L_{DF}/L_{NL} for the OA signals may vary over a broad range. If $L_{DF}/L_{NL} = \varepsilon\omega^2 a^2 v_a/2c_0^3 \gg 1$, nonlinear distortions to the wave profile will originate as in a plane wave. If, on the other hand, $L_{DF}/L_{NL} \ll 1$ the nonlinear effects may become noticeable only in the wave far field. In the crossover zone the transformation of the pulse profile is given by the parabolic diffraction equation

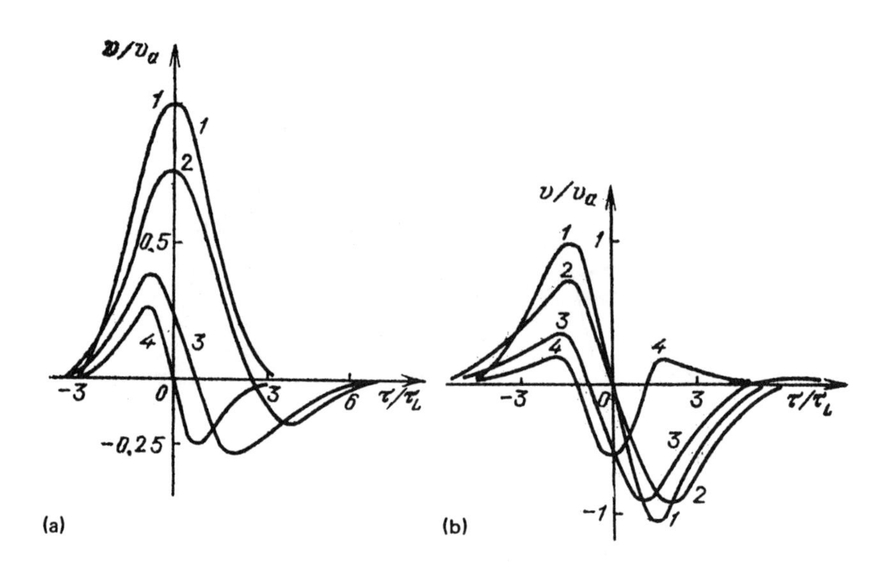

FIG. 2.12. OA-signal transformation in the crossover from the near field to the far field. The parameter z/L_{DF} is equal to 0 (1), 0.1 (2), 1 (3), 10 (4).

$$\frac{\partial}{\partial\tau}\left(\frac{\partial v}{\partial z}\right)=\frac{c_0}{2}\,\Delta_\perp v.$$

We consider the diffraction change in the OA signal spectrum. For definiteness assume the laser beam has a Gaussian intensity distribution in the cross section:

$$I(t,\mathbf{r}_\perp)=I_0f(t)H(\mathbf{r}_\perp)=I_0f(t)\exp[-(x^2+y^2)/a^2]. \tag{2.46}$$

Here $\mathbf{r}_\perp=\{x,y\}$ is the radius vector perpendicular to the laser beam; a is the beam radius. Then, the OA-signal spectrum can be represented as

$$\begin{aligned}\tilde{v}(\omega,z,\mathbf{r}_\perp)&=\pi a^2K(\omega)I_0\tilde{f}(\omega)(2\pi)^{-2}\iint_{-\infty}^{\infty}\exp\left[i(\mathbf{k}_\perp\mathbf{r}_\perp)-i\frac{k_\perp^2z}{2\omega/c_0}-\frac{k_\perp^2a^2}{4}\right]d\mathbf{k}_\perp\\&=K(\omega)I_0\tilde{f}(\omega)\left(1+i\frac{z}{L_{\mathrm{DF}}}\right)^{-1}\exp\left\{-r_\perp^2\left[a^2\left(1+i\frac{z}{L_{\mathrm{DF}}}\right)\right]^{-1}\right\},\end{aligned} \tag{2.47}$$

where $L_{\mathrm{DF}}=\omega a^2/2c_0$ is the diffraction length of frequency ω. The transverse distribution of each of the harmonics continues to be a Gaussian distribution. Field behavior on the axis $\mathbf{r}_\perp=0$ is of the most significant interest. In this case, the spectrum of the diffraction wave is the initial spectrum multiplied by the diffraction factor.

Figure 2.12 shows the change in pulse wave form as it travels from the near wave field to the far wave field for rigid (a) and free (b) boundaries. Equation (2.47) can be significantly simplified at large distances $z\gg\alpha a^2$:

$$(1+iz/L_{\mathrm{DF}})^{-1}=(1+i2c_0z/\omega a^2)^{-1}\approx-i\omega a^2/2c_0z.$$

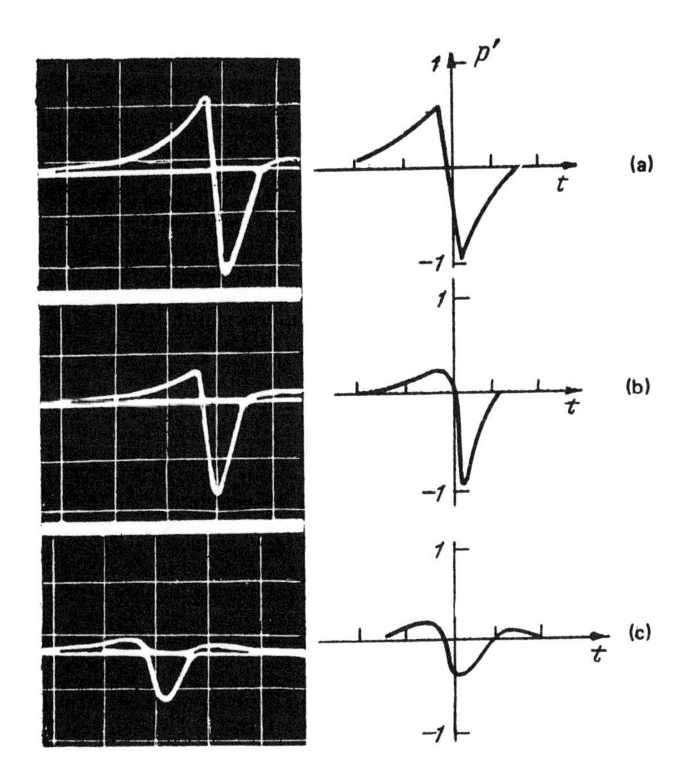

FIG. 2.13. OA-signal profile for a free boundary at various distances z from the boundary of 2 mm (a), 7 mm (b), and 70 mm (c) (left). Calculation of the wave form at these distances (right).

Hence axially

$$\tilde{v} = -i\omega K(\omega) I_0 \tilde{f}(\omega)(a^2/2c_0 z),$$

and, consequently, the wave profile in the wave zone will be proportional to the derivative of the profile at the boundary, i.e., the derivative of the solution of the one-dimensional problem. With a rigid boundary the compressional pulse becomes a bipolar pulse as it propagates, and with a free boundary the bipolar pulse becomes a triphase pulse with two compressional phases and a singular rarefactional phase.

Figure 2.13 shows the experimental[14] and theoretical profiles of an OA signal excited for the case of a free boundary in different wave propagation zones. The experimental data are confirmed by calculation (2.47).

A stage-by-stage approach is valid under the conditions $\alpha L_{NL}, \alpha L_{DF} \gg 1$. However, these conditions may be violated depending on the geometry of the heating zone and the exposure intensity. Sound excitation will be examined in Sec. 2.4 for the case where the condition $\alpha L_{DF} \gg 1$ is violated and we now consider the condition $\alpha L_{NL} \sim 1$. Since the acoustic nonlinearity parameter of homogeneous media lies in the range $\varepsilon = 2\text{–}14$, acoustic nonlinear effects in the generation zone may be manifest only at rather high Mach numbers of the excited wave $\varepsilon v_a / c_0 \lesssim 1$, which is achieved at the evaporation threshold of the media (see Table 1.2).

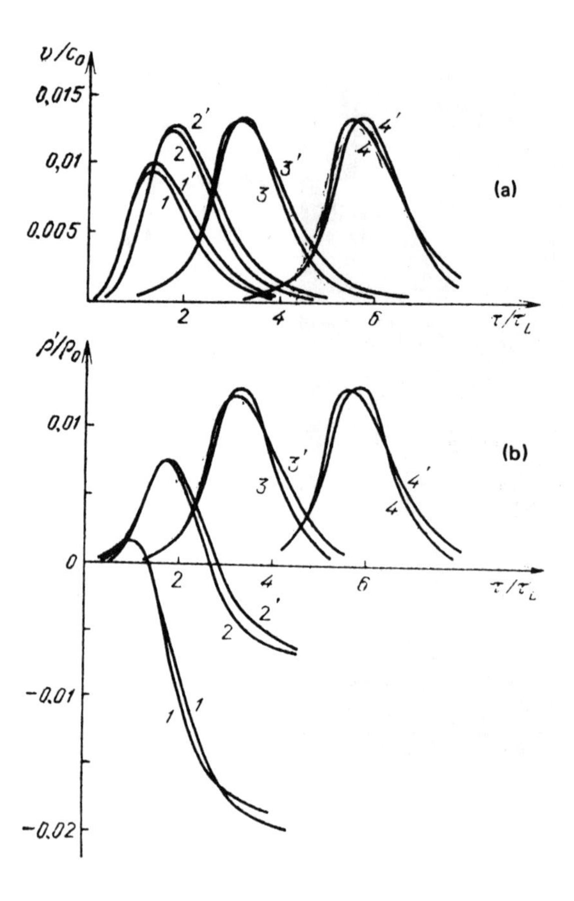

FIG. 2.14. Effect of an acoustic nonlinearity on the particle velocity (a) and density (b) profiles for the case of a rigid boundary ($\alpha c_0 \tau_L = 2$). The profiles are shown at distances αz of 1 (1), 3 (2), 5 (3), and 10 (4). A solution of the linear problem (curves 1′–4′) is given for comparison.

References 15 and 16 derived a numerical solution to the problem of thermo-optical excitation of sound accounting for the finite nature of the amplitude of the excited wave. Figure 2.14 shows the time dependences of the vibrational velocity and density in the heating zone accounting for and neglecting the nonlinearity. It is clear that there is only a minor difference in the waveform of the optoacoustic signal. Moreover, it is possible to describe the excitation and nonlinear evolution stage by means of a model inhomogeneous Burgers' equation

$$\frac{1}{c_0}\frac{\partial v}{\partial t} + \frac{\partial v}{\partial z} - \frac{\varepsilon}{c_0^2} v \frac{\partial v}{\partial t} - \frac{b}{2\rho_0 c_0^3}\frac{\partial^2 v}{\partial t^2} = -\frac{\beta I_0}{2\rho_0 c_p} f(t)\bar{g}(z),$$

where the source function $\bar{g}(z)$ [see Eq. (2.39)] is continued into the domain $z < 0$ evenly or oddly for the rigid and free boundaries, respectively. Figure 2.15 gives the OA-signal profiles obtained by numerical solution of complete, model, and linearized problems. The model inhomogeneous Burgers' equation provides a sufficiently accurate approximation.

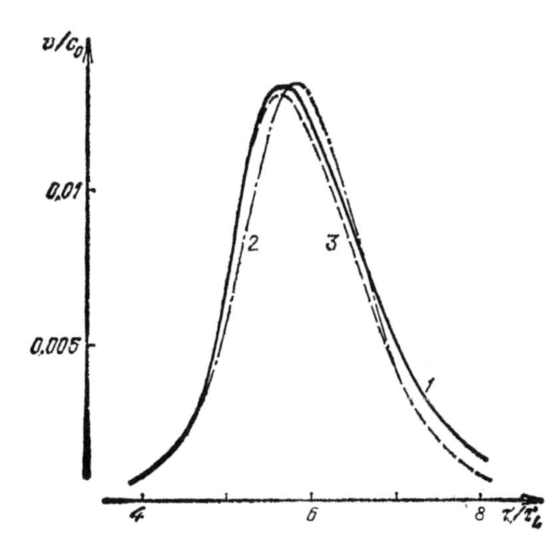

FIG. 2.15. OA-signal profiles obtained by solving complete (1), linearized (2), and model (3) problems for a rigid boundary.

2.4. Directional Patterns of Thermooptical Sound Sources

The finite size of the laser beam generally plays a significant role in thermooptical excitation of sound. This is due to both the broadband nature of the sources and therefore the fact that the spectrum contains rather low-frequency harmonics, as well as the significant distance traveled by the wave in the medium. The heating zone may have different geometries and wave diffraction may be significant as early as the wave excitation stage (see Refs. 17–19 and references cited therein). We return to Eq. (2.13). Let light distribution in the beam cross section be described by the function $H(x,y) = H(\mathbf{r}_\perp)$, while the depthwise distribution of the thermal sources is given by the function

$$g(z) = -\frac{d}{dz}\left[\exp\left(-\int_0^z \alpha(\xi)d\xi\right)\right].$$

Then

$$\operatorname{div}\langle\mathbf{S}\rangle = -I_0 f(t)H(\mathbf{r}_\perp)g(z) \tag{2.48}$$

and Eq. (2.13) takes the form

$$\frac{\partial^2\varphi}{\partial t^2} - c_0^2\,\Delta\varphi = -\frac{c_0^2\beta}{\rho_0 c_p} I_0 f(t)H(\mathbf{r}_\perp)g(z). \tag{2.49}$$

We solve this equation by the spectral method. Utilizing Eqs. (2.16) and (2.18) we obtain the inhomogeneous Helmholtz equation

$$\Delta\widetilde{\varphi} + \frac{\omega^2}{c_0^2}\widetilde{\varphi} = \frac{\beta I_0}{\rho_0 c_p} H(\mathbf{r}_\perp)g(z)\widetilde{f}(\omega). \tag{2.50}$$

Depending on the acoustic impedance ratio of the transparent and absorbing media this equation is supplemented by the boundary condition

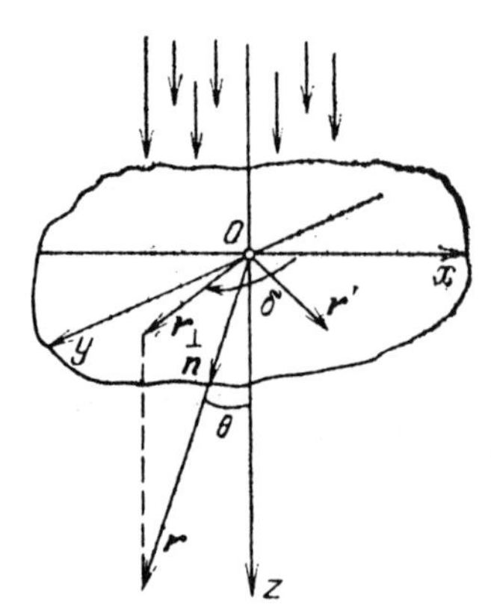

FIG. 2.16. Coordinate system used to analyze OA-signal excitation in the three-dimensional case.

$$\left(\frac{\partial\varphi}{\partial z}\right)\bigg|_0 = 0 \quad \text{(rigid boundary)}$$

or

$$\varphi|_0 = 0 \quad \text{(free boundary)}.$$

The Green's function of Helmholtz equation (2.50) in an unbounded medium takes the form

$$G(\mathbf{r},\mathbf{r}') = (4\pi)^{-1}\exp[i(\omega/c_0)|\mathbf{r}-\mathbf{r}'|]\,|\mathbf{r}-\mathbf{r}'|^{-1} \qquad (2.51)$$

(the sign on the exponent is based on the emission condition). Vectors $\mathbf{r}$ and $\mathbf{r}'$ represent the radii vectors of the observation point and the sources (Fig. 2.16), θ is the angle of arrival of the wave at the observation point. However, the solution of Eq. (2.50) with Green's function (2.51) will not satisfy the boundary conditions; it must be changed for this purpose;

$$\tilde{\varphi} = \frac{\beta I_0\tilde{f}(\omega)}{4\pi\rho_0 c_p}\int_0^\infty dz' \iint_{-\infty}^{\infty} dx'\,dy'\,H(x',y')g(z')$$
$$\times\left\{\frac{\exp\{i(\omega/c_0)[(x-x')^2+(y-y')^2+(z-z')^2]^{1/2}\}}{[(x-x')^2+(y-y')^2+(z-z')^2]^{1/2}}\right.$$
$$\left.\mp\frac{\exp\{i(\omega/c_0)[(x-x')^2+(y-y')^2+(z+z')^2]^{1/2}\}}{[(x-x')^2+(y-y')^2+(z+z')^2]^{1/2}}\right\}.$$

A minus sign between the two ratios is used in the braces in the case of a free boundary, while a plus sign is used in the case of a rigid boundary. The same result is achieved if sound excitation in an infinite medium is analyzed from the purely formal viewpoint, while the sources $g(z)$ are continued to the domain $z<0$ either evenly $[\bar{g}(z)=\bar{g}(-z)]$ for a rigid boundary or oddly $[\bar{g}(z) = -\bar{g}(-z)]$ for a free boundary. Denoting the corresponding continuation of the source function by $\bar{g}(z)$, we write the solution as

$$\tilde{\varphi}(\omega,r) = \frac{\beta I_0 \tilde{f}(\omega)}{4\pi\rho_0 c_p} \iiint_{-\infty}^{\infty} dx'\, dy'\, dz'\, H(x',y')\bar{g}(z')$$
$$\times \frac{\exp\{i(\omega/c_0)[(x-x')^2+(y-y')^2+(z-z')^2]^{1/2}\}}{[(x-x')^2+(y-y')^2+(z-z')^2]^{1/2}}. \tag{2.52}$$

This equation represents an expansion of the acoustic field in divergent spherical waves.

As in the case of plane waves, complex conditions of standing and traveling acoustic waves occur in the region where the thermal sources have an effect and the wave will become a purely traveling wave only outside the heating zone. We will be interested in the spectrum of this wave. We therefore set $|\mathbf{r}'| \ll |\mathbf{r}|$ in integral (2.52) and, as is usually the case, represent

$$G(\mathbf{r},\mathbf{r}) \approx (4\pi\mathbf{r}|)^{-1} \exp\left(i\frac{\omega}{c_0}|\mathbf{r}|\right)\exp\left[-i\frac{\omega}{c_0}\frac{(\mathbf{r}\mathbf{r}')}{|\mathbf{r}|}\right].$$

Taking this into account, Eq. (2.52) is written as

$$\tilde{\varphi}(\omega,\mathbf{r}) = (\beta I_0 \tilde{f}(\omega)/4\pi\rho_0 c_p)\exp[i(\omega/c_0|\mathbf{r}|)]|\mathbf{r}|^{-1}$$
$$\times \int\int_{-\infty}^{\infty}\int H(\mathbf{r}_\perp')\bar{g}(z')\exp\left[-i\left(\frac{\omega}{c_0}\right)(\mathbf{n}\mathbf{r}')\right]d\mathbf{r}_\perp'\, dz'. \tag{2.53}$$

Here $\mathbf{n} = \mathbf{r}/|\mathbf{r}|$ is the unit vector in the direction of the observation point. As we see from solution (2.53), the OA-signal spectrum is proportional to the light intensity spectrum $\tilde{f}(\omega)$. In this case the transfer function is no longer dependent solely on frequency, as in the one-dimensional case, but also on the direction of observation $\mathbf{n}$. This relation provides the directional pattern of the thermooptical sound sources. As before, the transfer function is the Fourier spectrum of the spatial distribution of the sources $\bar{g}(z)H(\mathbf{r}_\perp)$ (continued to $z<0$). In other words, the efficiency of sound excitation is proportional to the spectral component in the direction of observation $\mathbf{n}$ of the expansion of the sources in plane waves:

$$\tilde{\varphi}(\omega,\mathbf{r}) = |\mathbf{r}|^{-1}\exp[i(\omega/c_0)|\mathbf{r}|]K_\varphi(\omega,\mathbf{n})I_0\tilde{f}(\omega).$$

The expression for the transfer function K_φ is determined by comparison to Eq. (2.53):

$$K_\varphi(\omega,\mathbf{n}) = \frac{\beta}{4\pi\rho_0 c_p}\iiint_{-\infty}^{\infty} H(\mathbf{r}_\perp')\bar{g}(z')\exp\left[-i\frac{\omega}{c_0}(\mathbf{n}\mathbf{r}')\right]d\mathbf{r}_\perp'\, dz'.$$

However, the transfer function of the velocity potential φ should not be used, but rather the transfer function of the vibrational velocity or pressure should be used. Since the radial component of the velocity in the wave zone is related to the pressure increment in the same manner as in a plane wave, either of these quantities can be considered. Hence, as previously, we find the transfer function for the vibrational velocity. Since $\boldsymbol{v} = \nabla\varphi$, the principal velocity component in Eq. (2.53) will be the radial component $v_r = \partial\varphi/\partial r$. Since $\mathbf{kr} = \omega r/c_0 \gg 1$ in the wave zone, then

$$\tilde{v}(\omega,\mathbf{r}) = \frac{\beta I_0}{4\pi\rho_0 c_p}|\mathbf{r}|^{-1}\exp\left[i\frac{\omega}{c_0}|\mathbf{r}|\right]i\frac{\omega}{c_0}\tilde{f}(\omega)$$

$$\times\iint_{-\infty}^{\infty} H(\mathbf{r}'_\perp)\exp\left[-i\frac{\omega}{c_0}(\mathbf{n}_\perp\mathbf{r}'_\perp)\right]d\mathbf{r}'_\perp$$

$$\times\int_{-\infty}^{\infty}\bar{g}(z')\exp\left[i\frac{\omega}{c_0}z'\cos\theta\right]dz$$

$$= (2\pi)^{-1}|\mathbf{r}|^{-1}\exp\left[i\frac{\omega}{c_0}|\mathbf{r}|\right]K(\omega)I_0\tilde{f}(\omega). \tag{2.54}$$

The transfer function of the three-dimensional source distribution will therefore take the form

$$K(\omega) = \frac{\beta}{2\rho_0 c_p}i\frac{\omega}{c_0}\iint_{-\infty}^{\infty} H(\mathbf{r}'_\perp)\exp\left[-i\frac{\omega}{c_0}(\mathbf{n}_\perp\mathbf{r}'_\perp)\right]d\mathbf{r}'_\perp$$

$$\times\int_{-\infty}^{\infty}\bar{g}(z')\exp\left[-i\frac{\omega}{c_0}z'\cos\theta\right]dz'. \tag{2.55}$$

The physical meaning of each of the cofactors is rather evident. The last co-factor in Eq. (2.55) is the transfer function obtained by solving one-dimensional problem (2.39). We denote this function by

$$K_\parallel\left(\frac{\omega\cos\theta}{c_0}\right) = \left(\frac{\beta}{2\rho_0 c_p}\right)\int_{-\infty}^{\infty}\bar{g}(z')\exp\left[-i\left(\frac{\omega}{c_0}\right)z'\cos\theta\right]dz'. \tag{2.56}$$

Unlike the plane geometry case, Eq. (2.56) does not include the complete wave vector ω/c_0, but rather only its axial component. The other co-factor in transfer function (2.55) is determined from the intensity distribution in the beam cross section:

$$\iint_{-\infty}^{\infty} H(\mathbf{r}'_\perp)\exp\left[-i\frac{\omega}{c_0}(\mathbf{n}_\perp\mathbf{r}'_\perp)\right]d\mathbf{r}'_\perp = \tilde{H}\left(\frac{\omega}{c_0}\mathbf{n}_\perp\right), \tag{2.57}$$

which represents the spatial Fourier spectrum of this distribution. Finally, the multiplier $i\omega/c_0$ is related to wave diffraction and is manifested in the transition to the far field (see Sec. 2.3).

Therefore, the effect of the finite size of the laser beam is manifested as an additional spectral factor in the transfer function which is determined by diffraction and the transverse source distribution,

$$K(\omega,\mathbf{n}) = i\frac{\omega}{c_0}K_\parallel\left(\frac{\omega}{c_0}\cos\theta\right)\tilde{H}\left(\frac{\omega}{c_0}\mathbf{n}_\perp\right).$$

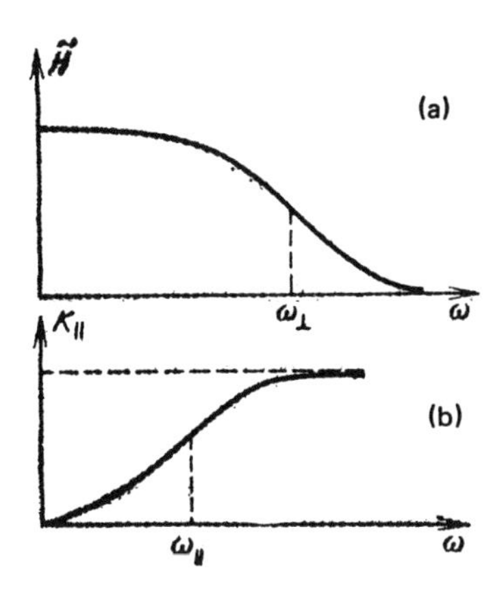

FIG. 2.17. Angular spectrum (a) and longitudinal component of transfer function (b).

The aperture factor $\widetilde{H}\ (\omega \mathbf{n}_\perp / c_0)$ imposes an upper limit on the frequency range of the excited waves of

$$\omega_\perp = 2c_0/(a \sin \theta),$$

while diffraction and the transfer function impose a lower limit of

$$\omega_\parallel = \alpha c_0 / \cos \theta = \omega_a / \cos \theta$$

(Fig. 2.17). Hence, sound excitation will be most efficient in the frequency band

$$\omega_\parallel = \alpha c_0/\cos \theta < \omega < 2c_0/a \sin \theta = \omega_\perp.$$

It follows that the thermooptical sources efficiently radiate sound within a cone with an apex angle θ:

$$\tan \theta < 2(\alpha a)^{-1}. \tag{2.58}$$

Therefore, under strong absorption of light ($\alpha a \gg 1$) the acoustic field will be concentrated at the beam axis (the heating zone appears as a disk). In the case of weak absorption ($\alpha a \lesssim 1$) a broad radiation directional pattern is obtained (Fig. 2.18).

Diagrams of the OA-excitation efficiency lines in the wave vector plane (with the wave frequency plotted radially and the direction identical to that of the vector $\mathbf{n}$)

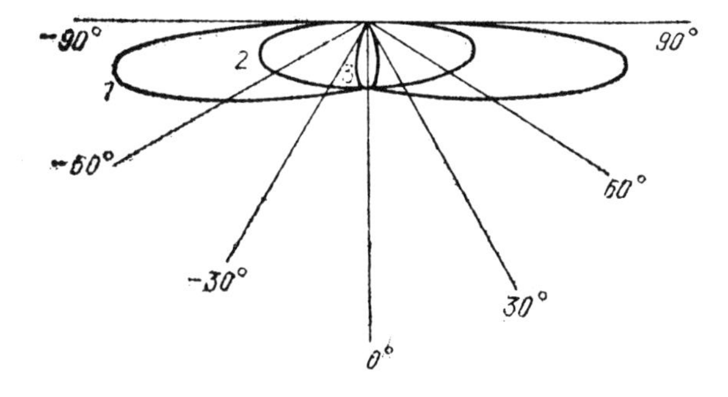

FIG. 2.18. Directional pattern of OA-signal radiation for different source geometries. The parameter αa is equal to 0.03 (1), 0.3 (2), and 3 (3).

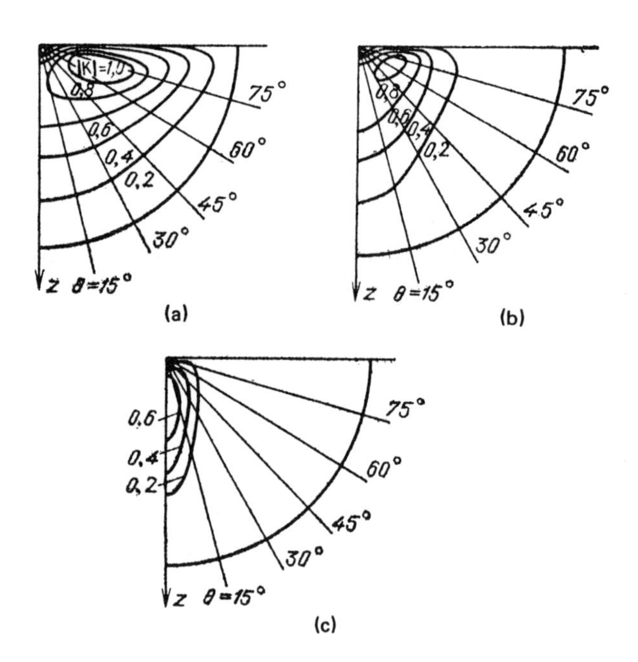

FIG. 2.19. OA-excitation efficiency levels for different source geometries. The parameter αa is equal to 0.15 (a), 0.45 (b), and 1.2 (c).

provide a convenient representation of the acoustic field distribution. These diagrams are shown in Fig. 2.19 for different values of αa.

In the general case, the OA-signal spectrum and wave form will be determined by the ratio of three frequencies: $\omega_{\parallel}$, $\omega_{\perp}$, and τ_L^{-1} [the latter is given by the spectral range of light intensity $\tilde{f}(\omega)$]. However, since efficient sound excitation is only possible for $\omega_{\parallel} < \omega_{\perp}$, we need only consider three cases:

$$\begin{cases} \textbf{1.}\ \omega_{\parallel}\tau_L > 1, \\ \textbf{2.}\ \omega_{\parallel}\tau_L < 1 < \omega_{\perp}\tau_L, \\ \textbf{3.}\ \omega_{\perp}\tau_L < 1. \end{cases}$$

1. The first case $\alpha c_0 \tau_L / \cos\theta > 1$ corresponds to long laser pulses or acoustic radiation along the interface surface $z = 0$ for $\alpha a \gg 1$. In this case, the spectral range of light intensity $\tilde{f}(\omega)$ is narrower than the ranges of the longitudinal (2.56) and transverse (2.57) components of the transfer function. Vibrational velocity spectrum (2.54) is given by

$$\tilde{v}_r(\omega,\mathbf{r}) = -i\frac{\omega}{c_0}\frac{\exp[i(\omega/c_0)|\mathbf{r}|]}{2\pi|\mathbf{r}|}\frac{\beta W_0}{\rho_0 c_p}\tilde{f}(\omega)$$

for a rigid boundary and

$$\tilde{v}_f(\omega,\mathbf{r}) = (-i\omega/\alpha c_0)\cos\theta\,\tilde{v}_r(\omega,\mathbf{r})$$

for a free boundary. The spectral density of the OA-signal in the low-frequency range is not determined by the intensity but rather by the power of light W_0. In this case, the temporal profile of the acoustic pulse will be the first derivative of the laser pulse (for a rigid boundary)

$$v_r\left(\mathbf{r},\tau=t-\frac{|\mathbf{r}|}{c_0}\right)=\frac{\beta W_0}{\rho_0 c_p}\frac{f'(\tau)}{2\pi c_0|\mathbf{r}|}$$

or its second derivative (for a free boundary)

$$v_f(\mathbf{r},\tau)=\frac{\beta W_0}{\rho_0 c_p c_0}\frac{\cos\theta f''(\tau)}{2\pi\xi c_0|\mathbf{r}|}. \tag{2.59}$$

In the case of a rigid boundary the thermooptical sources are monopole sources and in the case of a free boundary they are dipole sources (with the dipole oriented normal to the boundary).

2. The case $\omega_{\|}\tau_L<1<\omega_{\perp}\tau_L$ corresponds to sound excitation by a comparatively "thin" laser beam $a<2c_0\tau_L/\sin\theta$ with a moderate absorption coefficient $\alpha<\cos\theta/c_0\tau_L$. An analogous relation will hold when sound is radiated along the z axis. In this case, as in the preceding case, the intensity distribution in the beam cross section will have no effect on the OA-signal spectrum:

$$\widetilde{H}(\omega\sin\theta/c_0)\approx\pi a^2=\text{const.}$$

Therefore, the spectrum is determined solely by the longitudinal component of the transfer function:

$$\widetilde{v}(\omega,\mathbf{r})=\frac{\exp[i\omega|\mathbf{r}|/c_0]}{2\pi|\mathbf{r}|}\pi a^2K_{\|}\left(\frac{\omega}{c_0}\cos\theta\right)I_0\widetilde{f}(\omega)\left(i\frac{\omega}{c_0}\right).$$

Since $K_{\|}$ is the transfer function of the one-dimensional problem, in this case the pulse wave form will be the derivative of the pulse from the plane problem (see Sec. 2.2) compressed by a factor of $(\cos\theta)^{-1}$. The pulse will be a bipolar pulse for a rigid boundary and a tripolar pulse for a free boundary.

We consider in greater detail axial ($\mathbf{n}_{\perp}=0$) sound emission for the case of homogeneous absorption: $\alpha=\text{const}$ and $g(z)=\alpha\exp(-\alpha z)$. Here the "thin" beam condition $a<2c_0\tau_L/\sin\theta$ holds. Moreover, the solution of the one-dimensional problem will, due to the condition $\alpha c_0\tau_L<\cos\theta$, consist of two exponents $\exp(\mp\alpha c_0\tau)$, and the crossover region between these exponents with a duration of the order of τ_L (Fig. 2.4). The wave profile in the far field will therefore be the derivative of the solution of the plane problem v_0:

$$v\left(r,\tau=t-\frac{|\mathbf{r}|}{c_0}\right)=\frac{1}{2\pi|\mathbf{r}|}\frac{\beta W_0}{\rho_0 c_p c_0}\frac{d}{d\tau}v_0(\tau) \tag{2.60}$$

(with the same boundary type). Hence, by virtue of (2.28) for the rigid boundary case, the pulse wave form in the far field will correspond to the solution of the plane problem with a free boundary. However, with a free boundary the pulse in the far field will consist of two exponential compressional phases and a rarefactional phase: an "inverted" laser pulse (Fig. 2.20) where the amplitude of the rarefactional phase far exceeds the amplitude of the compressional phases [by a factor of $(\alpha c_0\tau_L)^{-1}$].

discussion pertains to f(t) and V_o given on p. 21.

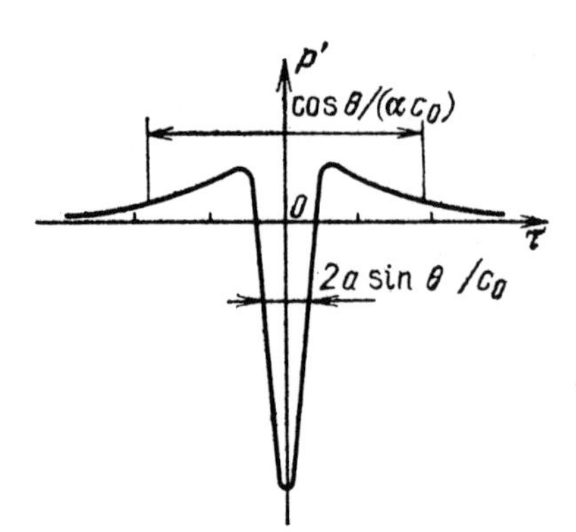

FIG. 2.20. OA-signal profile in the wave far field for a free boundary. The parameter $\alpha c_0 \tau_L = 0.3$.

3. Finally, if the pulse is sufficiently short

$$\alpha c_0 \tau_L / \cos\theta < 2 c_0 \tau_L / (a \sin\theta) < 1$$

(and the direction of observation is not too close to the axis), the OA-signal wave form will be determined solely by the thermooptical source distribution throughout the medium volume:

$$\tilde{v}(\omega,\mathbf{r}) = \frac{\beta \mathscr{E}_0}{\rho_0 c_p} \frac{\exp[i(\omega/c_0)|\mathbf{r}|]}{4\pi|\mathbf{r}|} i \frac{\omega}{c_0} \iint_{-\infty}^{\infty} H(\mathbf{r}_\perp') \exp\left[-i\frac{\omega}{c_0}(\mathbf{n}_\perp \mathbf{r}_\perp')\right] d\mathbf{r}_1' \times \int_{-\infty}^{\infty} \bar{g}(z') \exp\left(\frac{\omega}{c_0} z' \cos\theta\right) dz'. \quad (2.61)$$

The amplitude is proportional to the absorbed energy density $\alpha \mathscr{E}_0$.

We consider a Gaussian intensity distribution in the beam cross section

$$H(\mathbf{r}_\perp) = \exp(-\mathbf{r}_\perp^2/a^2)$$

and a homogeneously absorbing medium ($\alpha = \text{const}$). Then we write the solution of Eq. (2.61) as

$$\tilde{v}(\omega,\mathbf{r}) = \frac{\beta \mathscr{E}}{\rho_0 c_p} \frac{\exp[i(\omega/c_0)|\mathbf{r}|]}{2\pi|\mathbf{r}|} \left(-i\frac{\omega}{c_0}\right) \exp\left(-\frac{\omega^2 a^2}{4 c_0^2} \sin^2\theta\right) \times \begin{cases} \dfrac{\alpha^2}{(\alpha^2 + \omega^2 \cos^2\theta / c_0^2)} & \text{(rigid boundary)}, \\ \dfrac{-i(\alpha\omega/c_0)\cos\theta}{\alpha^2 + \omega^2\cos^2\theta/c_0^2} & \text{(free boundary)}. \end{cases} \quad (2.62)$$

Here $\mathscr{E} = \pi a^2 \mathscr{E}_0$ is the laser pulse energy. The last cofactor in Eq. (2.62) is an ordinary transfer function in the plane geometry [see Eqs. (2.24) and (2.25)], with an absorption coefficient $(\cos\theta)^{-1}$ times larger:

$$(\alpha/\cos\theta)^2[(\alpha/\cos\theta)^2 + (\omega/c_0)^2]^{-1} \quad \text{(rigid boundary)},$$

$$-i(\alpha/\cos\theta)(\omega/c_0)[(\alpha/\cos\theta)^2+(\omega/c_0)^2]^{-1} \quad \text{(free boundary)}.$$

Hence the angular spectrum $\widetilde{H}$ $(\omega/c_0 \sin\theta)$ in (2.62) can be treated as the temporal spectrum of an effective Gaussian light pulse

$$f(t)=\exp[-(c_0 t/a \sin\theta)^2]$$

of duration $\tau_L^*=a\sin\theta/c_0$. The diffraction factor $-i\omega/c_0$, as usual, yields time differentiation of the OA-signal wave form. Therefore, in the case of a rigid boundary, the acoustic pulse profile in the far field will be identical to the acoustic signal profile with a free boundary (see Fig. 2.4), where $\alpha a \tan\theta$ is used as the parameter $\alpha c_0 \tau_L$; in the case of a free boundary, it coincides with its derivative (see Fig. 2.20).

In summary, we can state that the OA-signal profile in the far field has a bipolar wave form in the case of a rigid boundary, and contains two compressional phases separated by a rarefactional phase in the case of a free boundary. An antenna effectively "radiates" through a cone with an angle $\theta \sim \arctan[(\alpha a)^{-1}]$ (sound excitation becomes inefficient for $\alpha a \tan\theta > 1$).

Supplementary information can be found in Refs. 20–25. Reference 17 provides a detailed survey of research on laser pulse-initiated thermooptical sound excitation in homogeneous and inhomogeneous media and on the effect of surface wave in liquid on the directional pattern of the acoustic radiation.

REFERENCES

1. V. Novatski, *Teoriya uprugosti* [*Elasticity Theory*] (Mir, Moscow, 1975).
2. L. D. Landau and E. M. Lifshitz, *Statisticheskaya fizika* [*Statistical Physics*] (Nauka, Moscow, 1990).
3. L. V. Burmistrova, A. A. Karabutov *et al.*, Akust. Zh. **24**, 655 (1978) [Sov. Phys. Acoust. (1978)].
4. M. A. Isakovich, *Obshchaya akustika* [*General Acoustics*] (Nauka, Moscow, 1973).
5. E. F. Carome, N. A. Clark, and C. E. Moeller, Appl. Phys. Lett. **4**, 95 (1964).
6. A. A. Karabutov, A. I. Portnyagin *et al.*, Pis'ma Zh. Tekh. Fiz. **5**, 328 (1979) [Sov. Tech. Phys. Lett. **5**, 131 (1979)].
7. A. A. Karabutov, N. N. Omel'chuk *et al.*, Vestn. Moscow State Univ. Phys. Astron. Ser. **26**, 62 (1985).
8. A. A. Karabutov and O. B. Ovchinnikov, *Sudostroitel'naya promyshlennost'* [*Shipbuilding Industry*], Acoustics Series ("RUMB," Leningrad, 1987), No. 2, p. 93.
9. S. A. Akhmanov and O. V. Rudenko, Pis'ma Zh. Tekh. Fiz. **1**, 725 (1975) [Sov. Tech. Phys. Lett. **1**, 318 (1975)].
10. N. S. Bakhvalov, Ya. M. Zhileykin, and E. A. Zabolotskaya, *Nelineynaya teoriya zvukovykh puchkov* [*Nonlinear Theory of Acoustic Beams*] (Nauka, Moscow, 1982).
11. O. V. Rudenko and S. I. Soluyan, *Teoreticheskie osnovy nelineynoy akustiki* [*Theoretical Foundations of Nonlinear Acoustics*] (Nauka, Moscow, 1975).
12. J. Wizen, *Linear and Nonlinear Waves* (Mir, Moscow, 1977).
13. O. A. Vasil'eva, A. A. Karabutov *et al.*, *Vzaimodeystvie odnomernykh voln v sredakh bez dispersii* [*Interaction of One-Dimensional Waves in Dispersionless Media*] (MSU, Moscow, 1983).
14. A. A. Karabutov, A. I. Portnygin *et al.*, Akust. Zh. **26**, 296 (1980) [Sov. Phys. Acoust. **26**, 162 (1980)].
15. A. A. Karabutov, E. A. Lapshin *et al.*, *Tr. IX Vses. akust. konf.* [*Conf. Proceedings of the Ninth All-Union Acoustical Conference*] (Nauka, Moscow, 1977), p. 25.
16. A. A. Karabutov, E. A. Lapshin *et al.*, *Vyshislitel'nye metody i programmirovanie* [*Computational Methods and Programming*] (Moscow State University, Moscow, 1979), Vol. 31, p. 174.

17. L. M. Lyamshev and L. V. Sedov, Akust. Zh. **27**, 5 (1981) [Sov. Phys. Acoust. **27**, 4 (1981)].
18. B. K. Novikov, O. B. Rudenko, and V. I. Timoshenko, *Nelineynaya gidroakustika* [*Nonlinear Hydroacoustics*] (Sudostroenie, Leningrad, 1981).
19. L. M. Lyamshev, Usp. Fiz. Nauk **135**, 634 (1981) [Sov. Phys. Usp. 24, 977 (1981)].
20. A. A. Karabutov, O. V. Rudenko, and E. B. Cherepetskaya, Akust. Zh. **25**, 383 (1979) [Sov. Phys. Acoust. **25**, 218 (1979)].
21. A. I. Bozhkov and F. V. Bunkin, Kvantovaya Elektron. **2**, 1763 (1975) [Sov. J. Quantum Electron. (1975)].
22. L. M. Lyamshev and L. V. Sedov, Akust. Zh. **23**, 91 (1977) [Sov. Phys. Acoust. **23**, 49 (1977)].
23. L. M. Lyamshev and L. V. Sedov, Akust. Zh. **23**, 788 (1977) [Sov. Phys. Acoust. **23**, 450 (1977)].
24. S. G. Kasoev and L. M. Lyamshev, Akust. Zh. **24**, 1978 (1978) [Sov. Phys. Acoust. (1978)].
25. K. A. Naugol'nykh, in *Nelineynye volny* [*Nonlinear Waves*], edited by A. V. Gapnovov (Nauka, Moscow, 1979), p. 324.

Chapter 3

Thermooptical Excitation of Sound in an Isotropic Solid

Thermooptical excitation of sound in liquids was examined in Chap. 1. The situation becomes more complex in solids due to the fact that both transverse and Rayleigh waves may be excited in addition to longitudinal waves. The basic equations must therefore be derived once again. The effect of thermal conduction, we recall, in liquids (with the possible exception of mercury) is manifested only at low frequencies, while it can be ignored in the ultrasonic range. The situation may be fundamentally different for solids and hence the thermal conduction must be accounted for from the very outset.

3.1. Thermooptical Equations in an Isotropic Solid

We utilize equations of elasticity theory[1,2] to describe thermooptical sound excitation in solids

$$\rho \frac{\partial^2 u_i}{\partial t^2} = \frac{\partial \sigma_{ik}}{\partial x_k}, \tag{3.1}$$

$$\rho c_V \frac{\partial T}{\partial t} - \rho (c_p - c_V) \beta^{-1} \operatorname{div} \frac{\partial \mathbf{u}}{\partial t} = \kappa \Delta T - \operatorname{div}\langle \mathbf{S} \rangle, \tag{3.2}$$

where $\mathbf{u} = \{u_1, u_2, u_3\}$ is the displacement vector of the particles in the medium. The stress tensor σ_{ik} must incorporate the thermoelastic terms

$$\sigma_{ik} = -K\beta T' \delta_{ik} + K u_{ll} \delta_{ik} + 2\mu (u_{ik} - \delta_{ik} u_{ll}/3). \tag{3.3}$$

Here K and μ represent the bulk compression and shear moduli, β is the thermal coefficient of volume expansion, while the linearized strain tensor

$$u_{ik} = \left(\frac{\partial u_i}{\partial x_k} + \frac{\partial u_k}{\partial x_i} \right) \Big/ 2. \tag{3.4}$$

Summation is assumed over all repeating indices. The equation system is used in a linear approximation and no differences between adiabatic and isothermal moduli are assumed. This is valid[2] at least over the frequency range $\omega \ll \omega_\chi = c_L^2 \rho_0 c_p / \kappa$. Here and below $c_{L,T}$ are the velocities of longitudinal and transverse acoustic waves, and κ is thermal conductivity.

Equation system (3.1)–(3.4) describes a coupled system of acoustic and thermal waves. However, acoustic and thermal mode coupling is weak in adiabatic sound propagation (for $\omega \ll \omega_\chi$) and separate equations can be used for these modes. The thermal problem can be independently solved. Scalar and vector potentials are convenient. We represent the particle velocity as follows:

$$\frac{\partial \mathbf{u}}{\partial t} = \operatorname{grad} \varphi + \operatorname{rot} \psi. \tag{3.5}$$

The equation system (3.1)–(3.4) reduces to

$$c_L^{-2} \frac{\partial^2 \varphi}{\partial t^2} - \Delta \varphi = -\beta \left(1 - 4 \frac{c_T^2}{3c_L^2}\right) \frac{\partial T}{\partial t} = -\beta^* \frac{\partial T}{\partial t}, \tag{3.6}$$

$$c_T^{-2} \frac{\partial^2 \psi}{\partial t^2} - \Delta \psi = 0, \tag{3.7}$$

$$\frac{\partial T}{\partial t} = \chi \, \Delta T - (\rho_0 c_p)^{-1} \operatorname{div}\langle \mathbf{S} \rangle \tag{3.8}$$

($\chi = \kappa/\rho_0 c_p$ is the thermal diffusivity, β^* is the "effective" thermal coefficient of volume expansion). Therefore thermal expansion will only excite the potential part of the particle velocity, while its solenoidal part is manifested strictly due to boundary conditions from longitudinal wave reflection off the boundary.

We consider the general case of an impedance boundary. This requires accounting for waves excited in a transparent medium. The equations describing these waves are entirely analogous to Eqs. (3.6)–(3.8) with the sound velocity and other parameters replaced by the corresponding parameters of the transparent medium. The boundary conditions for equivalent displacements and normal and tangential stresses

$$u_i|_{z=0} = u_i^{\text{tr}}|_{z=0}, \quad \sigma_{iz}|_{z=0} = \sigma_{iz}^{\text{tr}}|_{z=0}$$

(the superscript tr denotes the corresponding quantities in a transparent medium) are written as follows:

$$\rho_0 c_T^2 \left[2 \frac{\partial^2 \varphi}{\partial x \partial z} + \frac{\partial^2 \psi_z}{\partial y \partial z} - \frac{\partial^2 \psi_y}{\partial z^2} + \frac{\partial^2 \psi_x}{\partial x^2} - \frac{\partial^2 \psi_x}{\partial x \partial y} \right]_{z=0} = \rho_0^{\text{tr}} (c_T^{\text{tr}})^2 [\cdots]_{z=0}^{\text{tr}},$$

$$\rho_0 c_T^2 \left[2 \frac{\partial^2 \varphi}{\partial y \partial z} + \frac{\partial^2 \psi_x}{\partial z^2} - \frac{\partial^2 \psi_z}{\partial x \partial z} - \frac{\partial^2 \psi_x}{\partial y^2} + \frac{\partial^2 \psi_y}{\partial x \partial y} \right]_{z=0} = \rho_0^{\text{tr}} (c_T^{\text{tr}})^2 [\cdots]_{z=0}^{\text{tr}},$$

$$\rho_0 c_L^2 \left[\Delta \varphi - 2 \frac{c_T^2}{c_L^2} \Delta_\perp \varphi + 2 \frac{c_T^2}{c_L^2} \frac{\partial}{\partial z} \left(\frac{\partial \psi_y}{\partial x} - \frac{\partial \psi_x}{\partial y} \right) - \beta^* \frac{\partial T}{\partial t} \right]_{z=0} = \rho_0^{\text{tr}} (c_L^{\text{tr}})^2 [\cdots]_{z=0}^{\text{tr}},$$

$$\left[\frac{\partial \varphi}{\partial x} + \frac{\partial \psi_z}{\partial y} - \frac{\partial \psi_y}{\partial z} \right]_{z=0} = [\ldots]_{z=0}^{\text{tr}} \left[\frac{\partial \varphi}{\partial y} + \frac{\partial \psi_x}{\partial z} - \frac{\partial \psi_z}{\partial x} \right]_{z=0} = [\cdots]_{z=0}^{\text{tr}},$$

$$\left[\frac{\partial \varphi}{\partial z} + \frac{\partial \psi_y}{\partial x} - \frac{\partial \psi_x}{\partial y} \right]_{z=0} = [\cdots]_{z=0}^{\text{tr}}.$$

This system is rather cumbersome and it can be significantly simplified. It is advisable to introduce new variables

$$A = \frac{\partial \psi_y}{\partial x} - \frac{\partial \psi_x}{\partial y}, \quad B = \frac{\partial \psi_x}{\partial x} + \frac{\partial \psi_y}{\partial y} \tag{3.9}$$

(and analogous equations for a transparent medium). The quantity A represents the vortical displacement component $\partial u/\partial t$ normal to the interface surface. The quantities A and B satisfy a homogeneous wave equation analogous to Eq. (3.7):

$$\begin{aligned} & c_T^{-2} \frac{\partial^2 (A,B)}{\partial t^2} - \Delta (A,B) = 0, \\ & (c_T^{\mathrm{tr}})^{-2} \frac{\partial^2 (A^{\mathrm{tr}},B^{\mathrm{tr}})}{\partial t^2} - \Delta (A^{\mathrm{tr}},B^{\mathrm{tr}}) = 0. \end{aligned} \tag{3.10}$$

Expressing the first derivatives of ψ_z, ψ_z^{tr} based on auxiliary conditions (calibration conditions)

$$\operatorname{div} \boldsymbol{\psi} = \operatorname{div} \boldsymbol{\psi}^{\mathrm{tr}} = 0, \tag{3.11}$$

and the second derivatives with respect to z based on wave equations (3.7), after some simple yet cumbersome transforms, we obtain

$$\begin{aligned} & \left[\frac{\partial \varphi}{\partial z} + A\right]_{z=0} = [\cdots]_{z=0}^{\mathrm{tr}}, \\ & \left[\frac{\partial A}{\partial z} - \Delta_\perp \varphi\right]_{z=0} = [\cdots]_{z=0}^{\mathrm{tr}}, \\ & \left[\rho_0 c_T^2 \left(\frac{\partial}{\partial z}(\Delta_\perp \varphi) - \left(\frac{1}{2c_T^2}\frac{\partial^2 A}{\partial t^2} - \Delta_\perp A\right)\right)\right]_{z=0} = [\cdots]_{z=0}^{\mathrm{tr}}, \end{aligned} \tag{3.12}$$

$$\begin{aligned} & \left[\frac{\partial B}{\partial z} - \Delta_\perp \psi_z\right]_{z=0} = [\cdots]_{z=0}^{\mathrm{tr}}, \\ & \left[\frac{\partial}{\partial z}(\Delta_\perp \psi_z) - \left(\frac{1}{2c_T^2}\frac{\partial^2 B}{\partial t^2} - \Delta_\perp B\right)\right]_{z=0} = [\cdots]_{z=0}^{\mathrm{tr}}, \\ & \left[\frac{\partial B}{\partial z} + \left(\frac{1}{2c_T^2}\frac{\partial^2 \psi_z}{\partial t^2} - \Delta_\perp \psi_z\right)\right]_{z=0} = [\cdots]_{z=0}^{\mathrm{tr}} = 0. \end{aligned} \tag{3.13}$$

Therefore only the pair of variables $\varphi - A$ and $B - \psi_z$ will be coupled at the boundary. Since the equations for all these variables are independent, the general problem of thermooptical sound excitation (3.6), (3.7) in variables (3.9) is divided into two independent problems. The solution for the $B - \psi_z$ pair of variables is obvious, given by the homogeneity of the equations and the boundary conditions for B and ψ_z:

$$B = \psi_z = B^{\mathrm{tr}} = \psi_z^{\mathrm{tr}} = 0.$$

TABLE 3.1. Characteristic optoacoustic parameters.

Medium	ω_{χ}, c^{-1}	β^*, K^{-1}	c_L/χ, cm^{-1}	$\rho_0 c_L$, kg/(m^2×s)	$\beta^* \sqrt{\chi}$, cm/(K×s$^{1/2}$)	$\rho_0 c_p \sqrt{\chi}$, J/(cm^2×s$^{1/2}$)	$\beta^*/\rho_0 c_p$ cm^3/MJ
Aluminum	4.5×10^{11}	4.7×10^{-5}	7.2×10^{5}	1.7×10^{6}	4.4×10^{-5}	2.3	19
Water	1.5×10^{13}	1.8×10^{-4}	1.0×10^{8}	1.5×10^{5}	5.4×10^{-6}	0.16	43
Ethanol	1.6×10^{13}	1.1×10^{-3}	1.3×10^{8}	0.92×10^{5}	3.2×10^{-5}	0.058	560
Mercury	1.3×10^{11}	1.8×10^{-4}	9.1×10^{5}	2.0×10^{6}	7.2×10^{-5}	0.75	96
Quartz	6.2×10^{13}	0.7×10^{-6}	1.1×10^{8}	1.2×10^{6}	4.9×10^{-8}	0.10	0.47
Air	5.9×10^{9}	3.4×10^{-8}	1.7×10^{5}	4.3×10^{2}	1.5×10^{-3}	5.8×10^{-4}	2.6×10^{6}

Hence the acoustic part of the problem is significantly simplified.

Therefore thermooptical sound excitation at the boundary of a transparent medium and an absorbing medium is described by scalar wave equations (3.6), (3.10) and coupled boundary conditions (3.12). The excitation process is easily analyzed by the spectral method. This requires finding the spectral temperature density in the transparent and absorbing media.

3.2. Transfer Functions of Thermooptical Sound Excitation in a Heat-Conducting Medium

The change in the temperature of the media is determined by the absorbed energy, the thermal conductivity, and dilatation. However, thermal and acoustic wave coupling can be neglected for the majority of solids through frequencies $\omega \sim 10^{11}$ s^{-1} (Table 3.1).[2] In this frequency range sound propagates adiabatically: the heat does not diffuse over a distance of the order of the wavelength over the oscillation period. Consequently, the problem of determining the temperature field reduces to solving the equations

$$\frac{\partial T^{\text{tr}}}{\partial t} = \chi^{\text{tr}}\, \Delta T^{\text{tr}}; \quad z<0,$$

$$\frac{\partial T}{\partial t} = \chi\, \Delta T + \left(\frac{\alpha I_0}{\rho_0 c_p}\right) f(t) H(x,y) e^{-\alpha z}, \quad z>0. \tag{3.14}$$

We assume a constant light absorptivity α; I_0 is the absorbed part of the laser pulse intensity; its time dependence is given by $f(t)$, while its distribution over the laser spot is given by the function $H(x,y)$. Temperature equality and thermal flux continuity conditions will hold at the interface $z=0$:

$$T|_0 = T^{\text{tr}}|_0, \left[\rho_0 c_p \chi \frac{\partial T}{\partial z}\right]\bigg|_0 = [\cdots]^{\text{tr}}|_0, \tag{3.15}$$

We introduce the spectral density

$$\widetilde{T}(\omega,\mathbf{k}_\perp,z) = \iiint_{-\infty}^{\infty} T(t,\mathbf{r}_\perp z)\exp[i\omega t - i(\mathbf{k}_\perp \mathbf{r}_\perp)]dt\, d\mathbf{r}_\perp,$$

where $\mathbf{r}_\perp = \{x,y\}$ and $\mathbf{k}_\perp = \{k_x,k_y\}$. Then the solution of (3.14) and (3.15) can be given as

$$\widetilde{T}^{\text{tr}} = \frac{I_0}{\rho_0 c_p \chi} \frac{\alpha}{(i\omega/\chi - k_\perp^2 + \alpha^2)}$$
$$\times \frac{\alpha - (-i\omega/\chi + k_\perp^2)^{1/2}}{(-i\omega/\chi + k_\perp^2)^{1/2} + \rho^{\text{tr}} c_p^{\text{tr}} \chi^{\text{tr}}(-i\omega/\chi^{\text{tr}} + k_\perp^2)^{1/2}/\rho_0 c_p \chi}$$
$$\times \exp[(-i\omega/\chi^{\text{tr}} + k_\perp^2)^{1/2} z]\widetilde{f}(\omega)\widetilde{H}(\mathbf{k}_\perp), \tag{3.16}$$

$$\widetilde{T} = \frac{I_0}{\rho_0 c_p \chi} \frac{\alpha}{(i\omega/\chi - k_\perp^2 + \alpha^2)}$$
$$\times \left[\frac{\alpha + \rho^{\text{tr}} c_p^{\text{tr}} \chi^{\text{tr}}(-i\omega/\chi^{\text{tr}} + k_\perp^2)^{1/2}/\rho_0 c_p \chi}{(-i\omega/\chi + k_\perp^2)^{1/2} + \rho^{\text{tr}} c_p^{\text{tr}} \chi^{\text{tr}}(-i\omega/\chi^{\text{tr}} + k_\perp^2)^{1/2}/\rho_0 c_p \chi} \right.$$
$$\left. \times \exp[-(-i\omega/\chi + k_\perp^2)^{1/2} z] - \exp(-\alpha z) \right] \widetilde{f}(\omega) H(\mathbf{k}_\perp).$$

These very complex expressions can be significantly simplified by taking the following into account. Since the angular spectrum $\widetilde{H}(k_\perp)$ limits the transverse component of the wave vector of the field to

$$|\mathbf{k}_\perp| \lesssim a^{-1},$$

where a is the characteristic dimension (radius) of the laser beam, with the exception of a very low-frequency range of $\omega \lesssim \chi a^{-2}$ thermal diffusion transverse through the beam will not be significant. This approximation is valid in the ultrasonic frequency band even for metals and rather sharp beam focusing. If $\omega \gg \chi a^{-2}$, expressions (3.16) can be significantly simplified:

$$\widetilde{T}^{\text{tr}} = \frac{I_0}{\rho_0 c_p \chi} \frac{\alpha}{i\omega/\chi + \alpha^2} \frac{\alpha(-i\omega/\chi)^{-1/2} - 1}{1 + R_T}$$
$$\times \exp[(-i\omega/\chi^{\text{tr}})^{1/2} z]\widetilde{f}(\omega)\widetilde{H}(\mathbf{k}_\perp), \tag{3.17}$$

$$\widetilde{T} = \frac{I_0}{\rho_0 c_p \chi} \frac{\alpha}{i\omega/\chi + \alpha^2} \left[\frac{\alpha(-i\omega/\chi)^{-1/2} + R_T}{1 + R_T} \right.$$
$$\left. \times \exp[-(-i\omega/\chi)^{1/2} z] - \exp(-\alpha z) \right] \widetilde{f}(\omega) H(\mathbf{k}_\perp),$$

where $\mathrm{Re}(-i\omega/\chi)^{1/2} > 0$, while the parameter

$$R_T = \rho^{\text{tr}} c_p^{\text{tr}} (\chi^{\text{tr}})^{1/2} / \rho_0 c_p (\chi)^{1/2}$$

determines the ratio of thermal fluxes to the transparent and absorbing media. Solution (3.17) corresponds to the solution of the problem in the plane geometry case [for $\tilde{H}(\mathbf{k}_\perp) \sim \delta(\mathbf{k}_\perp)$]. The temperature field in the absorbing medium $\tilde{T}$ consists of the temperature distribution corresponding to the density of heating from absorbed light and the natural thermal wave which decays from the boundary into the medium. Evidently the transparent medium will only contain the natural thermal wave. The relative amplitude of the different Fourier harmonics of the temperature field is directly proportional to the intensity spectrum of the laser radiation $\tilde{f}(\omega)$ where the proportionality factor is dependent on the frequency, the thermal diffusivity, and the absorptivity of light.

At low frequencies $\omega \ll \omega_T = \alpha^2\chi$ the field in the depth of the absorbing medium is determined by the thermal conductivity:

$$\tilde{T} \approx \frac{I_0}{\rho_0 c_p} \frac{\exp[-(-i\omega/\chi)^{1/2} z]}{(-i\omega\chi)^{1/2}} \tilde{f}(\omega),$$

and has a universal character. This spectrum corresponds to the known solution of Danilovskaya's[2,3] surface heating problem. The length of the thermal wave is much less than the depth of penetration of light in the high-frequency range $\omega \gg \omega_T = \alpha^2\chi$ and the temperature field (with the exception of a narrow surface region) is determined by heating

$$\tilde{T} \approx \frac{I_0}{\rho_0 c_p} \frac{\alpha}{(-i\omega)} \exp(-\alpha z)\tilde{f}(\omega),$$

which corresponds to a thermally nonconducting medium (see Chap. 2).

The temperature distribution in a transparent medium is determined by the thermal conductivity across the entire frequency range due to the fundamental surface nature of such heating [see Eq. (3.17)].

We use solution (3.17) to analyze acoustic wave excitation. We limit the analysis to the case of a broad laser beam $\alpha a \gg 1$, which permits considering the one-dimensional problem [$\tilde{H}(\mathbf{k}_\perp) \sim \delta(\mathbf{k}_\perp)$] in the excitation zone. In this case a purely longitudinal wave will be excited, since the transverse field gradients are uniformly equal to zero [specifically, in boundary conditions (3.12)] and, consequently, $A = 0$. Therefore, we have the following system of equations for the spectral density of the scalar potential φ:

$$\frac{\omega^2}{c_L^2} \tilde{\varphi} + \frac{\partial^2 \tilde{\varphi}}{\partial z^2} = -i\omega\beta^* \tilde{T},$$

$$\frac{\omega^2}{(c_L^{tr})^2} \tilde{\varphi} + \frac{\partial^2 \tilde{\varphi}^{tr}}{\partial z^2} = -i\omega\beta^{*tr} \tilde{T}^{tr}, \tag{3.18}$$

$$\left[\frac{\partial \tilde{\varphi}}{\partial z}\right]\bigg|_0 = [\cdots]^{tr}|_0,$$

$$[\rho_0 \tilde{\varphi}]|_0 = [\cdots]^{tr}|_0. \tag{3.19}$$

The quantities $\tilde{T}$ and $\tilde{T}^{tr}$ are determined from Eq. (3.17). The auxiliary conditions include radiation conditions for φ as $z \to +\infty$ and for φ^{tr} as $z \to -\infty$. These correspond to the case of an absence of waves traveling from infinity to the boundary (such conditions are analogous to the vanishing of temperature at infinity). This yields a solution of the type

$$\widetilde{\varphi}^{\mathrm{tr}} = \varphi_0^{\mathrm{tr}} \exp\left(- \frac{i\omega z}{c_L^{\mathrm{tr}}} \right) - \frac{i\omega\beta^{*\mathrm{tr}} I_0 \widetilde{f}(\omega)}{[(\omega/c_L^{\mathrm{tr}})^2 - i\omega/\chi^{\mathrm{tr}}]\rho_0^{\mathrm{tr}} c_p^{\mathrm{tr}}}$$

$$\times \frac{\alpha/\chi}{\alpha^2 + i\omega/\chi} \frac{\alpha(- i\omega/\chi)^{-1/2} - 1}{1 + R_T} \exp\left[\left(- \frac{i\omega}{\chi^{\mathrm{tr}}} \right)^{1/2} z \right], \tag{3.20}$$

$$\widetilde{\varphi} = \varphi_0 \exp\left(\frac{i\omega z}{c_L} \right) - \frac{i\omega\alpha\beta^* I_0 \widetilde{f}(\omega)}{(\alpha^2 + i\omega/\chi)\rho_0 c_p \chi}$$

$$\times \left[\frac{R_T + \alpha(- i\omega/\chi)^{-1/2}}{1 + R_T} \frac{\exp[- (- i\omega/\chi)^{1/2} z]}{(\omega/c_L)^2 - i\omega/\chi} - \frac{\exp(- \alpha z)}{\alpha^2 + (\omega/c_L)^2} \right],$$

where φ_0^{tr} and φ_0 are determined from boundary conditions (3.19).

Only purely traveling acoustic waves of amplitudes φ_0 and φ_0^{tr} are found outside the heating zone

$$|z| > \max[\alpha^{-1}, (\chi/\omega)^{1/2}, (\chi^{\mathrm{tr}}/\omega)^{1/2}]$$

[since the exponential multipliers on the right-hand sides of Eq. (3.20) are small at these distances]. Hence, as in the case of liquids the particle velocity spectrum can be represented as

$$\widetilde{v} = \frac{\partial\widetilde{\varphi}}{\widetilde{\omega} z} = K(\omega) I_0 \widetilde{f}(\omega) \exp\left(- \frac{i\omega z}{c_L} \right),$$

$$\widetilde{v}^{tr} = \frac{\partial\widetilde{\varphi}^{\mathrm{tr}}}{\partial z} = K^{\mathrm{tr}}(\omega) I_0 \widetilde{f}(\omega) \exp\left(\frac{i\omega z}{c_L^{\mathrm{tr}}} \right). \tag{3.21}$$

The optoacoustic signal spectrum in both an absorbing and a transparent medium is proportional to the laser radiation intensity spectrum $\widetilde{f}(\omega)$ while the transfer functions $K(\omega)$ and $K^{\mathrm{tr}}(\omega)$ are determined by the light absorptivity and the physical parameters of the media.

The optoacoustic problem also reduces to determining the transfer function. In this connection we differentiate two versions of optoacoustic spectroscopy: an indirect version [signal recorded in a transparent medium $K^{\mathrm{tr}}(\omega)$] and a direct method [$K(\omega)$ is measured]. The transfer functions appear as follows in the case of a homogeneously absorbing medium:

$$K^{\mathrm{tr}}(\omega) = \frac{\beta^*}{\rho_0 c_p} \frac{N}{N+1} \frac{1}{1 + i\omega/\omega_T} \left[\frac{i\omega/\omega_T}{1 - i\omega/\omega_a} + \frac{1+b}{1+R_T} \right.$$

$$\left. + \left(- i\frac{\omega}{\omega_T} \right)^{1/2} \frac{R_T - b}{R_T + 1} \right], \tag{3.22}$$

$$K(\omega)=\frac{\beta^*}{\rho_0 c_p}\frac{1}{N+1}\frac{1}{1+i\omega/\omega_T}\left[i\frac{\omega}{\omega_T}\frac{1-iN\omega/\omega_a}{1+(\omega/\omega_a)^2}+\frac{1+b}{1+R_T}\right.$$
$$\left.+\left(-i\frac{\omega}{\omega_T}\right)^{1/2}\frac{R_T-b+Nm_\chi}{R_T}\right],$$

where $\omega_T\alpha^2\chi$; $\omega_a=\alpha c_L$; $N=\rho_0 c_L/\rho^{tr}c_L^{tr}$; $b=\beta^{*tr}\sqrt{\chi^{tr}}/\beta^*\sqrt{\chi}$; $m_x=\omega_T/\omega_a=\alpha\chi/c_L$. N is the ratio of the acoustic impedances of the absorbing and transparent media; the parameter b characterizes the relative contribution of these media to the optoacoustic signal.

There are several characteristic thermooptical sound excitation frequencies in heat-conducting media. One of these, ω_T, was introduced previously in the thermal problem: the light absorption layer is thermally "thin" at frequencies $\omega \ll \omega_T$, while at $\omega \gg \omega_T$ the situation is reversed: the layer is thermally "thick." Hence, thermal conduction will only be significant in the "low"-frequency range $\omega \lesssim \omega_T$. The second characteristic frequency $\omega_a=\alpha c_L$ defines the frequency range of efficient sound excitation (see Chap. 2). The light absorption layer is acoustically thin for $\omega \ll \omega_a$, and is acoustically thick for $\omega \gg \omega_a$. And, finally, the third frequency $\omega_\chi=c_L^2/\chi$ corresponds to the boundary of adiabatic sound propagation and defines the applicability range of this theory. The ratio of these three frequencies can be characterized by the single dimensionless parameter

$$m_\chi=\omega_T/\omega_a=\omega_a/\omega_\chi=\alpha\chi/c_L. \qquad (3.23)$$

As suggested by Table 3.1, the parameter m_χ, may be comparable to unity for strongly absorbing media only ($\alpha > 10^4$ cm^{-1}). This is possible for both metals and semiconductors (in the interband absorption region). The characteristic optoacoustic sound excitation frequencies are quite extensively spaced ($m_\chi \ll 1$) at moderately high absorption coefficients of the laser light ($\alpha < 10^4$ cm^{-1}).

An expression can be written in a form analogous to Eq. (3.21) for the vibrational velocity of boundary $z=0$, which is commonly measured in optoacoustic spectroscopy:

$$v=\int_{-\infty}^{\infty}K_0(\omega)I_0\tilde{f}(\omega)\exp(i\omega t)d\omega,$$

where

$$K_0(\omega)=\frac{\beta^*}{\rho_0 c_p}\frac{1}{N+1}\frac{1}{1+i\omega/\omega_T}\left[\frac{1+b}{1+R_T}+\frac{(1+N)/(1+R_T)}{1+i\omega/\omega_\chi}-\frac{iN\omega/\omega_\chi}{1-i\omega/\omega_a}\right.$$
$$\left.+\left(-i\frac{\omega}{\omega_T}\right)^{1/2}\frac{R_T-b+Nm_\chi}{R_T+1}+\frac{(-i\omega/\omega_T)^{1/2}}{1+i\omega/\omega_\chi}\frac{R_T(N+1)}{R_T+1}\right]. \qquad (3.24)$$

Clearly the transfer function in the case of an impedance boundary of the absorbing medium takes the form of a weighted sum of the transfer functions with rigid (K_r) and free (K_f) boundaries:

$$K=(K_r+NK_f)/(1+N). \qquad (3.25)$$

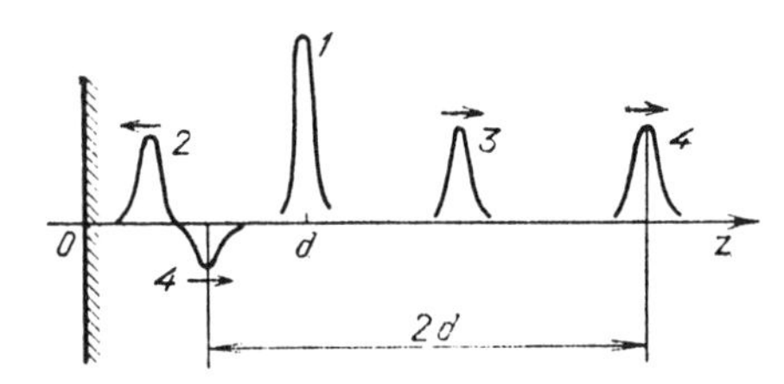

FIG. 3.1. Formation of the OA signal with an impedance boundary.

This fact can conveniently be represented (Fig. 3.1) by means of Green's functions (the system response functions to a δ perturbation). The source of the δ perturbation (1) at a distance d from the boundary generates two pulses (2 and 3) that travel along and counter to the z axis. Pulse 2 partially penetrates to the transparent medium upon reaching the impedance boundary, and is also partially reflected in antiphase (pulse 4). The last part follows pulse 3 with a time delay $\Delta\tau = 2d/c_L$. The resulting acoustic signal (the set of pulse 4) can be represented as the weighted sum of two signals (Fig. 3.2): the first signal is the system response with a rigid boundary ($N=0$) and pulse 2 will be reflected in phase from the acoustically more dense medium. The second signal is the system response with a free boundary ($N \gg 1$) and pulse 2 is reflected in antiphase of the acoustically less-dense medium. In other words, when the boundary is an impedance boundary, the Green's function is the weighted sum of the Green's functions in the case of rigid and free boundaries; the acoustic impedance ratio N is used as the weight coefficient. Since the spectral transfer function is the Fourier transform of the Green's function, this result is also valid for $K(\omega)$.

We note that $K_r^{tr} \equiv 0$. Since the transparent medium is acoustically far denser than the absorbing medium, thermal expansion of the absorbing medium can have no effect on it. The transparent medium will experience thermal expansion at the free boundary and, by virtue of the adiabaticity of sound propagation ($\omega \ll c_L^2/\chi$, length of the thermal wave is much less than the length of the acoustic wave) poor sound excitation results.

Consistent with Eq. (3.25) the temporal wave form of the acoustic pulses with an arbitrary impedance ratio N will be the weighted sum of the wave forms with rigid ($N=0$) and free ($N \gg 1$) boundaries (for the same light pulse).

We consider thermooptical sound excitation with strong light absorption and strong thermal conduction (as a rule, these two properties occur simultaneously;

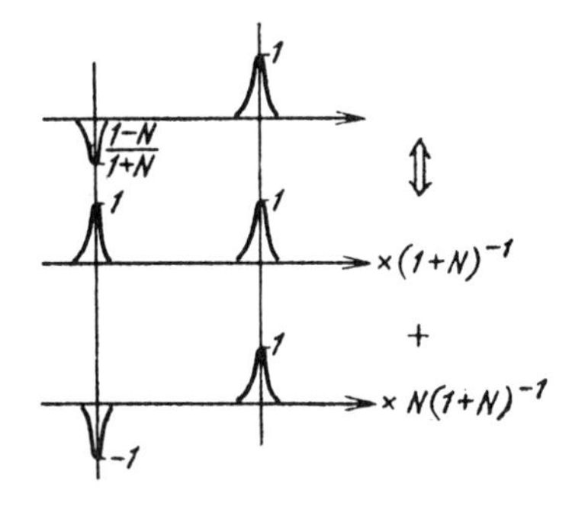

FIG. 3.2. Representation of the OA signal with an impedance boundary in the form of a weighted sum of signals for the rigid and free boundaries.

the absorbing medium is a metal or a semiconductor: $m_\chi \sim 1$). Then, the characteristic frequencies will not be appreciably separated $\omega_T \sim \omega_a \sim \omega_\chi \gg \omega$ and thermal conduction will be significant across the entire working frequency range. In this case the expressions for transfer functions (3.22)–(3.24) are significantly simplified:

$$K(\omega) = \frac{\beta^*}{\rho_0 c_p} \frac{1 + b + N(-i\omega/\omega_\chi)^{1/2}}{(N+1)(1+R_T)}. \tag{3.26}$$

With a rigid boundary of the absorbing medium ($N=0$) the transfer function of the vibrational velocity of sound in this medium will be independent of frequency:

$$K_r(m_\chi \sim 1) = \frac{\beta^*}{\rho_0 c_p} \frac{1+b}{1+R_T}. \tag{3.27}$$

In the time domain this means that the acoustic signal mimics the light pulse wave form. The second cofactor in the numerator determines the contribution of thermal expansion of the transparent medium to sound excitation in the absorbing medium (the larger the value b, the more significant this contribution). It follows from Table 3.1 that the transparent medium may significantly increase acoustic signal generation efficiency and in order to improve such efficiency it is necessary to use media with high values of β^* and $\beta^* \sqrt{\chi}$ (for example, aluminum as the absorbing medium and transformer oil as the transparent medium).

With a free acoustic boundary ($N \gg 1$)

$$K_f(m_\chi \sim 1) = (\beta^*/\rho_0 c_p)(-i\omega/\omega_\chi)^{1/2}/(1+R_T) \tag{3.28}$$

the transfer function corresponds to the known solution of the Danilovskaya's problem of sound excitation by surface heating in a heat-conducting medium. In this case, the transparent medium has no effect on the excited wave due to its low wave impedance.

It follows from Eqs. (3.27) and (3.28) that the sound excitation efficiency in strongly absorbing media with a free boundary is much less than the case with a rigid boundary:

$$\frac{|K_f(m_\chi \sim 1)|}{|K_r(m_\chi \sim 1)|} \sim \frac{1}{1+b} \sqrt{\frac{\omega}{\omega_\chi}} \ll 1. \tag{3.29}$$

The sound excitation efficiencies become comparable only at frequencies $\omega \sim \omega_\chi (1+b)^2/N^2$. The effect of the dipole sources will only become manifested with large N (a near-free boundary; gas as the transparent medium). Consequently, it is necessary to reduce the impedance ratio N of the absorbing and transparent media in order to improve the sound excitation efficiency in highly absorbing media. This apparently explains the experimentally observed high increase in the signal excited in metals when their surface is covered with a liquid (even a transparent liquid).[4]

The impedance ratio N also has a strong effect on the phase ψ of the transfer function in an absorbing medium:

$$\tan \psi = -N(\omega/2\omega_\chi)^{1/2}[(1+b) + N(\omega/2\omega_\chi)]^{-1/2}. \tag{3.30}$$

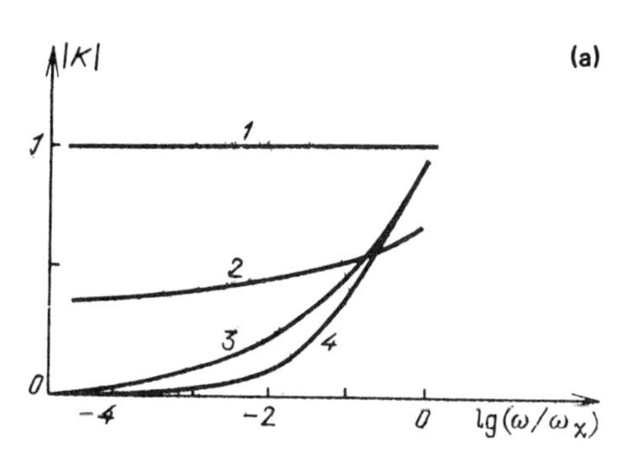

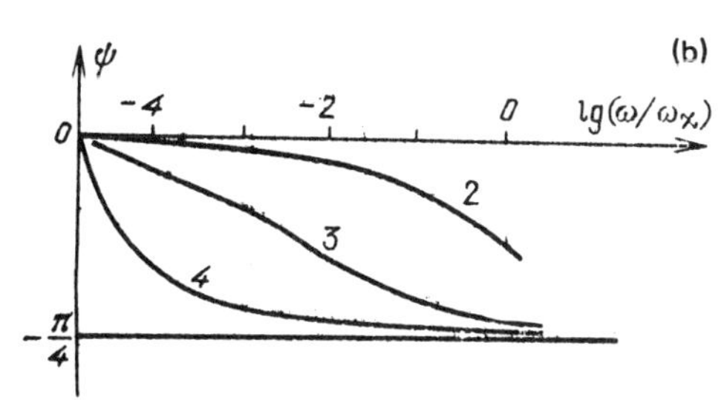

FIG. 3.3. Frequency dependence of the modulus (a) and phase (b) of the transfer function in a strongly absorbing medium for different boundary impedances N of 0 (1), 1 (2), 10 (3), and 100 (4).

The phase shift $\tan\psi \sim N$ in the low-frequency range $\omega < \omega_\chi(1+b)^2/N^2$. This makes it possible to determine the nature of the acoustic boundary of the absorbing medium based on the phase of the optoacoustic signal. The modulus and phase of the transfer function are shown in Fig 3.3 for a strongly absorbing medium ($m_\chi \sim 1$).

The transfer function is independent of frequency in a transparent medium in the case $m_\chi \sim 1$:

$$K^{\text{tr}}(\omega) = \frac{\beta^*}{\rho_0 c_p}\frac{N}{N+1}\frac{1+b}{1+R_T}. \tag{3.31}$$

This is evident, since the heating depth in a transparent medium will always be small compared to the wavelength of the acoustic wave for $\omega \ll \omega_\chi$. In this version of "indirect" optoacoustic spectroscopy the signal amplitude grows with increasing acoustic impedance ratio N, while the thermal expansion of a medium with a higher coefficient $\beta^*/\sqrt{\chi}$ makes the primary contribution to the signal. Specifically, if the transparent medium is a gas, no significant thermal expansion of the absorbing medium will occur (see Table 3.1).

The transfer function of the boundary vibrational velocity K_0 in the case of strong absorption $m_\chi \sim 1$ will also be independent of frequency:

$$K_0(\omega) = \frac{\beta^*}{\rho_0 c_p}\frac{2+b+N}{(N+1)(R_T+1)}. \tag{3.32}$$

It is weakly dependent on the impedance ratio N.

Figure 3.4 shows sample wave forms of optoacoustic signals in a heat-conducting medium for the case of a Gaussian laser pulse profile: $f(t) = \pi^{-1/2}\exp(-t^2/\tau_L^2)$ calculated by means of Eq. (3.26). In this case the wave profile is described by

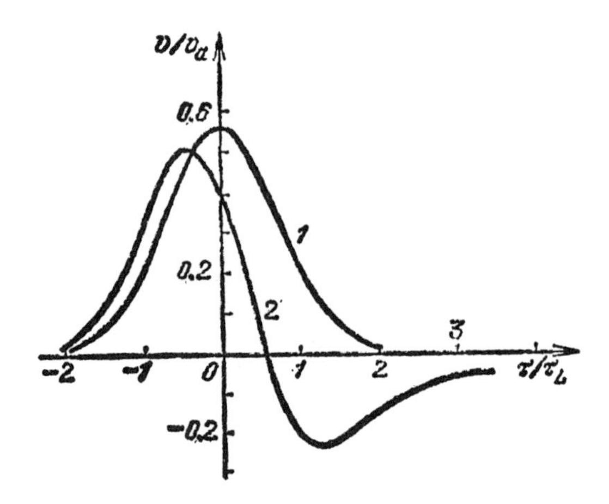

FIG. 3.4. OA-signal profiles in a heat-conducting strongly absorbing medium for the case of rigid (1) and free (2) boundaries. [The second signal is amplified by a factor of $(\omega_\chi\tau_L)^{1/2}$.]

$$v(\tau) = (\beta^* I_0/\rho_0 c_p)(1+R_T)^{-1}(1+N)^{-1}[(1+b)\pi^{-1/2}\exp(-t^2/\tau_L^2)$$
$$+N(\sqrt{2}/\omega_\chi\tau_L\pi)^{1/2}\exp(-t^2/2\tau_L^2)D_{1/2}(\tau\sqrt{2}/\tau_L)],$$

where $D_{1/2}(z)$ is the parabolic cylinder function.

With a rigid boundary $N=0$ the temporal dependences of the acoustic and laser pulses are identical. With a free boundary $N \gg 1$ the second term is the principal term and the acoustic pulse is a bipolar pulse. The asymptotics of the vibrational velocity for $\tau \gg \tau_L$ has universal representation $v \sim \tau^{-3/2}$ characteristic of a heat-conducting medium. The relative amplitude of the OA signals for both free and rigid (hard) boundaries is given by $(\omega_\chi\tau_L)^{-1/2} \ll 1$. They become comparable as the laser pulses become shorter. The normalized amplitude varies from $1+b$ (for $N=0$) to 1 (for $N \gg 1$). This version of optoacoustic spectroscopy yields less information in the case of strong thermal conduction.

We now consider the opposite case of $m_\chi \ll 1$ (moderate light absorptivities: $\alpha \lesssim 10^4\ \mathrm{cm}^{-1}$). In this case the characteristic frequencies ω_χ, ω_a, and ω_T are spaced out significantly across the band ($\omega_T \ll \omega_a \ll \omega_\chi$) and different terms in Eqs. (3.2)–(3.4) make a contribution to sound generation in the different frequency bands. Figure 3.5 shows the modulus and phase of the transfer function for an absorbing medium plotted as a function of frequency for different impedance ratios N of the media. The sound generation efficiency at moderately high frequencies $\omega \ll \omega_a$ diminishes as we go from a rigid ($N=0$) to a free ($N \gg 1$) boundary. The efficiency peak shifts towards higher frequencies $\omega \sim \omega_a$ with increasing N, and sound generation is most intense at these frequencies with a free boundary [Fig. 3.5(a)].

Heat conduction of the media in the case $m_\chi \ll 1$ will only be significant at very low frequencies $\omega \ll \omega_T$. The transfer functions in this frequency range take the form

$$K(\omega) = \frac{\beta^*}{\rho_0 c_p}\frac{1}{N+1}\left[\frac{1+b}{1+R_T} + N\frac{(\omega/2\omega_\chi)^{1/2}}{1+R_T}(1-i)\right],$$

$$K^{\mathrm{tr}}(\omega) = \frac{\beta^*}{\rho_0 c_p}\frac{N}{N+1}\frac{1+b}{1+R_T}, \qquad (3.33)$$

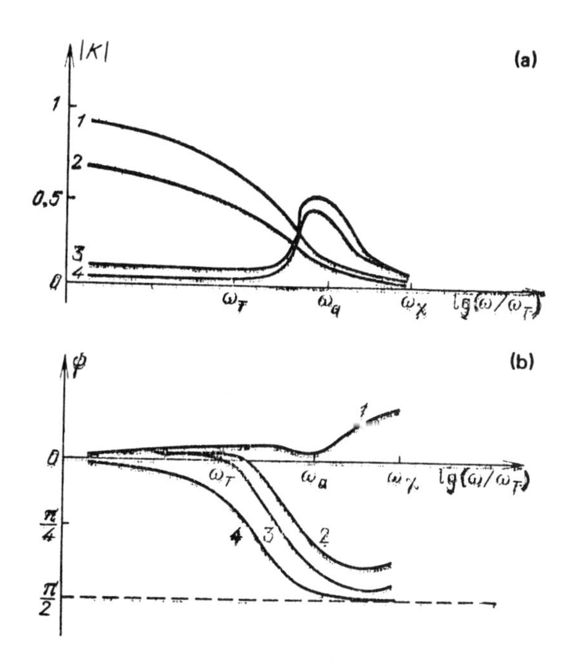

FIG. 3.5. Frequency dependence of the modulus (a) and phase (b) of the transfer function for weak heat conduction ($m_\chi = 10^{-2} \ll 1$) and different boundary impedances of 0.1 (1), 1 (2), 10 (3), and 100 (4).

$$K_0(\omega) = \frac{\beta^*}{\rho_0 c_p} \frac{1}{N+1} \left[\frac{2+b+N}{1+R_T} + N \frac{m_\chi + R_T}{1+R_T} \left(\frac{\omega}{2\omega_\chi} \right)^{1/2} (1-i) \right].$$

The acoustic wave excitation efficiency in a transparent medium will be independent of frequency although it will depend on the impedance ratio N and grows with an increasing ratio. The transfer function K_0 of the boundary vibrational velocity behaves analogously. Its frequency dependence will only be manifested with large N at frequencies $\omega \gtrsim \omega_\chi / N^2$.

The modulus of the transfer function is weakly dependent on frequency in an absorbing medium for moderate acoustic impedance ratios N: $N m_\chi (\omega/\omega_\chi)^{1/2} \ll 1$. If the density of the absorbing medium far exceeds that of the transparent medium ($N \gg 1$) a frequency dependence of sound generation efficiency will appear. The phase shift of the transfer function is proportional to $\sqrt{\omega}$.

The transfer functions at moderately low frequencies $\omega \ll \omega_\chi$ in the case of $m_\chi \ll 1$ will behave in a manner analogous to the case of highly thermally conductive media [see Eqs. (3.16)–(3.22)]. This analogy can be attributed to the fact that heat conduction plays a more significant role in thermooptical sound excitation in both cases.

The transfer functions over the frequency range $\omega_T \ll \omega \ll \omega_\chi$ (for $m_\chi \ll 1$) reduce to the form obtained previously for the case of thermally nonconducting media:

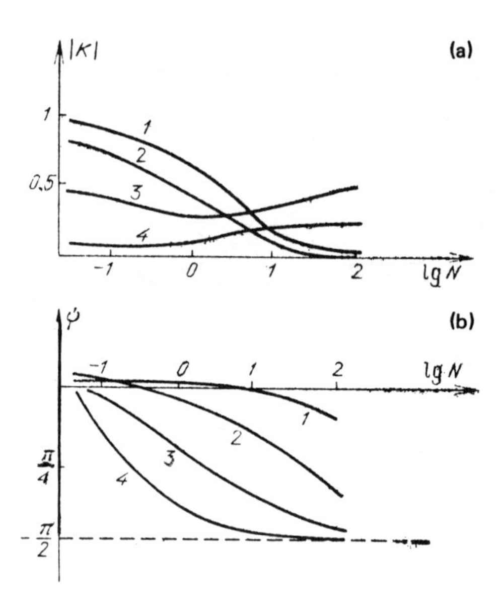

FIG. 3.6. The dependence of the modulus (a) and phase (b) of the transfer function for weak heat conduction ($m_\chi = 10^{-2}$) on the boundary impedance: 1—ω_T; 2—$0.1\omega_a$; 3—ω_a; 4—$10\omega_a$.

$$K(\omega) = \frac{\beta^*}{\rho_0 c_p} \frac{1}{N+1} \frac{1 - iN\omega/\omega_a}{1 + (\omega/\omega_a)^2} ,$$

$$K^{\text{tr}}(\omega) = \frac{\beta^*}{\rho_0 c_p} \frac{N}{N+1} \frac{1 + i\omega/\omega_a}{1 + (\omega/\omega_a)^2} , \tag{3.34}$$

$$K_0(\omega) = -\frac{\beta^*}{\rho_0 c_p} \frac{N}{N+1} \frac{1 + i\omega/\omega_a}{1 + (\omega/\omega_a)^2} .$$

The heat released from laser pulse absorption within this frequency range will not experience heat conduction loss to a significant degree, since the heat conduction length is far less than the size of the absorption region as $\omega \gg \omega_T$.

We note that the phase of transfer function K and the transfer function moduli K^{tr} and K_0 are the most sensitive to N.

The modulus and phase of the transfer function of the absorbing medium are shown in Fig. 3.6 in a plot against the acoustic impedance ratio N for different frequencies. Clearly the highest N gradient in the relation of the transfer function modulus is observed at low and medium frequencies $\omega \lesssim \omega_a$ for values of $N \lesssim 10$. The relation of the transfer function phase has a significant N gradient at medium frequencies $\omega_T < \omega \lesssim \omega_a$ across the entire possible range of N. Sound excitation efficiency is high in this frequency range (see Fig. 3.5) and, hence, it is best to employ the phase of the vibrational velocity of the particles in the absorbing medium over the frequency range $\omega_T < \omega \lesssim \omega_a$ for diagnostics of the boundary between the absorbing and transparent media and to measure the impedance ratio.

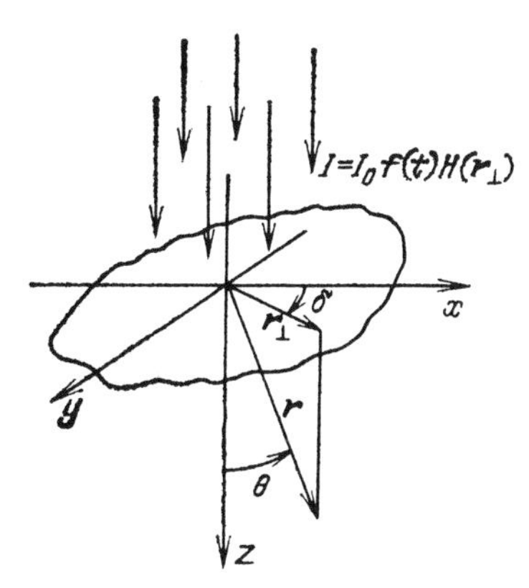

FIG. 3.7. Coordinate system used to analyze laser excitation of surface acoustical waves.

This analysis indicates the role of heat conduction in the thermooptical excitation of sound for the case of an arbitrary type of acoustic boundary (an arbitrary acoustic impedance ratio) in the case of broad laser beams.

3.3. Laser Excitation of Rayleigh Waves

Laser radiation may excite not only bulk waves in solids but also surface acoustic waves (SAW's). This process is described by the same system of wave equations connected through the boundary conditions for the scalar potential and the vortical velocity component normal to the surface. It is necessary in the boundary conditions to account for the fact that a gas is the transparent medium in this case and it is therefore sufficient to consider the last two relations from (3.12); these relations take the form

$$\frac{\partial(\Delta_\perp \varphi)}{\partial z} - \left((2c_T^2)^{-1} \frac{\partial^2 A}{\partial t^2} - \Delta_\perp A \right) = 0,$$

$$\frac{\partial A}{\partial z} + \left[(2c_T^2)^{-1} \frac{\partial^2 \varphi}{\partial t^2} - \Delta_\perp \varphi \right] = 0. \tag{3.35}$$

Wave equations (3.6) and (3.10) maintain their form:

$$c_L^{-2} \frac{\partial^2 \varphi}{\partial t^2} - \Delta \varphi = -\beta^* \frac{\partial T}{\partial t},$$

$$c_T^{-2} \frac{\partial^2 A}{\partial t^2} - \Delta A = 0 \tag{3.36}$$

(it is not necessary to solve the problem in a transparent medium).

We introduce a Fourier transform with respect to time and the transverse coordinates $\mathbf{r}_\perp = \{x,y\}$ (Fig. 3.7) (the transformed functions will be denoted by the "tilde") and a Laplace transform with respect to the coordinate z:

$$\hat{T}(\omega,\mathbf{k}_\perp,p) = \int_0^\infty dz \iiint_{-\infty}^{\infty} dt\, d\mathbf{r}_\perp \exp[i\omega t - i(\mathbf{k}_\perp \mathbf{r}_\perp) - pz]\, T(t,\mathbf{r}_\perp,z)$$

$$= \int_0^\infty \tilde{T}(\omega,\mathbf{k}_\perp,z)\exp(-pz)dz.$$

Here (as in the case of liquids; see Chap. 2) system (3.36) reduces to the system of algebraic equations

$$(k_L^2 - \mathbf{k}_\perp^2 + p^2)\hat{\varphi} - \left(\frac{d\tilde{\varphi}}{dz}\right)\Big|_0 - p\tilde{\varphi}|_0 = -i\omega\beta^* \hat{T},$$

$$(k_T^2 - \mathbf{k}_\perp^2 + p^2)\hat{A} - \left(\frac{d\tilde{A}}{dz}\right)\Big|_0 - p\tilde{A}|_0 = 0, \tag{3.37}$$

where $k_L = \omega/c_L$, $k_T = \omega/c_T$ are the wave vectors of the longitudinal and shear waves, while $\tilde{\varphi}|_0$, $\tilde{A}|_0$, $(d\tilde{\varphi}/dz)|_0$, $(d\tilde{A}/dz)|_0$ are the values of the functions and their derivatives on the boundary $z = 0$. The boundedness of $\tilde{\varphi}$ and $\tilde{A}$ as $z \to \infty$ dictates the absence of poles of their transforms $\hat{\varphi}$ and $\hat{A}$ in the right semiplane $\mathrm{Re}\, p > 0$. The following relations are therefore obtained from Eqs. (3.37) and these radiation conditions:

$$\left(\frac{d\tilde{\varphi}}{dz}\right)\Big|_0 + (\mathbf{k}_\perp^2 - k_L^2)^{1/2}\tilde{\varphi}|_0 = i\omega\beta^* \hat{T}(\omega,\mathbf{k}_\perp,(\mathbf{k}_\perp^2 - k_L^2)^{1/2}),$$

$$\left(\frac{d\tilde{A}}{dz}\right)\Big|_0 + (\mathbf{k}_\perp^2 - k_T^2)^{1/2}\tilde{A}|_0 = 0. \tag{3.38}$$

Boundary conditions (3.35) appear as follows in spectral form:

$$\mathbf{k}_\perp^2\left(\frac{d\tilde{\varphi}}{dz}\right)\Big|_0 = \left(\frac{k_T^2}{2} - \mathbf{k}_\perp^2\right)\tilde{A}|_0,$$

$$\left(\frac{d\tilde{A}}{dz}\right)\Big|_0 = \left(\frac{k_T^2}{2} - \mathbf{k}_\perp^2\right)\tilde{\varphi}|_0. \tag{3.39}$$

The potentials φ and A separately have no physical meaning for a surface acoustic wave. The normal velocity component on the surface of the medium $w(t,\mathbf{r}_\perp, z = 0)$ is of principal interest. Its spectrum $\tilde{w}(\omega,\mathbf{k}_\perp)$ can be expressed through the potentials as follows:

$$\tilde{w}(\omega,\mathbf{k}_\perp) = \left(\frac{d\tilde{\varphi}}{dz}\right)\Big|_0 + \tilde{A}|_0. \tag{3.40}$$

Hence, it is only necessary to solve Eqs. (3.38) and (3.39) and substitute these into Eq. (3.40):

$$\tilde{w}(\omega,\mathbf{k}_\perp) = i\omega\beta^* \frac{[k_T^2(\mathbf{k}_\perp^2 - k_T^2/2)/2]\hat{T}(\omega,\mathbf{k}_\perp,(\mathbf{k}_\perp^2 - k_L^2)^{1/2})}{\mathbf{k}_\perp^2[(\mathbf{k}_\perp^2 - k_L^2)(\mathbf{k}_\perp^2 - k_T^2)]^{1/2} - (\mathbf{k}_\perp^2 - k_T^2/2)^2}\,. \tag{3.41}$$

Therefore, Eq. (3.41) provides a solution to the problem of SAW excitation by a nonstationary, inhomogeneous temperature field.

The surface acoustic waves in Eq. (3.41) correspond to the poles

$$k_\perp = \mp k_R = \mp \omega/c_R,$$

that in turn correspond to the roots of the Rayleigh determinant

$$k_R^2(k_R^2 - k_L^2)^{1/2}(k_R^2 - k_T^2)^{1/2} - (k_R^2 - k_T^2/2)^2 = 0.$$

Carrying out an inverse Fourier transform with respect to frequency and wave vectors, and evaluating the integrals by the calculus of residues, we can obtain the dependence of the surface vibrational velocity on coordinates and time:

$$\begin{aligned} w(t,\mathbf{r}_\perp) = & -(2\pi^2)^{-1}\mathscr{F}\beta^* \int_{-\infty}^{\infty} d\omega \int_0^{2\pi} \exp c_R^{-2}\omega^3 \\ & \times[-i\omega(t - r_\perp \cos(\delta - \xi)/c_R)]\hat{T} \\ & \times(\omega,\omega/c_R,\xi,|\omega|(c_R^{-2} - c_L^{-2})^{1/2})d\xi, \end{aligned} \tag{3.42}$$

where the dimensionless factor

$$\mathscr{F} = \left(\frac{c_R}{2c_T}\right)^2 \left\{\left[2 - \left(\frac{c_R}{c_T}\right)^2\right]\left[2 + \left(1 - \frac{c_R^2}{c_T^2}\right)^{-1} + \left(1 - \frac{c_R^2}{c_L^2}\right)^{-1}\right] - 8\right\}^{-1}$$

is determined solely by the elastic properties of the medium (and ordinarily is of the order of 0.1). Cylindrical coordinates (see Fig. 3.7) were used in Eq. (3.42). The physical meaning of solution (3.42) is simple: this expression is the expansion of the SAW field in plane traveling monochromatic waves. The spectral density of the field is determined by the spectral component of the temperature field at frequency ω and its corresponding spatial frequency ω/c_R and the attenuation coefficient of the surface acoustic waves in the medium bulk $|\omega|(c_R^{-2} - c_L^{-2})^{1/2}$.

The structure of the temperature field is determined by the light intensity distribution in the beam cross section and the dependence of the light absorptivity on depth. We assume a laser beam of intensity

$$I = I_S H(t,\mathbf{r}_\perp)$$

is incident perpendicular to the surface of the body. In this case the temperature increment is described by a type (3.14) heat conduction equation

$$\frac{\partial T}{\partial t} - \chi\,\Delta T = +\left(\frac{I_0}{\rho_0 c_p}\right)H(t,\mathbf{r}_\perp)g(z), \quad I_0 = AI_S, \tag{3.43}$$

where the function $g(z)$ describes the depthwise light intensity distribution:

$$g(z) = -\frac{d}{dz}\exp\left[-\int_0^z \alpha(\zeta)d\zeta\right].$$

$\alpha(\zeta)$ is the distribution of light absorptivity. Since the absorbing medium bounds a vacuum, the thermal flux through the boundary will be equal to zero:

TABLE 3.2. Wave vector moduli in different media.

$\omega/2\pi$, 10^9 s^{-1}	μm^{-1}	Al	Cu	SiO_2	Si
	P	0.2	0.26	0.16	0.1
0.1	k_R	0.22	0.30	0.19	0.13
	$(\omega/\chi)^{1/12}$	2.7	2.3	9.6	3.0
	p	2.0	2.6	1.6	1.0
1.0	k_R	2.2	3.0	1.9	1.3
	$(\omega/\chi)^{1/2}$	8.6	7.4	30	9.4
	p	20	26	16	10
10	k_R	22	30	19	13
	$(\omega/\chi)^{1/2}$	27	23	96	30

$$\left(\frac{\partial T}{\partial z}\right)\Big|_{z=0}=0 \tag{3.44}$$

[compare to Eq. (3.15)].

The solution of system of equations (3.43), (3.44) in spectral form (accounting for the vanishing of the temperature field at infinity):

$$T(z\to\infty)\to 0$$

can be written as

$$\widehat{T}(\omega,\mathbf{k}_\perp,p)=\frac{I_0}{\rho_0 c_p \chi}\widetilde{H}(\omega,\mathbf{k}_\perp)\frac{(\mathbf{k}_\perp^2-i\omega/\chi)^{1/2}\widehat{g}(p)-p\widehat{g}[(\mathbf{k}_\perp^2-i\omega/\chi)^{1/2}]}{(\mathbf{k}_\perp^2-i\omega/\chi)^{1/2}(p^2-\mathbf{k}_\perp^2+i\omega/\chi)}, \tag{3.45}$$

where

$$\widetilde{H}(\omega,\mathbf{k}_\perp)=\iiint_{-\infty}^{\infty}\exp[i\omega t-i(\mathbf{k}_\perp\mathbf{r}_\perp)]H(t,\mathbf{r}_\perp)dt\,d\mathbf{r}_\perp,$$

$$\widehat{g}(p)=\int_0^\infty g(z)\exp(-pz)dz$$

and $\mathrm{Re}[(\mathbf{k}_\perp^2-i\omega/\chi)^{1/2}]>0$. Expression (3.45) is rather cumbersome. However, the wave vector of the thermal wave far exceeds the wave vector of the Rayleigh wave over a frequency band through a few gigahertz $\omega\ll c_R^2/\chi$ (Table 3.2). An assumption of $\mathbf{k}_\perp^2\ll\omega/\chi$ and $p^2\ll\omega/\chi$ in Eq. (3.45) is therefore possible. Equation (3.45) then reduces to

$$\widehat{T}(\omega,\mathbf{k}_\perp,p)=(i\omega)^{-1}(I_0/\rho_0 c_p)\widetilde{H}(\omega,\mathbf{k}_\perp)\widehat{g}(p), \tag{3.46}$$

which corresponds to an approximation of a heat-nonconducting medium.

Equations (3.42) and (3.46) provide a solution to the problem of laser excitation of surface acoustic waves:

$$w(t,\mathbf{r}_\perp) = \frac{\beta^* I_0}{\rho_0 c_p} \mathscr{F} \int_{-\infty}^{\infty} d\omega \int_0^{2\pi} \frac{i\omega^2}{c_R^2} \exp\left[-i\omega\left(t - \frac{r_\perp}{c_R}\cos(\delta - \xi)\right)\right]$$

$$\times \widetilde{H}\left(\omega, \frac{\omega}{c_R}, \xi\right) \hat{g}[|\omega|(c_R^{-2} - c_L^{-2})^{1/2}]d\xi. \qquad (3.47)$$

Equation (3.47) relates the temporal spectrum of the Rayleigh wave to the space-time spectrum of laser radiation intensity. In principle, this permits determining the depthwise distribution of the absorption coefficient of light when the intensity distribution in the beam cross section and the temporal wave form of the laser pulse are known (and vice versa, when absorption is known, to determine the radiation intensity distribution in the beam cross section).

It is advisable to consider two different cases: a fixed and a moving laser beam (the case of a fixed beam corresponds to beam velocities far less than the velocity of the Rayleigh wave). We first consider the case of a fixed beam. Then the spatial and temporal spectra of the sources are factorized:

$$\widetilde{H}(\omega,\mathbf{k}_\perp) = \widetilde{H}(\mathbf{k}_\perp)\widetilde{f}(\omega). \qquad (3.48)$$

This permits introducing the concept of the transfer function by analogy to laser excitation of bulk waves (see Chap. 2 and Sec. 3.2).

If the wave sources are one-dimensional sources—the beam becomes a strip extended along the y axis,

$$H(\mathbf{r}_\perp) = H(x),$$

plane Rayleigh waves will be excited and Eq. (3.47) is significantly simplified. Substituting

$$\widetilde{H}(\mathbf{k}_\perp) = 2\pi\delta(k_\perp \sin \xi)\cdot\widetilde{H}(k_\perp \cos \xi)$$

into Eq. (3.47) and retaining only the wave traveling in the direction of increasing x (an analogous wave is also traveling in the opposite direction) we obtain for the surface vibrational velocity

$$w_+\left(\tau = t - \frac{x}{c_R}\right)$$

$$= \frac{\beta^* I_0}{\pi\rho_0 c_p} \mathscr{F} \int_{-\infty}^{\infty} -i\frac{\omega}{c_R}\exp(-i\omega\tau)$$

$$\times \widetilde{H}\left(\frac{\omega}{c_R}\right)\hat{g}[|\omega|(c_R^{-2} - c_L^{-2})^{1/2}]\widetilde{f}(\omega)d\omega. \qquad (3.49)$$

The function

$$K_1(\omega) = -i\frac{\omega}{c_R}\widetilde{H}\left(\frac{\omega}{c_R}\right)\hat{g}[|\omega|(c_R^{-2} - c_L^{-2})^{1/2}]\frac{\beta^*}{\pi\rho_0 c_p}\mathscr{F} \qquad (3.50)$$

can be called the spectral transfer function of laser SAW sources for the case of a two-dimensional geometry, since it relates the spectrum of a traveling Rayleigh wave to the temporal spectrum of laser radiation. It is clear on this basis that efficient excitation of a Rayleigh wave of frequency ω requires the spectral component of the intensity distribution (corresponding to this frequency) to be significant. In other words, the laser beam must be narrower than the wavelength of the Rayleigh wave $k_R a \lesssim 1$, and moreover, the light absorptivity must be sufficiently large, $\alpha \gtrsim k_R$ (here, the thickness of the absorption layer must be less than the wavelength of the Rayleigh wave). The multiplier $-i\omega/c_R$ in Eq. (3.50) demonstrates that the SAW pulse is proportional to the derivative of the envelope of the intensity distribution at the laser spot $H'(x)$ in a case of a short laser pulse and surface absorption. The difference between SAW excitation transfer function (3.50) and bulk wave transfer function (3.22) is that the SAW spectrum is determined by the Laplace transform of the depthwise source distribution $\hat{g}[|\omega|(c_R^{-2}-c_L^{-2})^{1/2}]$, while the spectrum of the spatial waves is determined by its Fourier transform $\tilde{g}(\omega/c_L)$. Therefore, sources in the bulk of the medium will have no effect on laser excitation of surface acoustic waves (unlike volume wave excitation).

The transfer function of laser sources of surface acoustic waves can be introduced in a more general form. This requires isolation of the traveling wave in solution (3.42). This wave is formed outside the heating zone and hence for $\omega r_\perp/c_R \gg 1$ Eq. (3.42) in the far field reduces to

$$w(t,\mathbf{r}_\perp,\delta) = -2\pi \frac{\beta^* I_0}{\rho_0 c_p} \mathscr{F} \int_{-\infty}^{\infty} \frac{\omega^2}{c_R^2} \left(\frac{c_R}{\pi|\omega|r_\perp}\right)^{1/2} \exp\left[-i\omega\left(t-\frac{r_\perp}{c_R}\right)-\frac{i\pi}{4}\right]$$

$$\times \tilde{H}\left(\frac{\omega}{c_R},\delta\right)\hat{g}[|\omega|(c_R^{-2}-c_L^{-2})^{1/2}]\tilde{f}(\omega)d\omega. \qquad (3.51)$$

Equation (3.51) in the far wave field yields the expansion of the surface acoustic waves in the diverging plane waves. As expected, the angular dependence of Rayleigh wave excitation efficiency is determined solely by the light intensity distribution in the beam cross section. Specifically, for an axially symmetrical laser beam, the SAW will also be axially symmetric. Its amplitude diminishes in proportion to $r^{-1/2}$ with distance, as would be expected for a cylindrical wave. The function

$$K_2(\omega,\delta) = -|\omega|^{3/2}\tilde{H}\left(\frac{\omega}{c_R},\delta\right)\hat{g}[|\omega|(c_R^{-2}-c_L^{-2})^{1/2}] \qquad (3.52)$$

can also be referred to as a transfer function. Unlike the plane-wave case (as in the three-dimensional bulk wave case) it is dependent on the angle of arrival of the waves.

We estimate the amplitude of the vibrational velocity in the laser-excited surface wave. The amplitude for a monochromatic wave excited by laser intensity modulation at frequency ω in a case of a beam narrow compared to the wavelength of the Rayleigh wave ($\omega a/c_R \lesssim 1$) and strong absorption ($\omega/c_R \lesssim \alpha$) will be

$$w_a \sim 2\mathscr{F}\beta^* I_0 a/\rho_0 c_p c_R$$

for a plane beam and

$$w_a \sim \mathscr{F}\beta^* \frac{W}{\rho_0 c_p} \frac{\omega^2}{c_R^2} \left(\frac{c_R}{\pi r_\perp \omega}\right)^{1/2}$$

in the cylindrically symmetrical case (W is the variable part of the laser radiation power). For aluminum this yields in the planar case

$$w_a \sim (3.8\times 10^{-6}\ \mathrm{cm}^3/\mathrm{J}) I_0$$

and a quantity that is $(a/r_\perp)^{1/2}$ times smaller in the second case. These estimates for the intensity conversion efficiencies yield

$$I_{\mathrm{SAW}}/I_0 \sim (7\times 10^{-13}\ \mathrm{cm}^2/\mathrm{W}) I_0.$$

The amplitude of a thermooptically excited SAW is directly proportional to the absorbed laser radiation intensity. The energy conversion efficiency is therefore also directly proportional to the light intensity. These mechanisms which are valid for optical excitation of bulk waves also hold in the case of SAW excitation. As in the case of volume wave excitation, the optoacoustic sources of Rayleigh waves are broadband sources. Hence, they can be used to generate short and powerful SAW pulses. If short ($c_R\tau_L \ll a$) laser pulses are used in a medium that strongly absorbs light ($\alpha a \gg 1$) for thermooptical excitation of SAW's, the SAW spectrum in Eq. (3.49) will be determined by the radiation distribution over the laser spot $\widetilde{H}(\omega/c_R)$. The amplitude of the surface displacement velocity can be determined by the formula

$$w_a \sim 2\mathscr{F}\beta^* \frac{\mathscr{E}_0 c_R}{\rho_0 c_p a}$$

in the planar case and the amplitude is $(a/r_\perp)^{1/2}$ times smaller in the case of a cylindrical beam geometry; $\mathscr{E}_0$ is the absorbed energy density at the surface. In a planar geometry these estimates yield for aluminum

$$w_a \sim (1\ \mathrm{cm/s})(\mathscr{E}_0/a\ \mathrm{cm}^3/\mathrm{J}).$$

If the depth of penetration of light is significant $\alpha c_R\tau_L \ll 1$, $\alpha a \ll 1$, the wave form of the SAW pulse will be determined solely by the depthwise distribution of light absorption. In this case a spectral representation of the solution is most convenient for analysis.

We now consider Rayleigh wave excitation by a moving beam:

$$H(t,\mathbf{r}_\perp) = H(\mathbf{r}_\perp - \mathbf{V}_0 t) f(t).$$

In this case [unlike Eq. (3.48)] the spatial and temporal spectra of light intensity are related:

$$\widetilde{H}(\omega,\mathbf{k}_\perp) = \widetilde{H}(\mathbf{k}_\perp)\widetilde{f}(\omega - (\mathbf{k}_\perp\mathbf{V}_0)).$$

Without loss of generality it is possible to direct the beam velocity $\mathbf{V}_0$ along the x axis. Then, Eq. (3.47) becomes

$$w(t,\mathbf{r}_\perp) = -\frac{\beta^* I_0}{\rho_0 c_p}\mathscr{F}\int_{-\infty}^{\infty} d\omega \int_0^{2\pi} i\frac{\omega^2}{c_R^2}\exp\left[-i\omega\left(t - \frac{r_\perp}{c_R}\cos(\delta-\xi) + t\right)\right]$$

$$\times \widetilde{H}\left(\frac{\omega}{c_R},\xi\right)\widetilde{f}\left(\omega - \frac{\omega}{c_R}V_0\cos\xi\right)\hat{g}[|\omega|(c_R^{-2} - c_L^{-2})^{1/2}]d\xi.$$

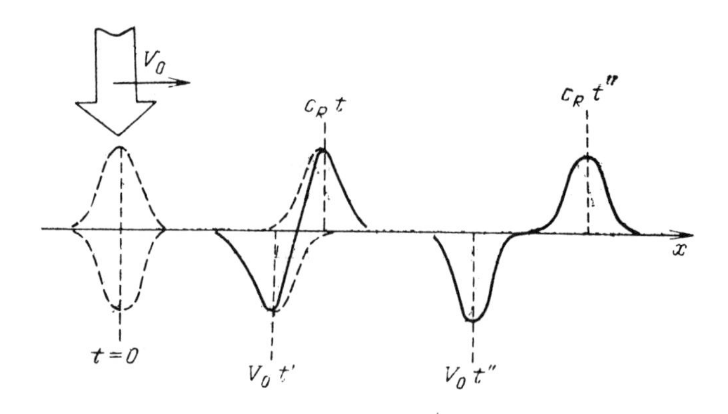

FIG. 3.8. Excitation of a Rayleigh wave by a moving laser beam ($t'' > t' > 0$).

We consider a one-dimensional source distribution $H(\mathbf{r}_\perp) = H(x)$:

$$\widetilde{H}(\mathbf{k}_\perp) = 2\pi\delta(k_\perp \sin \xi)\widetilde{H}(k_\perp \cos \xi)L$$

(L is beam length on the y axis) and isolate the wave traveling in the positive direction on the x axis:

$$w_+\left(\tau = t - \frac{x}{c_R}\right) = -\frac{\beta^* I_0}{\rho_0 c_p}\mathcal{F}\int_{-\infty}^{\infty} i\frac{\omega}{c_R}\exp(-i\omega\tau)\widetilde{H}\left(\frac{\omega}{c_R}\right)\widetilde{f}\left[\omega\left(1-\frac{V_0}{c_R}\right)\right]$$

$$\times\widehat{g}[|\omega|(c_R^{-2} - c_L^{-2})^{1/2}]d\omega. \tag{3.53}$$

Solution (3.53) for a moving beam differs from Eq. (3.49) in the case of a fixed beam by the additional Doppler factor $(1 - V_0/c_R)$ in the spectrum of the temporal dependence of light intensity, i.e., a spatial harmonic with the wave vector ω/c_R is excited by the spectral component at the frequency $\omega(1 - V_0/c_R)$.

The case of cw radiation is of primary interest in the laser excitation of SAW's by a moving beam (no sound excitation occurs in this case with a fixed beam). Let

$$f(t) = \theta(t),$$

where θ is the Heaviside function. Then, with surface absorption ($\alpha a \gg 1$) wave (3.53) is written as

$$w_+ = \frac{\beta^* I_0}{\rho_0 c_p}\mathcal{F} L c_R \frac{H(x - c_R t)}{c_R - V_0}\theta(t).$$

Accounting for the stimulated wave traveling with the beam [and, naturally, neglected in solution (3.49)] the vibrational surface velocity can be given as

$$w(t,x) = \frac{\beta^* I_0}{\rho_0 c_p}\mathcal{F} L c_R \theta(t)\,\frac{H(x - c_R t) - H(x - V_0 t)}{c_R - V_0}.$$

This wave consists of two diverging pulses (Fig. 3.8) whose amplitude grows as $|V_0 - c_R|^{-1}$ as the beam scanning velocity V_0 approaches the velocity of the Rayleigh wave c_R. With precise coincidence $V_0 = c_R$ the wave profile is proportional to the derivative of the intensity distribution. The surface displacement mimics this distribution correspondingly:

$$w(t,x) = -\frac{\beta^* I_0}{\rho_0 c_p}\mathscr{F} L c_R \theta(t) t \frac{\partial}{\partial x} H(x - c_R t).$$

Naturally this increase may be limited by diffraction (at distances $x \sim L^2/a$) or a nonlinearity. We note that a stage-by-stage analysis (see Sec. 2.3) is not suitable in the case of a moving beam.

This analysis is easily generalized to the case of sources of finite size.

Additional information on laser excitation of sound in isotropic solids can be found in Refs. 5–7.

3.4. Optical Generation of Acoustic Waves in Anisotropic Solids

We consider a general methodological approach to the description of acoustic wave generation by electromagnetic irradiation of crystals. In the final analysis this will substantially enhance our understanding of the physical mechanism of thermoelastic excitation of sound.

According to elasticity theory[1] a force **f** acting on a crystalline lattice results from the spatial inhomogeneity of a certain tensor σ_{ij}. In this case the equation of motion takes the form

$$\rho_0 \frac{\partial^2 u_i}{\partial t^2} = f_i = \frac{\partial \sigma_{ij}}{\partial x_j}. \tag{3.54}$$

The work of this force directed on crystal deformation is related to the change in the free-energy density F of the crystal-electromagnetic field system by the thermodynamic identity

$$d\tilde{F} = d(F - E_i D_i/4\pi) = s\,dT - (4\pi)^{-1} D_i\,dE_i + \sigma_{ij}\,d\xi_{ij}, \tag{3.55}$$

where

$$\xi_{ij} = \frac{\partial u_i}{\partial x_j}. \tag{3.56}$$

The effect of the magnetic field is neglected in energy conservation law (3.55) and the analysis below. The free-energy density $\tilde{F}$ introduced here is dependent on the electrical field. This quantity is more easily used than F which is a function of the electrical induction.[8] Relation (3.55) makes it possible to determine the density of forces acting on the body if the dependence of $\tilde{F}$ on the tensor ξ_{ij} is known:

$$f_i = \frac{\partial}{\partial x_j}\left(\frac{\partial \tilde{F}}{\partial \xi_{ij}}\right)_{T,E_i}. \tag{3.57}$$

Equations (3.54)–(3.57) can be used to describe sound propagation in crystals in zero electromagnetic field. Traditionally, the elastic energy density of the crystal is expanded in powers of the strain tensor U_{ij}:[9]

$$U_{ij} = (\xi_{ij} + \xi_{ji} + \xi_{ki}\xi_{kj})/2. \tag{3.58}$$

Using the first terms of the expansion

$$\widetilde{F}_{\text{elas.}} = F_{\text{elas.}} = F(T) + c_{ijkl}U_{ij}U_{kl}/2 \tag{3.59}$$

and linearized strain tensor (3.4), we obtain equations for the acoustic eigenwaves (of infinitely small amplitude) of an anisotropic solid:

$$\rho_0 \frac{\partial^2 u_i}{\partial t^2} - c_{ijkl} \frac{\partial^2 u_l}{\partial x_j \, \partial x_k} = 0. \tag{3.60}$$

Here c_{ijkl} are the second-order isothermal elasticity moduli. We note that the tensor σ_{ij} is identical to the elastic stress tensor from Ref. 1 in this approximation. As we know[9] Eq. (3.60) determines the possibility for propagation of three acoustic waves in an arbitrary direction of the crystal: a longitudinal wave L and two transverse waves (a fast wave FT and a slow wave ST).

The nonlinearity of the strain-displacement relation (3.58) will cause the wave equation to be nonlinear even within the framework of approximation (3.59). However, terms of the same order also appear when the possible dependence of the second-order elasticity moduli on strain are accounted for. Therefore, in order to derive a wave equation accurate to the quadratic terms it is also necessary to use the following terms of the expansion of the free-energy density $F_{\text{elas.}}$:[9]

$$\widetilde{F}_{\text{elas.}} = F(T) + c_{ijkl}U_{ij}U_{kl}/2 + c_{ijklmn}U_{ij}U_{kl}U_{mn}, \tag{3.61}$$

where c_{ijklmn} is the tensor of the third-order elasticity moduli. Cubic terms obtained in Eq. (3.61) by means of Eqs. (3.56) and (3.58) represent interaction between elastic waves. We provide as an example an equation that describes the propagation of a plane longitudinal acoustic wave along the axis $x_1 \equiv z$:

$$\frac{\partial^2 u_1}{\partial t^2} - c_L^2 \frac{\partial^2 u_1}{\partial z^2} = -c_L^2 \varepsilon \frac{\partial}{dz} \left(\frac{\partial u_1}{\partial z} \right)^2. \tag{3.62}$$

Physically in Eq. (3.62), $c_L^2 \equiv c_{11}/\rho_0$ is the longitudinal sound velocity in direction x_1, while $\varepsilon = -3(1 + 2c_{111}/c_{11})/2$ is the acoustic nonlinearity parameter.[10,11] The number of subscripts on the elasticity modulus tensors was reduced by commonly used principles.[9] Using the slowly varying profile method,[10] Eq. (3.62) can be reduced to simple wave equation (2.40).

In order to describe acoustic wave generation processes in accordance with Eqs. (3.54)–(3.57) it is necessary to determine the crystalline-lattice–electromagnetic-field interaction energy. The first (quadratic) terms of the expansion of $\widetilde{F}$ in powers of u_{ij} (3.4) and E_k are represented as[8,9]

$$\widetilde{F} = \widetilde{F}_0 + \frac{1}{2} c_{ijkl}u_{ij}u_{kl} - \frac{\varepsilon_{ij}}{8\pi} E_i E_j - e_{kij}E_k u_{ij}. \tag{3.63}$$

Here ε_{ij} and e_{kij} are the permittivity and piezoelectric tensors. Relation (3.63) leads to the following equation describing sound excitation in a piezodielectric:

$$\rho_0 \frac{\partial^2 u_i}{\partial t^2} - c_{ijkl} \frac{\partial^2 u_l}{\partial x_j \, \partial x_k} = -\frac{\partial}{\partial x_j} (e_{kij}E_k). \tag{3.64}$$

We note that an elastic wave in piezocrystals according to Eq. (3.55) is accompanied by the electrical induction field $\mathbf{D}_k = -4\pi(\partial\tilde{F}/\partial E_k)_{T,\xi ij} = 4\pi e_{kij}u_{ij}$ even in zero external electrical field. Hence, in order to describe acoustic wave propagation in piezocrystals it is necessary to solve a coupled (through the piezoeffect) system of equations for the electrical field and deformations.[9] In describing the sound generation processes it is natural to assume in a first approximation that the external electrical field E_k far exceeds the field E'_e coupled to the excited acoustic wave:

$$E'_e \sim 4\pi\varepsilon_{ke}^{-1} e_{kij}\xi_{ij} \sim 4\pi\varepsilon_{ke}^{-1} e_{kij} c_{ijmn}^{-1} e_{pmn} E_p \cong (4\pi e^2/\varepsilon c) E_e.$$

The electrical field in Eq. (3.64) in the sound generation region can therefore be assumed to be known and identical to the external field in the case of a small electromechanical coupling factor $(4\pi e^2/\varepsilon c)^{1/2}$.[9]

When necessary it is possible to account for weak renormalization of the elastic moduli and frequency dispersion of the acoustic waves in the piezocrystals.[9] Specifically, Eq. (3.64) predicts the possibility for direct conversion (without a change in frequency) of electromagnetic radiation into coherent acoustic waves by transmitting light through an ideal vacuum-piezocrystal boundary.[9] In this case, the acoustic wave sources are, according to Eq. (3.64), δ localized at the crystal surface.

Reference 12 reported the experimental generation of coherent transverse acoustic wave at frequencies of 0.891 and 2.53 THz at helium temperatures using this method. Modulated laser radiation in the far-IR range (light wavelengths of 337 and 118 μm) was used for this purpose. However, subsequently several groups of researchers were not able to reproduce the experiment from Ref. 12 even using more advanced signal storage and processing techniques.[13] In addition to strong scattering of terahertz acoustic waves during propagation, an additional reason for these may be, as reported in Ref. 13, the nonideal nature of the piezocrystal surface.

We employ a simple calculation to demonstrate this assumption. We consider the generation of longitudinal sound by an electromagnetic wave polarized and traveling along the z axis:

$$\mathbf{E} = E_0 f(t)\mathbf{n}_z \exp[-i\omega_L(t - z/c_n)] \equiv E_z\mathbf{n}_z.$$

Excitation of a plane longitudinal wave will be described by the inhomogeneous equation

$$\rho_0\left(\frac{\partial^2 v}{\partial t^2} - c_L^2\frac{\partial^2 v}{\partial z^2}\right) = -\frac{\partial^2(e_{11}E_z)}{\partial z\,\partial t}.$$

We model damage of the near-surface crystal layer, assuming that the piezoproperties are not manifested suddenly (at the boundary $z = 0$), but rather are gradually manifested with distance from the surface:

$$e_{11} \equiv e_0[1 - \exp(-z/l)].$$

Here l represents the depth of the damaged layer.

We utilize relations (2.33) and (2.35) to calculate the vibrational velocity spectrum. In this case $G(t,z) = e_{11}E_Z$, and for a free surface $z = 0$ we obtain

$$\hat{G}(\omega,p) = e_0 E_0 \tilde{f}(\omega - \omega_L)[(i\omega_L/c_n - p - l^{-1})^{-1} - (i\omega_L/c_n - p)^{-1}],$$

$$\tilde{v}(\omega) = -\frac{e_0 E_0}{\rho_0 c_L} \tilde{f}(\omega - \omega_L)\left[\frac{1}{1-(\omega_L c_L/\omega c_n)^2} - \frac{1}{1-(\omega_L c_L/\omega c_n + i c_L/\omega l)^2}\right].$$

The last relation in the frequency range $\omega \sim \omega_L$ can be significantly simplified due to the significant difference in the speed of light c_n and speed of sound c_L:

$$\tilde{v}(\omega) = -(e_0 E_0/\rho_0 c_L)\tilde{f}(\omega - \omega_L)[1 + (l\omega/c_L)^2]^{-1}. \tag{3.65}$$

The multiplier $\tilde{f}(\omega - \omega_L)$ in representation (3.65) describes the spectrum of the acoustic waves in the case of an ideal boundary ($l = 0$). The sound is quasimonochromatic ($\omega \simeq \omega_L$) for $\omega_L \gg \tau_L^{-1}$. However, according to Eq. (3.65), the nonideal nature of the boundary will effectively suppress high-frequency acoustic waves of frequencies $\omega \gtrsim c_L/l$. Physically, this is due to the fact that sound is not synchronously generated at different points in the surface layer and as a result is quenched by interference. The sound generation efficiency diminishes with increasing l. The thickness of the damaged layer cannot exceed the wavelength of the acoustic wave (~1–10 nm) for efficient sound generation in the terahertz range. It is difficult from the engineering standpoint to fabricate a piezocrystal with such a surface quality.[13]

Lasers in the visible range cannot be used to generate acoustic waves by the surface piezoeffect, since acoustic waves of such high frequencies ($\omega \sim 10^{15}\ \mathrm{s}^{-1}$) will not travel through the crystals. Moreover, there will be no piezoeffect in any crystals having a center of symmetry. In this case, it is necessary to account for the subsequent terms in the expansion of the interaction energy density. From the physical viewpoint the cubic terms in $\tilde{F}$ are related to the possible dependence of the tensors c_{ijkl}, ε_{ij}, and e_{kij} on strain and the electrical field.[14] For example, the dependence of the permittivity on u_{ij} and the piezomoduli on E_l makes a contribution to energy proportional to the combination $u_{ij}E_kE_l$. This nonlinear interaction determines electrostrictive generation of sound

$$f_i \sim \frac{\partial}{\partial x_j}(a_{ijkl}E_kE_l),$$

where a_{ijkl} is the photoelastic, or electrostriction, tensor. Such a sound generation mechanism may be quite active in the field of two convergent light beams, if the optical dispersion of the medium permits synchronous generation of sound at the difference frequency ($\omega = \omega_{L1} - \omega_{L2}$, $\omega_{L1,2}$ are the laser radiation frequencies).

The wave equation describing electrostriction sound excitation in a crystal takes the form

$$\rho_0 \frac{\partial^2 u_i}{\partial t^2} - c_{ijkl}\frac{\partial^2 u_l}{\partial x_j \partial x_k} = -(8\pi)^{-1}\frac{\partial}{\partial x_j}(a_{ijkl}E_kE_l).$$

Let two plane light waves polarized along the x axis converge along z axis:

$$\mathbf{E} = E_1\mathbf{n}_x \sin[\omega_{L1}(t - z/c_{n1})] + E_2\mathbf{n}_x \sin[\omega_{L2}(t - z/c_{n2}) + \varphi_0].$$

Here φ_0 is a fixed phase shift.

Assuming that the special selection of the geometry of the problem and a crystal of specific symmetry makes it possible to describe[9] interaction by a single electrostriction constant Y, we obtain the following equation for the longitudinal acoustic waves:

$$\rho_0\left(\frac{\partial^2 v}{\partial t^2} - c_L^2 \frac{\partial^2 v}{\partial z^2}\right) = -\frac{\partial^2}{\partial z\, \partial t}\left[\frac{YE_1E_2}{8\pi}\cos(\omega t - k_a z + \Delta kz - \varphi_0)\right]. \tag{3.66}$$

Harmonic sources at frequencies $2\omega_{L1}$, $2\omega_{L2}$, $\omega_{L1}+\omega_{L2}$ that do not fall within the spectral range of acoustic waves are dropped on the right-hand side of Eq. (3.66). The designation Δk is used for the difference of the wave numbers of the acoustic and light waves:

$$\Delta k = k_a - k_{L1} + k_{L2} \equiv c_L^{-1}\omega - c_{n1}^{-1}\omega_{L1} - c_{n2}^{-1}\omega_{L2}.$$

The quantity $\omega/(k_a - \Delta k)$ represents the velocity V_0 of the acoustic sources. With phase matching ($\Delta k = 0$), the velocity V_0 will match the sound velocity: $V_0 = \omega/k_a = c_L$. In this case we expect efficient (synchronous) generation of acoustic waves in accordance with Sec. 1.3.

General solution (2.36) can be used to describe the profile of the vibrational velocity v in an acoustic wave if we additionally assume that the laser beam interaction region is spatially limited ($0 \leqslant z \leqslant L$). Assuming in Eq. (2.36)

$$G(z,t) = \frac{YE_1E_2}{8\pi}\cos[\omega t - (k_a - \Delta k)z - \varphi_0],$$

we obtain in the case of a rigid boundary $z=0$ a representation that will be valid in the domain $z \geqslant L$:

$$v(\tau) = -\frac{YE_1E_2\omega}{16\pi\rho_0 c_L^2}\left\{\frac{\sin(\Delta kL/2)}{\Delta k/2}\sin\left(\omega\tau + L\frac{\Delta k}{2} - \varphi_0\right)\right.$$
$$\left. + \frac{\sin[L(\Delta k/2 - k_a)]}{\Delta k/2 - k_a}\sin\left[\omega\tau + L\left(\frac{\Delta k}{2} - k_a\right) - \varphi_0\right]\right\}.$$

The second term here describes acoustic waves reflected by the boundary $z=0$. During generation these waves travel in the opposite direction of the source motion. If $|\Delta k| \ll k_a$, their generation will not be efficient compared to excitation of the comoving wave. The amplitude of the wave traveling from the boundary $z=0$ will be highly dependent on the shift of the wave vectors Δk and the length of the interaction region of the counterpropagating electromagnetic waves:

$$|v|_{\max} = \frac{|Y|E_1E_2\omega}{16\pi\rho_0 c_L^2}\left|\frac{\sin(L\Delta k/2)}{\Delta k/2}\right|.$$

Consequently, for a fixed value of Δk the acoustic wave amplitude grows with the length L if the last one does not exceed the coherence length $L_c \equiv \pi/|\Delta k|$ (Fig. 3.9). A further increase in L will cause the wave to shift towards the domain of the antiphase sources (due to the difference between the sound and the source velocities) and to decay. For $L > 2L_c$, the wave begins to rise again (however, with an

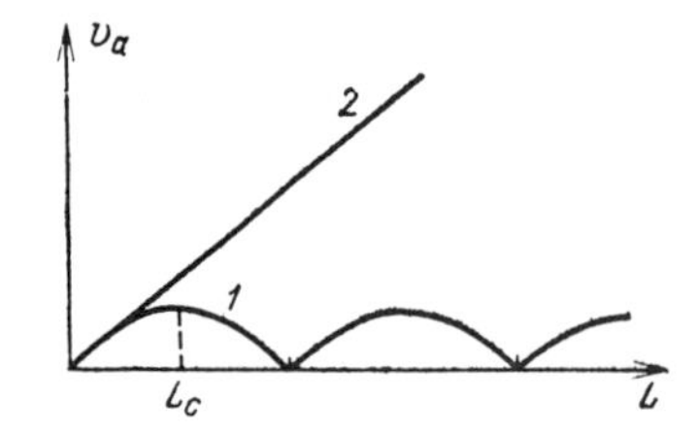

FIG. 3.9. Acoustic wave amplitude plotted as a function of the wave interaction length for $\Delta k \neq 0$ (1) and $\Delta k = 0$ (2).

opposite phase), etc. In the case of exact phase matching ($\Delta k = 0$) the coherence length becomes infinite, while the amplitude of the acoustic wave will rise in proportion to the length L. In this situation it may be necessary to account for the linear and (or) nonlinear absorption of acoustic waves (Sec. 6.3) to describe limiting of the sound amplitude.

The use of an expansion of the energy in powers of the strain tensor u_{ij} and the electrical field E_k of the type in (3.63) effectively assumes that the electromagnetic field has a direct effect on the motion of the crystal atoms. Field interaction with the charges bound to the atoms is implied. Where crystals may contain or support free charge carriers (metals, semiconductors) these carriers can be directly exposed to electromagnetic radiation. The subsequent interaction of free carriers with the crystal lattice may produce sound generation.

Underlying the description of this interaction is the assumption that an acoustic wave modulating the interatomic distances may thereby modulate the band structure of the crystal. It is assumed in a first approximation that shifts of the electron energy levels in the conduction band and holes in the valence band are proportional to the strain tensor of the crystal u_{ij}. Consequently, the free energy of the charge carriers and the lattice in the acoustic wave field will shift by[9]

$$\Delta F = n_e d_{ij}^{(e)} u_{ij} + n_h d_{ij}^{(h)} u_{ij}. \tag{3.67}$$

Here, n and d_{ij} represent the concentrations and deformation potential tensors of the charge carriers; the indices e and h are used for the electrons and holes, respectively. In the general case Eq. (3.67) must also include a sum over all remaining energy bands. However, it is qualitatively evident that the acoustic wave will have an insignificant effect on shell electrons strongly coupled to the atoms (the internal electrons). We also note that the notation for the interaction energy in the form of (3.67) assumes the change in the carrier energy is independent of its position within the given energy band, i.e., its wave vector. In the subsequent approximation accounting for this relation is equivalent to altering the effective masses of the carriers in the strain field.[9]

Relation (3.57) suggests that interaction (3.67) will produce a force on the crystal lattice of

$$f_i = \frac{\partial[n_e d_{ij}^{(e)} + n_h d_{ij}^{(h)}]}{\partial x_j}. \tag{3.68}$$

According to (3.68) sound may be generated by optical radiation-induced spatial modulation of the free charge-carrier concentration.[15–17]

At high conduction-electron concentrations (such as in metals) Coulomb interaction effectively hinders the electron spatial rearrangement, since modulation of the concentration of carriers of only a single type will unavoidably result in charge modulation. In this sense it is possible to achieve significantly higher modulation of carrier concentration by means of interband light absorption in semiconductors. Indeed, in this case, simultaneous electron and hole creation will not violate quasineutrality. The equality $n'_e = n'_h = n'$ holds for the photoexcited electron and hole concentrations, while the acoustic sources in the wave equation take the form

$$f_i = d_{ij} \frac{\partial n'}{\partial x_j}, \tag{3.69}$$

where d_{ij} is the sum deformation potential tensor of electrons and holes $(d_{ij} \equiv d_{ij}^{(e)} + d_{ij}^{(h)})$.

Henceforth, we will use the deformation potential constant d (I.9) to describe sound generation by the concentration-deformation mechanism (3.69) in the model of an isotropic solid. It is clearly evident that $d_{ij} = d\delta_{ij}$. Hence, the effect of shear deformations on carrier energy is ignored, which is valid for spherical energy surfaces.[9,18] We emphasize that description (3.69) which treats tensor d_{ij} as a constant evidently cannot be used when photogeneration of electron-hole pairs causes a substantial rearrangement of the band structure in the material. However, the issue of the behavior of $d_{ij} = d_{ij}(n')$ in such essentially nonequilibrium conditions has been virtually ignored to date.

There is one additional possibility for sound excitation by optical generation of spatially inhomogeneous charge-carrier distributions in piezoelectric semiconductors. A spatially inhomogeneous electrical field (a Dember field) will arise in this case even during ambipolar diffusion of a quasineutral electron-hole plasma. This field which is caused by the local charge separation is in fact responsible for the joint motion of electrons and holes that have different mobilities. In piezoelectric semiconductors the onset of such a field is accompanied by sound generation by the piezoeffect (3.63), (3.64).[19] The simultaneous application of a constant external electrical field[20–22] will facilitate charge separation and generation of inhomogeneous electrical fields in the spatially inhomogeneous photoexcitation range.

When electromagnetic irradiation heats the crystal, this approach makes it possible to describe thermoelastic sound generation by formally introducing the contribution associated with crystal deformation from changing temperature into the free energy of the system:[23]

$$F = F_0 + \tfrac{1}{2} c_{ijkl} u_{ij} u_{kl} - c_{ijkl} \alpha_{ij} u_{kl} T', \tag{3.70}$$

where T' is the temperature increment. Physically α_{ij} is the tensor of the coefficients of thermal expansion.[23] The wave equation corresponding to representation (3.70) takes the form

$$\rho_0 \frac{\partial^2 u_i}{\partial t^2} - c_{ijkl} \frac{\partial^2 u_l}{\partial x_j \partial x_k} = -c_{ijkl} \alpha_{kl} \frac{\partial T'}{\partial x_j}. \tag{3.71}$$

The following issues remain open. What physical interaction makes a contribution to energy proportional to temperature and strain? What crystal subsystem do the deformation waves interact with? In order to answer these questions we need only recall that the temperature characterizes thermal vibrations of the lattice, i.e., chaotic atomic motion near the equilibrium. Thermoelastic sound generation is therefore the result of the elastic interaction of regular acoustic waves with random acoustic waves describing thermal motion of the crystal lattice. We will provide a more detailed description of this process.

3.5. Laser Generation of Nonequilibrium Acoustic Phonons

We recall the method of describing thermal vibrations in a solid. An analysis of a number of simple physical situations reveals that pulsed optical irradiation can induce essentially nonequilibrium random acoustic fields in crystals.

Arbitrary motion of a solid can be represented as a superposition of plane harmonic waves of different polarizations within the framework of elasticity theory of a continuous medium. Each of the waves is characterized by an amplitude and a phase. The wave vector **k** is the continuous parameter that identifies the plane acoustic wave of a given polarization. Its direction indicates the direction of the phase velocity. The specific nature of a particular crystal is manifested in the dispersion law of acoustic waves defined by Eq. (3.60): the dependence of frequency ω on wave vector **k**: $\omega_{\mathrm{ph}} = \omega_{\mathrm{ph}}(\mathbf{k})$. The subscript ph $= L, FT, ST$ labels the acoustic modes. We note that there are no optical modes of vibrations associated with the possibility of a relative shift of the crystal sublattices within the framework of the continuous medium model.

Both an acoustic wideband pulse and thermal lattice vibrations can be resolved into plane waves. The specific nature of the latter lies in the fact that thermal vibrations are chaotic vibrations. In this case chaotic refers to the fact that with a large number of waves responsible for thermal motion of the medium, we cannot identify their phase variations. This is due to the fact that even weak interactions between isolated waves (when there is a large number of waves) causes rapid initiation of chaos of the phases of the individual vibrations, although there may be slow changes in vibration amplitude.[24] It is therefore possible to attempt to simplify the description of lattice thermal vibrations by using statistical averaging (averaging over the phases). Indeed, such an approach makes it possible to obtain a kinetic equation for acoustic noise[25] without employing the apparatus of quantum mechanics.

From the viewpoint of quantum mechanics the energy of vibrations (which is proportional to the square of their amplitude and is independent of phase) cannot be measured simultaneously with the phase. Hence, information on the phase of the acoustic waves vanishes upon quantization of the acoustic vibrations of the continuous medium. After quantum-mechanical averaging (which qualitatively can be compared to statistical phase averaging) the vibrational state of the solid is characterized by the phonon distribution function $N_{\mathrm{ph}}(\mathbf{k})$. This function is equal to the number of elastic energy quanta (number of phonons) in the wave vector interval $\mathbf{k} - \mathbf{k} + d\mathbf{k}$ per unit of volume of the matter. Phonon energy is equal to $\hbar\omega_{\mathrm{ph}}(\mathbf{k})$, and all phases are assumed to be equally probable.[23] At the same time acoustic pulses excited from thermal expansion of the medium are coherent in the sense that the phase is an integral characteristic of pulse propagation. Such a crystal state cannot be described by the function $N_{\mathrm{ph}}(\mathbf{k})$.

It is possible to treat the phonon distribution function $N_{\rm ph}(\mathbf{k})$ as dependent on coordinates and time: in describing nonstationary spatially inhomogeneous chaotic crystal vibrations $N_{\rm ph}(\mathbf{k},\mathbf{r},t)$. The basis for this approach is that plane waves with similar wave vectors form upon superposition of the wave packets localized in space.[9,23] However, it is important to recall in this case that the spatial length of the wave packet $|d\mathbf{r}|$ cannot, by the uncertainty relation, be less than a quantity of the order of $|d\mathbf{k}|^{-1}$ ($|d\mathbf{k}|$ is an estimate of the width of the packet in the wave-vector space). Consequently, it makes sense to analyze the spatial variations of the function $N_{\rm ph}(\mathbf{k})$ only at distances far exceeding the lengths of acoustic waves producing the packet. Analogously the change in $N_{\rm ph}(\mathbf{k})$ over time can only be analyzed at times dt exceeding the acoustic vibration periods. Otherwise the wave-packet energy becomes undetermined: $d\omega\, dt \sim 1$. Given these circumstances the phonons are conveniently represented as particles transporting energy in the solid. In this case $N_{\rm ph}(\mathbf{k},\mathbf{r},t)$ is the phonon concentration in $\mathbf{k}$ space per unit of volume of the material.

In the absence of phonon-creation or -destruction processes, the complete time derivative of the distribution function is equal to zero:

$$\frac{dN_{\rm ph}}{dt} = \frac{\partial N_{\rm ph}}{\partial t} + \frac{\partial N_{\rm ph}}{\partial \mathbf{k}}\frac{\partial \mathbf{k}}{\partial t} + \frac{\partial N_{\rm ph}}{\partial \mathbf{r}}\frac{\partial \mathbf{r}}{\partial t} = 0. \tag{3.72}$$

According to wave-packet propagation theory[23] such packets travel at a group velocity $\partial \mathbf{r}/\partial t = \partial\omega/\partial\mathbf{k}$. The change in the wave vector is determined by the possible spatial inhomogeneity of the medium or the effect of external fields which modulate the dispersion law: $\partial\mathbf{k}/\partial t = -\partial\omega/\partial\mathbf{r}$. Then Eq. (3.72) describing phonon propagation takes the form

$$\frac{\partial N_{\rm ph}}{\partial t} + \frac{\partial\omega}{\partial\mathbf{k}}\frac{\partial N_{\rm ph}}{\partial\mathbf{r}} - \frac{\partial\omega}{\partial\mathbf{r}}\frac{\partial N_{\rm ph}}{\partial\mathbf{k}} = 0. \tag{3.73}$$

The phonons convey energy within the crystal without scattering in the case $\partial\omega/\partial\mathbf{r} = 0$. The direction of phonon propagation can be altered only from reflection off the crystal boundaries. Such vibrational energy-transfer conditions are called ballistic phonon heat conduction.[26]

Ballistic phonon propagation breaks down, for example, upon phonon scattering by defects, impurities, and dislocations. The number of phonons in mode $\mathbf{k}$ changes from interaction with other crystal subsystems; specifically, electrons in metals and semiconductors. Moreover, when the anharmonism of lattice vibrations (3.58), (3.61) is taken into account, phonon interaction will cause their spectral and polarization redistribution. The corresponding terms describing these processes are called collision integrals and are designated St (Ref. 27) and are placed on the right-hand side of Eq. (3.73):

$$\frac{dN_{\rm ph}}{dt} = St_{\text{ph-ph}} + St_{\text{ph-}e,h} + St_{\text{ph-}i} + St_{\text{ph-}L} + \cdots. \tag{3.74}$$

We henceforth use the subscripts ph, i, and L to designate phonons, impurities, and light. In these designations $St_{\text{ph-}i}$ denotes phonon scattering by impurities.

Phonon-phonon interactions that conserve total energy of the phonon system lead to an equilibrium phonon spectral distribution:

$$N^{B}_{\mathrm{ph}}(\mathbf{k}) = [\exp(\hbar\omega_{\mathrm{ph}}(\mathbf{k})/k_B T) - 1]^{-1}. \tag{3.75}$$

The Bose-Einstein distribution function (3.75), whose parameter is the thermodynamic temperature T of the crystal lattice (k_B is Boltzmann's constant), cancels the collision integral $St_{\mathrm{ph\text{-}ph}}$.[9,27] Phonon-electron interactions will also set up acoustic vibration spectrum (3.75) and lead to a Fermi electron distribution with an electron temperature equal to the lattice temperature ($T_e = T$).[9,27] Therefore, the free- energy density of the elastic vibrations of the lattice in equilibrium is determined by the thermodynamic temperature

$$F_0(T) = \sum_{\mathrm{ph}} \int (2\pi)^{-3}\hbar\omega_{\mathrm{ph}}(\mathbf{k})N_{\mathrm{ph}}(\mathbf{k},T)d\mathbf{k}.$$

When the evolution of thermal vibrations in the crystal can be represented as a transition through local equilibrium states (3.75) with a temperature dependent on the coordinate and time: $T = T(\mathbf{r},\mathbf{t})$, the energy-transfer equation adopts the characteristic form of the diffusion equation (3.8). The diffusive heat conduction equation (3.8) can be derived from kinetic phonon equation (3.74) (in the case of dielectrics) or from an electron kinetic equation (in the case of metals[27]).

Absorption of electromagnetic radiation will elevate crystal temperature in accordance with Eq. (3.8). We can now state that when such a description is used it is implicitly assumed that the phonons created from absorption of light thermalize over time periods substantially shorter than other characteristic times. The free path length of the phonons will therefore be significantly less than the characteristic spatial scales in Eq. (3.8). In fact, the ph-ph and ph-e interaction times at room temperature are so small that Eq. (3.8) adequately describes heating, even on a subnanosecond time scale. Nonetheless, nonequilibrium phonon distributions can be generated by means of picosecond and femtosecond pulses at high temperatures as well.

A significant increase in the phonon free path length at low (helium) temperatures[26] makes it necessary to directly employ kinetic equation (3.74) to describe energy transfer. A description of phonon sources in the energy-transfer equation is also significantly complicated. It is now necessary to represent the spectral distribution of phonons contributed to the lattice from external action, i.e., it is necessary to be able to calculate when and at what point in space absorption of light at point $\mathbf{r}$ at time t will produce what number of phonons, with what wave vectors and polarizations.

It is extraordinarily difficult to describe exactly the optical energy-to-chaotic acoustic vibration energy conversion process in a crystal. There are a wide variety of thermalization channels for the absorbed energy. We provide a few examples. The term $St_{\mathrm{ph}\text{-}L}$ on the right-hand side of Eq. (3.74) describes direct phonon creation from absorption of light. For example, light will be strongly absorbed in polar dielectrics when the light energy and the optical phonon energy are identical. At low temperatures, the excited optical phonons rapidly decay into low-energy acoustic phonons.

Phonons participate in electron absorption of electromagnetic radiation in metals and optical excitation of electron-hole (EH) pairs in indirect-band-gap semiconductors in accordance with the laws of conservation of energy and momentum.

In semiconductors, this will include short-wavelength intervalley phonons that rapidly decay into acoustic phonons. However, in the last two cases, $St_{\text{ph-}L}$ can be neglected, since the electron (or EH pair) acquires the predominant fraction of light energy and a significantly larger number of phonons is radiated from subsequent electron scattering (nonradiative EH pair recombination), i.e., $|St_{\text{ph-}L}| \ll |St_{\text{ph-}e,h}|$. As $St_{\text{ph-}e,h}$ is dependent on the distribution functions of the nonequilibrium charge carriers, it is necessary to investigate the effect of optical radiation on the electron and hole subsystems of the crystal. The kinetic equation for phonons must be analyzed in conjunction with kinetic equations for the charged carriers. The problem is further complicated by accounting for the reverse effect of the nonequilibrium carrier and phonon distributions on light absorption.

In order to derive qualitative concepts of ballistic phonon transport induced by spatially inhomogeneous pulsed light absorption, we will model the phonon sources on the right-hand side of Eq. (3.74):

$$\frac{\partial N_{\text{ph}}}{\partial t} + \frac{\partial \omega_{\text{ph}}}{\partial \mathbf{k}} \frac{\partial N_{\text{ph}}}{\partial \mathbf{r}} = \left(\frac{\partial N_{\text{ph}}(\mathbf{k})}{\partial t} \right)_g \Phi\left(\frac{x}{x_0}, \frac{y}{y_0}, \frac{z}{z_0}, \frac{t}{t_0} \right). \tag{3.76}$$

Here the function $[\partial N_{\text{ph}}(\mathbf{k})/\partial t]_g$ describes the spectrum and the polarization composition of the radiated phonons, Φ represents spatial and temporal modulation of the sources, $\Phi(0) = 1$. Notation (3.76) assumes spatial homogeneity of the crystal, $\partial\omega/\partial\mathbf{r} = 0$, a constant composition of the radiated phonons in space and time and no effect of the sources on the phonon propagation. In Eq. (3.76), $N_{\text{ph}}(\mathbf{k})$ represents the increment of the phonon distribution function relative to an equilibrium distribution function determined by the initial conditions of experiment [the initial crystal temperature T_0: $N_{\text{ph}}(\mathbf{k}) \gg N^B_{\text{ph}}(\mathbf{k},T_0)$].

Reference 28 reports a simple physical case where the assumptions underlying model (3.76) hold. Light is absorbed by a metallic film sprayed onto a transparent dielectric crystal. Absorption of intense light sets up a quasistationary electron energy distribution over times significantly shorter than the optical excitation time. In this case the energy absorbed by the electron subsystem ceases to grow, and all luminous energy absorbed by the electrons is radiated as phonons. The spectrum of emitted phonons in this case is nearly constant during the entire optical irradiation period. The metallic film must be sufficiently thin to avoid electron reabsorption of the radiated phonons. Therefore the phonons are ballistically distributed throughout the entire system. Even lower reabsorption of phonons by charge carriers can be expected in semiconductors.

We utilize the Debye model of a solid[29] to further simplify the problem. Specifically, this model assumes an isotropy of the medium and zero dispersion of the acoustic wave velocity ($\omega_{\text{ph}} = c^{\text{ph}}\mathbf{k}$, c^{ph} is independent of $k \equiv |\mathbf{k}|$, $c_{FT} = c_{ST}$). Here $\partial\omega_{\text{ph}}(\mathbf{k})/\partial\mathbf{k} = c^{\text{ph}}\mathbf{k}/k$. Then, in the case of uniform irradiation of the crystal surface $z = 0$ ($y_0, x_0 \to \infty$) Eq. (3.76) reduces to

$$\frac{\partial N_{\text{ph}}(k,\xi)}{\partial t} + c^{\text{ph}}\xi \frac{\partial N_{\text{ph}}(k,\xi)}{\partial z} = \left(\frac{\partial N_{\text{ph}}(\mathbf{k})}{\partial t} \right)_g \Phi\left(\frac{z}{z_0}, \frac{t}{t_0} \right), \tag{3.77}$$

where ξ is the cosine of the angle between the direction of phonon propagation (the vector $\mathbf{k}/k$) and the z axis.

If the phonons are isotropically emitted in all the directions $[\partial N_{ph}(\mathbf{k})/\partial t]_g \equiv [\partial N_{ph}(k)/dt]_g$, an equation is easily derived for

$$F'_{ph}(\xi) = \int (2\pi)^{-2} k^2 \hbar c^{ph} k N_{ph}(k,\xi) dk.$$

Physically, $F'_{ph}(\xi)$ is the free-energy density of ph-polarized phonons traveling in direction ξ–$\xi + d\xi$:

$$F = \sum_{ph} F_{ph} = \sum_{ph} \int_{-1}^{1} F'_{ph}(\xi) d\xi.$$

Integrating Eq. (3.77) with respect to the wave numbers and the polar angle, we obtain

$$\frac{\partial F'_{ph}}{\partial t} + c^{ph}\xi \frac{\partial F'_{ph}}{\partial z} = \frac{1}{2}\left(\frac{\partial F_{ph}}{\partial t}\right)_g \Phi\left(\frac{z}{z_0}, \frac{t}{t_0}\right). \qquad (3.78)$$

In this case the process is therefore independent of the spectrum of the emitted phonons. Equation (3.78) suggests that propagation of phonons of different polarizations can be studied independently, which is related to our neglect of $St_{ph\text{-}ph}$ and $St_{ph\text{-}i}$. We also do not account for a possible change in phonon polarization from reflection by the crystal boundaries. Then the phonon specular reflection condition off the boundary $z = 0$

$$N_{ph}(\xi,\mathbf{k},z=0,t) = N_{ph}(-\xi,\mathbf{k},z=0,t)$$

can, using Eq. (3.77), lead to $(\partial N_{ph}/\partial z)|_0 = 0$, and, consequently, $(\partial F'_{ph}/\partial z)|_0 = 0$. The specular reflection conditions can be automatically satisfied if the phonon sources are evenly continued into the domain $z < 0$.

The solution of Eq. (3.78) takes the characteristic form of a "retarded potential":

$$F'_{ph} = \frac{1}{2}\left(\frac{\partial F_{ph}}{\partial t}\right)_g \int_{-\infty}^{t} \Phi\left(\frac{|z - c^{ph}\xi(t-t')|}{z_0}, \frac{t'}{t_0}\right) dt'. \qquad (3.79)$$

In actual experiments, the ballistic phonons are ordinarily recorded outside of their generation region and after a time period that far exceeds the optical irradiation time. This in fact comprises the physical meaning of the ballistic state: ballistic phonons can be used for "bleaching" of solids. It is therefore important to identify the asymptotic properties of solution (3.79). If the phonons are generated in a limited time interval t_0 near $t = 0$, then integration in Eq. (3.79) is dropped for $t \gg t_0$:

$$F'_{ph} \approx \left(\frac{\partial F_{ph}}{\partial t}\right)_g t_0 \Phi\left(\frac{|z - c^{ph}\xi t|}{z_0}, 0\right). \qquad (3.80)$$

We obtain for the free-energy density

$$F_{ph} \approx \frac{1}{2}\left(\frac{\partial F_{ph}}{\partial t}\right)_g t_0 \int_{-1}^{1} \Phi\left(\frac{|z - c^{ph}\xi t|}{z_0}, 0\right) d\xi. \qquad (3.81)$$

If the generation region is localized within distances of the order of z_0 near the surface $z=0$, the integrand will be nonzero only for $(z-z_0)/c^{\mathrm{ph}}t \lesssim \xi \lesssim (z+z_0)/c^{\mathrm{ph}}t$. Then, for $t \gg z_0/c^{\mathrm{ph}}$ we have in the range $z \ll c^{\mathrm{ph}}t$

$$F_{\mathrm{ph}} \approx \left(\frac{\partial F_{\mathrm{ph}}}{\partial t}\right)_g \frac{t_0 z_0}{c^{\mathrm{ph}}t}. \tag{3.82}$$

According to asymptote (3.82) information on the type of space-time modulation of the sources will vanish after time $t \gg \max\{t_0, z_0/c^{\mathrm{ph}}\}$ in the range $z \ll c^{\mathrm{ph}}t$. The energy density of the phonon field will be spatially homogeneous within this range and will decay (as $1/t$) with time. The leading edge of this region travels at a constant velocity c^{ph} into the crystal bulk, which corresponds to conservation of total energy of the phonon field after termination of source irradiation.

According to Eq. (3.82) the decay of the energy density of the phonon field at the given point in space z is observed roughly after the time period corresponding to the transit time of the sound from the crystal boundary to this point ($t \gg z/c^{\mathrm{ph}}$). The first phonons from the generation region may reach the observation point prior to a time of the order of $z_0/c^{\mathrm{ph}} + t_0$. Therefore, the variation of F_{ph} at a given point in space as a function of time will manifest a pulsed behavior. Such pulses are called thermal pulses.[26] In our one-dimensional geometry of ballistic phonon energy transfer, the duration of the leading edge of the thermal pulse will be determined either by the phonon transit time through the irradiation region or by the irradiation time ($\sim \max\{z_0/c^{\mathrm{ph}}, t_0\}$). Here we have an analogy to the generation of coherent acoustic pulses (2.27), (2.28). However, the peak energy density of the phonon field at any instant in time is located on the surface $z=0$ (here we assume that the sources become weaker with distance from the boundary: $\partial\Phi/\partial z \leqslant 0$), i.e., in a one-dimensional geometry the thermal pulse does not separate from the boundary. This is due to the slow drift from the boundary of phonons traveling along it.

It is interesting to note that in a simple analysis of the operation of a classical thermal pulse generator (a heated metallic disk on a dielectric crystal) homogeneous kinetic equation (3.73) is traditionally utilized and a time-modulated phonon energy flux at the heater/crystal boundary is represented:[26,29]

$$\int (2\pi)^{-3} \hbar c^{\mathrm{ph}} k c^{\mathrm{ph}} \xi N_{\mathrm{ph}}(k,\xi)\,|_{z=0}\, dk = J_0 f\left(\frac{t}{t_0}\right). \tag{3.83}$$

If it is assumed that equilibrium phonon distribution (3.75) with a temperature T_H exceeding the initial crystal temperature T_0, is set up in the heater, then qualitatively

$$N_{\mathrm{ph}}(k,\xi)\,|_{z=0} \equiv N^B_{\mathrm{ph}}(k,\xi_{-\xi}, T_H) - N^B_{\mathrm{ph}}(k,\xi_{-\xi}, T_0).$$

At temperatures below the Debye temperature, such an assumption leads to $J_0 \sim T_H^4 - T_0^4$. Under pulsed optical excitation the heater temperature will manifest pulsed variations, which justifies the introduction of temporal modulation in boundary condition (3.83).[29] With an isotropic directional distribution of hot phonons and

$T_H^4 \gg T_0^4$ condition (3.83) is equivalent to a boundary representation of F'_{ph}:

$$F'_{\mathrm{ph}}|_0 = 2J_0 f(t/t_0)/c^{\mathrm{ph}}.$$

The corresponding solution of the kinetic equation makes it possible to determine the phonon free-energy density distribution:[29]

$$F_{ph} = \frac{2J_0}{c^{ph}} \int_0^1 f\left(\frac{t - z/c^{ph}\xi}{t_0}\right) d\xi.$$

The asymptotics of the derived solution for $t \gg t_0$ in the range $z \lesssim c^{ph}t$

$$F_{ph} \approx 2J_0 t_0 z/(c^{ph}t)^2$$

unlike Eq. (3.82) describes the growth of F_{ph} with distance from the radiator (in the region $z \lesssim c^{ph}t$). Therefore, the phonons also produce a thermal pulse in space in this case at a fixed instant in time. Analysis reveals that this is due to the assumption of an equilibrium phonon distribution within the heater and at the heater boundary. It is in fact implied that phonon-phonon or phonon-electron interactions will be able to supplement the shortage of phonons propagating perpendicular to the boundary; such phonons will escape the film more rapidly. This results in a reduction in the number of phonons traveling along the boundary. It is these phonons that are responsible for the fact that the maximum of F_{ph} does not separate from the boundary of the sample in the case of ballistic phonon transport within the film.[28]

We again emphasize that the concept of temperature cannot be used to describe this energy-transfer process, since the crystal is in an essentially nonequilibrium state. Generally, there will be no phonons traveling towards the crystal boundary outside the generation region, while introducing temperature into the Debye model assumes an isotropic directional phonon distribution.

We consider the limiting case $t_0 \ll z_0/c^{ph}$ for a graphical representation of these conclusions. In this case the phonons are not able to move appreciably over their generation time and instantaneous photoexcitation can be assumed. Solution (3.81) holds and this solution is easily transformed by going over to dimensionless variables $z = z/z_0$, $t = c^{ph}t/z_0$, $F_{ph} = F_{ph}/(\partial F_{ph}/\partial t)_g t_0$ and introducing the antiderivative of $\Phi(\Psi_z = \Phi)$:

$$F_{ph} = (2t)^{-1} \begin{cases} \Psi(z+t) + \Psi(|z-t|) - 2\Psi(0), & 0 \leqslant z \leqslant t, \\ \Psi(z+t) - \Psi(z-t), & z > t. \end{cases} \tag{3.84}$$

Since $\Phi(z \to \infty) \to 0$, then $\Psi(z \to \infty) \to$ const. Hence according to Eq. (3.84) for $z,t \gg 1$ the phonon energy density distribution in the comoving coordinate system (one traveling at the unit velocity in dimensionless variables) assumes a stationary profile with an amplitude decaying as t^{-1}:

$$F_{ph}(z,t \gg 1) \approx (2t)^{-1} \begin{cases} \Psi(\infty) + \Psi(t-z) - 2\Psi(0), & z - t \leqslant 0, \\ \Psi(\infty) - \Psi(z-t), & z - t > 0. \end{cases} \tag{3.85}$$

Asymptotic result (3.85) will be useful for analyzing the generation of coherent acoustic waves in the process of nonstationary ballistic phonon heat conduction.

Figures 3.10 and 3.11 reflect the evolution of the free-energy density of the phonon field in the specific case $\Phi = -\Psi = \exp(-z)$. Here, of course, z_0^{-1} represents the light intensity absorption coefficient ($z_0 \equiv \alpha^{-1}$). The energy density of the phonon field is shown in Fig. 3.10 at successive instants in time. The prop-

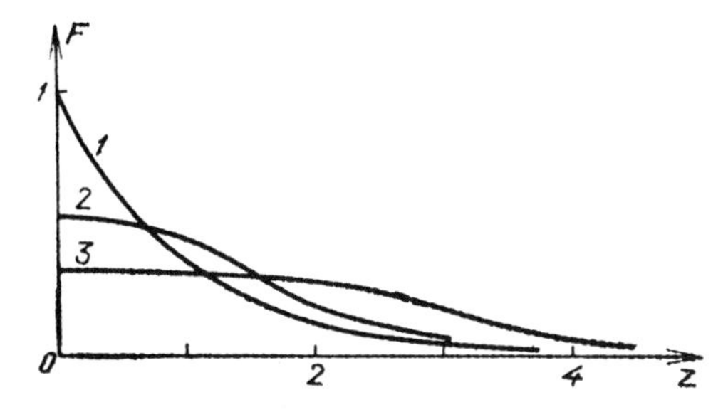

FIG. 3.10. Energy density distribution of the phonon field in space at different instants in time t: 1—0; 2—1.5; 3—3.

agation of the leading edge of the phonon flux into the crystal bulk is clearly visible. The change in the thermal pulse profile is reflected in Fig. 3.11 on selected axes as the phonon generation and detection regions are moving farther apart. Time in Fig. 3.11 is measured in units of the phonon transit time from the boundary to the detector. The curves are labeled in increasing order of dimensionless distance whose rise may be attributable to both increasing distance between the detector and the boundary and an increase in light absorption (reduction of z_0). The thermal signal manifests increasingly pulsed behavior with increasing z.

We cite the results of experimental study Ref. 30 as an example of the explicit manifestation of the kinematic properties of ballistic phonon transport examined here. This study analyzed nonstationary phonon heat conduction in the [111] direction (the z axis) in pure crystalline silicon. We note that the velocities of the variously polarized transverse phonons coincide in this direction ($c_{FT} = c_{ST} \equiv c_T$). The thermal pulses were recorded by means of a bolometer fabricated from granular aluminum sprayed on a crystal end opposite the photoexcited surface ($z = 0$). The sample was placed in a helium cryostat where its temperature was maintained near 1.8 K.

A temperature-tunable $Ba_2NaNb_5O_{15}$ parametric light generator pumped by the second harmonic of a Q-switched YAG:Nd^{3+} laser (where YAG denotes yttrium aluminum garnet) was employed as the radiation source. The parametric light generator (PLG) produced 10–12 ns pulses at a pulse energy of 1 μJ; the PLG was tunable from 1.03–0.917 μm. The light quantum energy in this case varies near the edge of the one-photon absorption band of silicon, which causes a significant change in the absorptivity of light (the absorption length α^{-1} varies over the range 4.7–0.02 cm). Therefore the phonon detector (mounted at a fixed distance $z \equiv L \approx 1$ cm from the front edge) gradually moves away from the phonon generation region with increasing optical radiation energy.

Figure 3.12 shows oscillograms of the response of a superconducting bolometer for different pump radiation wavelengths. Labels I and II correspond to the smallest

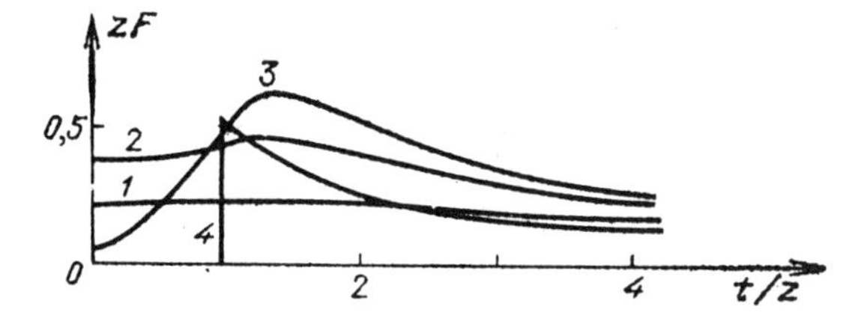

FIG. 3.11. Thermal pulse profile at different distances from the irradiated surface. The parameter z is equal to 0.3 (1), 1 (2), 3 (3), and ∞ (4).

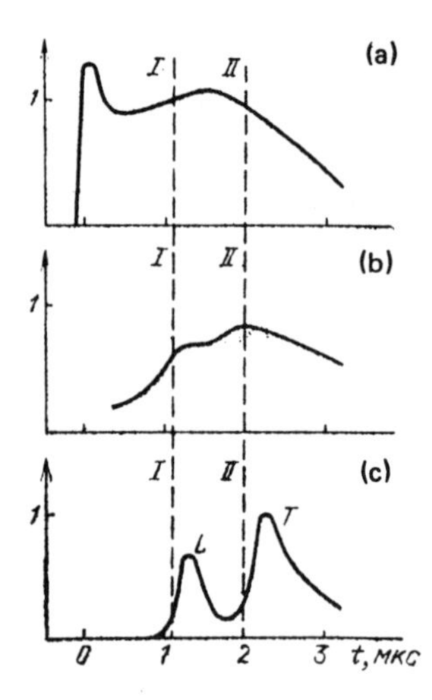

FIG. 3.12. Oscillograms of thermal pulses for different pump radiation wavelengths and absorption layer thicknesses α^{-1}: (a) $\lambda = 1.024\ \mu m$, $\alpha^{-1} = 3.36$ cm; (b) $1.008\ \mu m$, 0.25 cm; (c) $0.997\ \mu m$, 0.02 cm.

propagation time of the L and T phonons from the front edge of the crystal to the bolometer. The significant changes in thermal pulse wave form with increasing excitation energy observed in experiment are primarily related to the transformation of the energy liberation region resulting from the change in the depth of penetration of the optical radiation. In the case of strong absorption when the light penetrates the crystal to an insignificant depth ($\alpha^{-1} \ll L$), the time of the peak arrivals of thermal pulses of different phonon polarizations t_{ph}^{peak} is determined by the longitudinal size of the sample: $t_{ph}^{peak} \sim L/c^{ph}$ [Fig. 3.12(c)]. In the case of weakly homogeneous irradiation of the sample ($\alpha^{-1} > L$) the detector registers a single thermal pulse [Fig. 3.12(a)] (the first peak on this oscillogram is due to direct laser light action on striking the bolometer).

It is possible to estimate the wavelength of the pump light λ_0 at which the peaks of the thermal pulses of different polarizations will be separated out in time. The delay of the L- signal peak t_L^{peak} relative to the excitation pulse ought to be less than the time to the leading edge of the T signal t_T^l. However, consistent with the simplest model (3.84), t_T^l cannot be less than the direct transit time of the T phonons to the detector from the bulk of the absorption region: $t_T^l \gtrsim (L - \alpha^{-1})c_T^{-1}$. At the same time, the L-signal peak will appear earlier than phonons created in the bulk of the absorption region and reflected by the front crystal face: $t_L^{peak} \lesssim (L + \alpha^{-1})c_L^{-1}$. The peaks will then separate out when $(L + \alpha^{-1})c_L^{-1} \sim (L - \alpha^{-1})c_T^{-1}$. This relation makes it possible to estimate the depth of absorption at $\alpha^{-1} \sim L(c_L - c_T)(c_L + c_T)^{-1} \sim 0.3$ cm in silicon, corresponding to a wavelength $\lambda_0 \sim 1.02\ \mu m$ at $T_0 \sim 1.8$ K. This estimate is confirmed experimentally.

Analysis reveals that a quantitative description of the experimental results of Ref. 30 requires that Eq. (3.84) be significantly complicated taking account, specifically, of the three-dimensionality of the phonon generation region and the significant lifetime of the photoexcited EH pairs. However, the following qualitative result demonstrated by Ref. 30 is the only important result to us: the duration of the leading edge of the thermal pulse shrinks with diminishing depth of penetration of the laser light. This is in agreement with the results of the preceding theoretical analysis. With significant light absorptivities [Fig. 3.12(c)] the duration of the leading edge of the thermal pulses is determined by the nonradiative recombination time of the EH pair, which far exceeds the laser irradiation duration at the pump energies employed here. In this case $t_0 \gg z_0 c^{\text{ph}}$ and the natural subject of analysis is the regime of phonon generation at the surface $[\Phi(z/z_0,t/t_0)\rightarrow z_0\delta(z)f(t/t_0)]$.

We have therefore analyzed how the energy from thermal oscillations is transferred in a solid when the traditional concept of temperature cannot be employed. It is now important to answer the question of how to describe the excitation of coherent acoustic waves in this case.

3.6. Generation of Coherent Acoustic Waves From Interaction of Thermal Vibrations (Phonons) in the Crystalline Lattice

An analysis of the interaction of phonons and deformation waves makes it possible to identify the physical mechanism of thermoelastic generation of sound. We consider the generation of acoustic pulses in nonstationary ballistic phonon heat conduction.

A coherent acoustic wave can be treated as an external field acting on phonons consistent with the concept of the deformation potential of vibrations of the crystal lattice.[23,31] It can easily be demonstrated that an acoustic wave alters the elastic properties of the medium and thereby modulates the frequencies of thermal vibrations. This assumption can be written as

$$\widetilde{\omega}_{\text{ph}}(\mathbf{k}) = \omega_{\text{ph}}(\mathbf{k})(1 + m_{il}^{\text{ph},\mathbf{k}} u_{il}), \tag{3.86}$$

where $m_{il}^{\text{ph},\mathbf{k}}$ is the tensor of the ph,**k** phonon deformation potential.[23] Taking account of the change in frequencies of the corresponding lattice vibrations (3.86) leads to the following expression for the elastic free-energy density of the crystal:

$$F = \frac{1}{2} c_{ijkl}^{(0)} u_{ij} u_{kl} + \sum_{\text{ph}} \int (2\pi)^{-3} \hbar\omega_{ph}(\mathbf{k}) N_{ph}(\mathbf{k}) d\mathbf{k}$$
$$+ \sum_{\text{ph}} \int (2\pi)^{-3} \hbar\omega_{\text{ph}}(\mathbf{k}) m_{il}^{\text{ph},\mathbf{k}} N_{\text{ph}}(\mathbf{k}) u_{il}\, d\mathbf{k}, \tag{3.87}$$

where $c_{ijkl}^{(0)}$ are the elasticity moduli at absolute zero temperature, $N_{\text{ph}}(\mathbf{k})$ are phonons that do not take part in zero-temperature lattice vibrations. The second term in Eq. (3.87) in fact represents the phonon-deformation wave interaction energy. According to the methodological approach outlined in Sec. 3.4, such interaction may lead to the appearance of sources in the wave equation for sound.

Using Eqs. (3.54) and (3.55) we obtain

$$\rho_0 \frac{\partial^2 u_i}{\partial t^2} - c_{ijkl}^{(0)} \frac{\partial^2 u_l}{\partial x_j \partial x_k} = \sum_{\text{ph}} \int (2\pi)^{-3} \hbar\omega_{\text{ph}}(\mathbf{k}) m_{il}^{\text{ph},\mathbf{k}} \frac{\partial N_{\text{ph}}(\mathbf{k})}{\partial x_l} d\mathbf{k}. \tag{3.88}$$

Using wave equation (3.88) and kinetic equation (3.73) it can be demonstrated that total energy (3.87) of the phonons and coherent waves is conserved.[23] Equation (3.88) demonstrates that the spatial inhomogeneity of the phonon distribution function generates regular lattice displacements. This equation describes the generation of deformation waves in the process of an arbitrary (yet known) evolution of the phonon fields $N_{\text{ph}}(\mathbf{k},\mathbf{r},t)$.

When changes in the occupation numbers of the phonon modes can be described by equilibrium functions (3.75) with a space- and time-dependent temperature, the right-hand side of Eq. (3.88) becomes

$$f_i = \sum_{\text{ph}} \int (2\pi)^{-3}\hbar\omega_{\text{ph}}^2(\mathbf{k}) \frac{\partial N_{\text{ph}}}{\partial \omega_{\text{ph}}}(\mathbf{k}) m_{il}^{\text{ph},\mathbf{k}} m_{jk}^{\text{ph},\mathbf{k}} \frac{\partial u_{jk}}{\partial x_l} d\mathbf{k}$$

$$+ \sum_{\text{ph}} \int (2\pi)^{-3}\hbar\omega_{\text{ph}}(\mathbf{k}) m_{il}^{\text{ph},\mathbf{k}} \left(\partial N_{\text{ph}}^B \frac{(\omega_{\text{ph}}(\mathbf{k}),T)}{\partial T} \right) \frac{\partial T}{\partial x_l} d\mathbf{k}. \quad (3.89)$$

The first term defines the temperature dependence of the elastic moduli (the isothermal moduli):

$$c_{ijkl} = c_{ijkl}^{(0)} + \sum_{\text{ph}} \int (2\pi)^{-3}\hbar\omega_{\text{ph}}^2(\mathbf{k}) \left(\frac{\partial N_{\text{ph}}^B}{\partial \omega_{ph}(\mathbf{k})} \right) m_{il}^{\text{ph},\mathbf{k}} m_{jk}^{\text{ph},\mathbf{k}} d\mathbf{k}.$$

The change in temperature in the second term of the last expression may be due to the presence of the acoustic wave and external heating. Specifically, if a new temperature is established in the acoustic wave, this will cause a substitution of isothermal moduli by adiabatic moduli.

External heating in accordance with Eq. (3.89) generates coherent waves. A comparison of the second term of force f_i to Eq. (3.71) indicates the relation of tensors m_{il}, c_{ijkl}, and α_{kl}. Therefore according to Eq. (3.89) thermoelastic sound generation is the result of the interaction of phonons and deformation waves.

In order to present the mechanism of such interaction in a convenient form, we introduce a different method of deriving a type (3.88) equation.[28] We proceed from an equation for waves in a solid accounting for the quadratic nonlinearity, which is derived with conservation of cubic terms in Eq. (3.61) for the free-energy density. In this case force f_i (3.57) in addition to the linear terms contains nonlinear terms of the type

$$\frac{\partial}{\partial x_j} \left[\left(\frac{\partial u_m}{\partial x_l} \right) \left(\frac{\partial u_k}{\partial x_n} \right) \right].$$

In the physical case of interest to us the field is naturally represented as a superposition of a regular field $\mathbf{u}$ (coherent acoustic waves) and a random field $\mathbf{u}'$ (thermal noise):

$$\mathbf{u} = \mathbf{u} + \mathbf{u}', \quad \langle \mathbf{u}' \rangle = 0,$$

where the angular brackets denote statistical averaging. Henceforth we analyze the excitation of plane longitudinal deformation waves. We therefore assume one-dimensionality of the regular longitudinal waves and the averaged characteristics of the noise field. Substituting the last representation into the nonlinear wave equation and carrying out averaging, we obtain

$$\frac{\partial^2 u_1}{\partial t^2} - c_L^2 \frac{\partial^2 u_1}{\partial x_1^2} = -c_L^2 \varepsilon \frac{\partial}{\partial x_1} \left(\frac{\partial u_1}{\partial x_1} \right)^2 + \rho_0^{-1} \langle f_i \rangle, \quad (3.90)$$

where the expression $\langle f_i \rangle$ will only include terms of the type

$$\frac{\partial}{\partial x_1}\left\langle \frac{\partial u'_m}{\partial x_l}\frac{\partial u'_k}{\partial x_n}\right\rangle .$$

Therefore, only a random field contributes to $\langle f_i \rangle$. Equation (3.90) demonstrates that the coherent component of the wave field is created from nonlinear interaction of the noise components. Regular crystal deformation results from autodetection of random fields at a lattice quadratic nonlinearity.

In order to reduce Eq. (3.90) to a form analogous to (3.88) we need only neglect the self-action of the regular wave and, by using a quantum description of the thermal lattice vibrations, carry out quantum-mechanical averaging. Reference 26 derived for the case of an isotropic solid and longitudinal thermal lattice vibrations only ($z \equiv x_1$)

$$\frac{\partial^2 u}{\partial t^2} - c_L^2 \frac{\partial^2 u}{\partial z^2} = -\rho_0^{-1}\int (2\pi)^{-3}\hbar c_L k(\varepsilon_2 + \varepsilon_1 \xi^2)\frac{\partial N(\mathbf{k})}{\partial z}\, d\mathbf{k}. \quad (3.91)$$

The dimensionless parameters ε_1 and ε_2 ($\varepsilon_1 + \varepsilon_2 = \varepsilon$) are related to the second- and third-order elasticity moduli:

$$\varepsilon_1 = (c_{12} - 24c_{155})/2c_{11} - 3/2,$$
$$\varepsilon_2 = (24c_{155} - 6c_{111} - c_{12})/2c_{11}. \quad (3.92)$$

The form of the expression obtained on the right-hand side of Eq. (3.91) is consistent with the fact that in an isotropic solid the general expression for the deformation potential tensor of longitudinal phonons takes the form[23] $m_{il} = m_1 k_i k_l / k^2 + m_2 \delta_{il}$. A comparison of Eqs. (3.88), (3.91), and (3.92) indicates the relation between the phonon deformation potential and the anharmonic coefficients of solids, which can also be identified in a more general case.[23]

The concept of phonon deformation potential in which the coherent acoustic field is treated as an external field assumes that the wavelength of regular sound far exceeds the wavelengths of the phonons. Otherwise, consistent with Sec. 3.5, it is not possible to use the concept of phonon wave packets. The acoustic system of the crystal breaks down into high-frequency (thermal noise) and low-frequency (coherent sound) components. It is important that the concept of deformation potential is effectively satisfied automatically during noise generation of sound. Indeed, according to Eqs. (3.88) and (3.91) derived here, the frequency spectrum of the excited acoustic pulses is not determined by the phonon frequencies, but rather by the characteristic scales of the variation of the phonon distribution function $N_{\mathrm{ph}}(\mathbf{k})$ in both space and time. Therefore, the limits on the applicability of the concept of deformation potential in this case correlates with the possibility for using a kinetic equation for phonons (Sec. 3.5).

We now analyze acoustic pulse generation from optical initiation of ballistic phonon heat conduction. In order to identify the fundamental differences from the thermoelastic sound excitation regime attributable to diffusive heat conduction (Sec. 3.1), we limit the analysis to the generation of plane longitudinal acoustic waves resulting from creation and propagation of longitudinal phonons. In a first approximation (neglecting the self-action of the excited sound and its respective effect on phonon distribution) the process in a one-dimensional geometry is described by Eqs. (3.91) and (3.77). In the case of isotropic phonon emission, kinetic equation (3.77) becomes Eq. (3.78). Carrying out integration over the polar angle and with respect to the wave numbers of the phonons on the right-hand side of Eq. (3.91), we obtain

$$\frac{\partial^2 u}{\partial t^2} - c_L \frac{\partial^2 u}{\partial z^2} = -\rho_0^{-1} \int_{-1}^{1} [\varepsilon + \varepsilon_1(\xi^2 - 1)] \frac{\partial F'}{\partial z} d\xi. \tag{3.93}$$

According to Eq. (3.93) the absence of forces acting on the free surface is given by

$$\left.\frac{\partial u}{\partial z}\right|_0 = (\rho_0 c_L^2)^{-1} \int_{-1}^{1} [\varepsilon + \varepsilon_1(\xi^2 - 1)] F'|_{z=0} \, d\xi. \tag{3.94}$$

Relation (3.94) defines the deformation to the crystal free surface and is a boundary condition for Eq. (3.93). Of course it is possible, as before (Sec. 3.2), to go to an equation for the sound velocity potential φ which is equal to zero at the free boundary. However, in this case it is more convenient to consider problem (3.93) and (3.94) in terms of the deformations $\mathscr{D} \equiv \partial u/\partial z$:

$$\frac{\partial^2 \mathscr{D}}{\partial t^2} - c_L^2 \frac{\partial^2 \mathscr{D}}{\partial z^2} = -\rho_0^{-1} \int_{-1}^{1} [\varepsilon + \varepsilon_1(\xi^2 - 1)] \frac{\partial^2 F'}{\partial z^2} d\xi, \tag{3.95}$$

$$\mathscr{D}|_0 = (\rho_0 c_L^2)^{-1} \int_{-1}^{1} [\varepsilon + \varepsilon_1(\xi^2 - 1)] F'|_0 \, d\xi. \tag{3.96}$$

The longitudinal deformation $\mathscr{D} = \partial u/\partial z$ describes here the relative change in crystal volume: $\mathscr{D} > 0$ which corresponds to expansion and $\mathscr{D} < 0$ corresponds to compression.

The specified fundamental feature of the sound generation process from ballistic phonon heat conduction is that the excited acoustic wave does not separate from the thermal wave. Indeed, according to Sec. 3.5, the leading edge of the thermal pulse travels at the sound speed and, consequently, does not lag the coherent wave. This means that we cannot accurately speak of a stationary form of the acoustic wave outside the generation region in this case, since the coherent signal is amplified at any distance from the boundary.

We define the asymptotic form of the deformation wave sources for $t \gg \max\{t_0, z_0/c_L\}$ as a confirmation of this argument. Using solution (3.79), we obtain the following asymptotic representation for the right-hand side of Eq. (3.95) in the case of instantaneous $[\Phi(z/z_0,t/t_0) \to t_0\delta(t)\Phi(z/z_0)]$ phonon generation:

$$\frac{\partial^2 \mathscr{D}}{\partial t^2} - \frac{\partial^2 \mathscr{D}}{\partial z^2} \cong -\frac{2}{z}\frac{\partial}{\partial \tau}\Phi(|\tau|). \tag{3.97}$$

The dimensionless variables $z = z/z_0$, $t = c_L t/z_0$, $\mathscr{D} = \mathscr{D}/\mathscr{D}_0$, $\mathscr{D}_0 = (\partial F/\partial t)_g t_0 \varepsilon / 4\rho_0 c_L^2$, $\tau = t - z$ is the dimensionless comoving coordinate, are introduced in the notation of Eq. (3.97).

An analysis reveals that the acoustic wave sources under the conditions of phonon generation at the surface $[\Phi(z/z_0,t/t_0) \to z_0\delta(z)f(t/t_0)]$ and phonon flux emission from the boundary (3.83) can be represented in an analogous form. It is sufficient to carry out the substitution $\Phi(|\tau|) \to f(\tau)$ in Eq. (3.97) for this purpose.[29] Of course, different normalization scales are selected in the last cases. For example, for a phonon flux (3.83): $\mathscr{D}_0 = \varepsilon J_0/\rho_0 c_L^3$, $t = t/t_0$, $z = z/c_L t_0$.[29]

According to Eq. (3.97) coherent sound continues to rise outside the phonon generation region for an extended period after termination of phonon generation ($z \gg 1$, $t \gg 1$), where the deformation wave is excited near the leading edge of the phonon flux ($|\tau| \sim |t - z| \sim 1$), i.e., the sources are separated from the boundary. The acoustic wave sources travel at the sound speed and, consequently, induce synchronous amplification of an acoustic wave propagating from the surface $z = 0$ into the crystal bulk. Assuming that the profile $\mathscr{D}_M$ of this wave changes slowly with distance from the boundary $\mathscr{D}_M = \mathscr{D}_M(\mu z, \tau)$ ($\mu \ll 1$ is the small parameter), we obtain a simplified equation for describing its transformation

$$\frac{\partial \mathscr{D}_M}{\partial z} \approx -z^{-1}\Phi(|\tau|).$$

According to this equality, the synchronously excited acoustic wave component grows logarithmically at large distances:

$$\mathscr{D}_M \approx -\Phi(|\tau|)\ln z. \tag{3.98}$$

The profile of the synchronously amplified wave mimics the spatial distribution of the phonon sources in the case of instantaneous phonon generation (3.86) or the time dependence of the phonon sources $\mathscr{D}_M \sim -f(\tau)\ln z$ in the case of surface-localized phonon sources.[29] In the general case the duration of the synchronous component of the deformation wave will be determined either by the phonon generation time or the sound transit time through the phonon generation region ($\sim \max\{t_0, z_0/c_L\}$).

We have determined the optoacoustic signal component that varies with distance from the boundary. The sound sources in wave equation (3.93) differing from (3.97) are localized near the boundary and excite a deformation $\mathscr{D}_S$ whose profile is stationary for $z \gg 1$: $\mathscr{D}_S = \mathscr{D}_S(\tau)$. The complete coherent acoustic signal can be given as ($z \gg 1$)

$$\mathscr{D} = \mathscr{D}_S(\tau) - \Phi(\tau)\ln z. \tag{3.99}$$

Since $\mathscr{D}_S \sim 1$ in dimensionless variables, the synchronously amplified wave will become dominant only at distances $\ln z \gg 1$ (in fact, $z \gtrsim 10^3$–10^4). This is due to its very slow (logarithmic) growth resulting from the decay in amplitude of the sources with distance (as $1/z$).

The profile of the synchronously excited wave component is independent of the parameter ε_1. The synchronous component of the signal $\mathscr{D}_M$ will always be a compressional wave ($\mathscr{D}_M < 0$) for $\varepsilon > 0$. The reason for this is, in fact, that the leading edge of the thermal pulse for $z \gg 1$ is generated by phonons traveling quasicollinearly to the axis ($\xi \approx 1$). Hence, the contribution to the right-hand side of Eq. (3.95) which is proportional to ε_1 is insignificant. Therefore, the synchronous sources are determined by space-time variations in the energy density of the phonon field.

In order to describe the profile of the deformation pulse in the region where the synchronous and asynchronous contributions may be comparable in magnitude it is necessary to use an exact solution of problems (3.95) and (3.96). If in dimensionless variables (3.95) and (3.96) take the form

$$\frac{\partial^2 \mathscr{D}}{\partial t^2} - \frac{\partial^2 \mathscr{D}}{\partial z^2} = -\frac{\partial^2}{\partial z^2} G(t,z), \quad \mathscr{D}|_{z=0} = G(t,0),$$

the solution is easily represented in the form

$$\mathscr{D} = \frac{1}{2}\int_{-\infty}^{t} dt \int_{z+t-t'}^{|z-t+t'|} \frac{\partial^2}{\partial z'^2} G(t',z')dz' + G(\tau,0). \qquad (3.100)$$

Here the integral defines the solution of the inhomogeneous equation with zero boundary conditions, while the second term defines the free-wave excited from deformation of the boundary $z=0$. We note that the limit process $z \to \infty$ cannot be carried out in Eq. (3.100) for the integral conversion to a form similar to Eq. (2.36) since the acoustic wave does not separate from the sources.

With use of Eq. (3.100), it is possible to obtain the following analytical description of the deformation wave profile for $z \gg 1$, when ballistic phonon heat conduction is initiated by instantaneous sources that decay exponentially with distance from the boundary $[\Phi = \exp(-z)]$:

$$\mathscr{D}|_{\tau<0} = \exp\tau \ln z - \exp\tau[\ln 2 + \gamma - 2\varepsilon_1/\varepsilon],$$

$$\mathscr{D}|_{\tau>0} = -\exp(-\tau)\ln z + \exp(-\tau)[\ln 2 + \gamma - 2\varepsilon_1/\epsilon] + \exp(-\tau) \qquad (3.101)$$

$$\times[2\ln\tau + Ei(\tau)] + \exp\tau[2E_1(2\tau) - E_1(\tau)] + \frac{4}{\tau}[1 - e^{-\tau}]$$

$$-\frac{4\varepsilon_1}{\varepsilon}\frac{1}{\tau}\left[e^{-\tau} + \frac{2}{\tau}\left(e^{-\tau} - \frac{1-e^{-\tau}}{\tau}\right)\right].$$

Here Ei and E_1 are integral exponential functions; $\gamma \approx 0.577$ is Euler's constant. The form of the solution confirms our hypothesis (3.99) of the possibility of selecting a logarithmically rising signal component.

Figure 3.13 shows the deformation wave profiles at various distances from the boundary calculated by means of (3.101) for the following specific values: (a)$\varepsilon_1/\varepsilon = 2$ and (b) $\varepsilon_1/\varepsilon = \frac{1}{2}$. We note that profiles 1 in Fig. 3.13 (for $z=1$) are profiles of the pulse component we have denoted by $\mathscr{D}_S(\tau)$. This component is not actually observable, since solution (3.99) is not valid for $z=1$. Consistent with this description the acoustic signal begins with the rarefactional phase for $z=10$ and $\varepsilon_1/\varepsilon = 2$, and in the case $\varepsilon_1/\varepsilon = \frac{1}{2}$ the acoustic signal begins with the compressional phase. This is consistent with the fact that in the first case the crystal surface compresses from photogeneration of nonequilibrium phonons, while in the second case it expands. Indeed, with an isotropic phonon directional distribution the sign of the surface deformation at the boundary $z=0$ ($F'|_{z=0}$ is independent of ξ) in accordance with Eq. (3.96) is identical to the sign labeling $1 - 2\varepsilon_1/3\varepsilon$. The growth of the synchronous wave component with increasing distance will cause the deformation wave to begin with the compressional phase in both cases (Fig. 3.13).

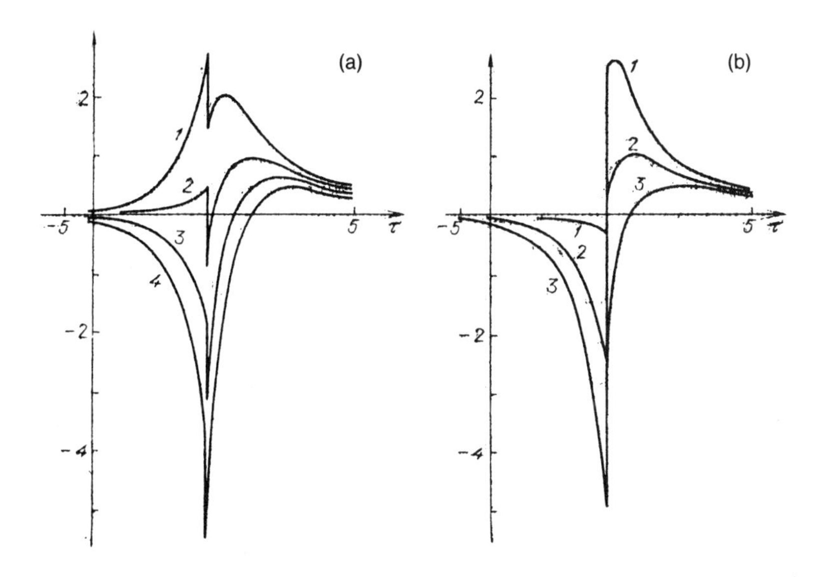

FIG. 3.13. Deformation pulse profiles for the case of instantaneous bulk laser generation of phonons at distances z from the boundary of (a) 1 (1), 10 (2), 10^2 (3), 10^3 (4); (b) 1 (1), 10 (2), 10^2 (3).

When phonons are injected through the free boundary of the crystal, the acoustic pulse profile in the region $z \gg 1$ [for a model of Lorentzian modulation of the phonon flux $f(t) = (1+t^2)^{-1}$] takes the form

$$\mathscr{D} = (1+\tau^2)^{-1}\{-\ln z + \ln[(1+\tau^2)^{1/2}/2] + \tau[\arctan \tau + \pi/2] + 2 + 2\varepsilon_1/3\varepsilon\}.$$

In Fig. 3.14 the profile is shown for the case where the surface $z = 0$ is undeformed ($\varepsilon_1/\varepsilon = 3/2$). The plots in Figs. 3.13 and 3.14 confirm that for $z \geqslant 10^3$ a unipolar

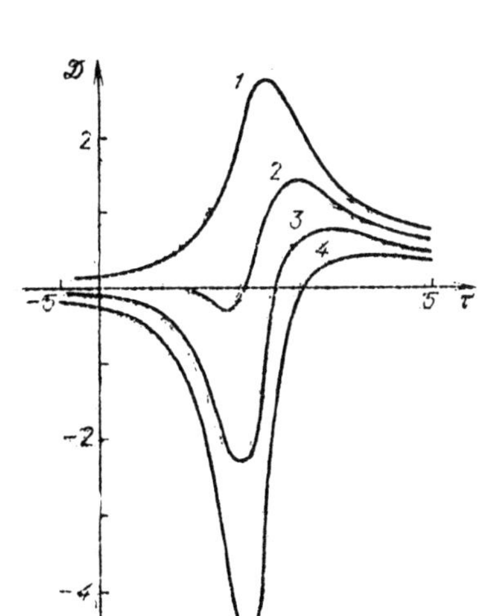

FIG. 3.14. Deformation pulse profiles for the case of optical initiation of a nonequilibrium phonon flux through the crystal boundary. The parameter z is equal to 1 (1), 10 (2), 10^2 (3), and 10^3 (4).

deformation pulse that is synchronously amplified at the leading edge of the non-equilibrium phonon flux will in fact be observed. Calculations[28] of a simple model situation where the phonon sources are not instantaneous and are spatially inhomogeneous: $\Phi(z/z_0,t/t_0)=[\theta(z)-\theta(z-z_0)][\theta(t)-\theta(t-t_0)]$ lead to similar results.

In conclusion, we emphasize that the possibility for the predominance of a synchronously enhanced component in an acoustic wave identified here justifies the rather coarse assumptions used to simplify the mathematical description of the process. For example, we have established (3.90) that coherent wave generation results from the interaction of random waves. However, random wave interaction will also result in energy redistribution across the spectrum, i.e., it is necessary to account for phonon-phonon interactions at the same times and distances, which we have not done. Finally, we generally neglected the contribution to the signal from the ballistic transverse phonon flux. The basis for this neglection is the separation of the leading edge of longitudinal phonon flux from the leading edge of the transverse phonon flux, which corresponds to arrival of thermal pulses at the detector at different points in time when $z \gg \max\{z_0,c^{ph}t_0\}$ [Fig. 3.12(c)]. Here, interaction of only longitudinal phonons at the leading edge of the flux conserves their total energy. Since only quasicollinear interactions of longitudinal phonons are possible with zero dispersion of phonon velocity (the Debye model),[9,10] phonon-phonon interactions do not effectively alter the directivity of the phonons composing the leading edge of the thermal pulse. Hence, the phonon-phonon collision integral $St_{\text{ph-ph}}$ makes no contribution to the description by Eq. (3.78) of the leading edge of the ballistic longitudinal phonon flux in the absence of umklapp processes.[23,29] Sound generation at the leading edge of the transverse phonon thermal pulse is not a synchronous process ($c_T \sim 0.5c_L$). Hence, the contribution of transverse phonons to a coherent acoustic wave at large distances ($\ln z \gg 1$) can be ignored. Therefore, the simple model examined here appears to be applicable for rough estimates in a plane experimental geometry. At the same time, it is not advisable to complicate the theory further without carrying out low-temperature optoacoustic experiments.

REFERENCES

1. L. D. Landau and E. M. Lifshitz, *Teoriya uprugosti* [*Elasticity Theory*] (Nauka, Moscow, 1987).
2. V. Novatski, *Teoriya uprugosti* [*Elasticity Theory*] (Mir, Moscow, 1975).
3. V. I. Danilovskaya, Prikl. Mat. Mekh. **14**, 422 (1950).
4. V. R. J. Gutfeld and R. L. Melchner, Appl. Phys. Lett. **30**, 257 (1977).
5. L. R. F. Rose, J. Acoust. Soc. Am. **75** (3), 723 (1984).
6. L. M. Lyamshev and B. I. Chelnokov, Akust. Zh. **29** (3), 372 (1983) [Sov. Phys. Acoust. **29**, 220 (1983)].
7. A. A. Kolomenskiy, Akust. Zh. **34**, 871 (1988) [Sov. Phys. Acoust. **34**, 504 (1988)].
8. L. D. Landau and E. M. Lifshitz, *Elektrodinamika sploshnykh sred* [*Electrodynamics of Continuous Media*] (Nauka, Moscow, 1982).
9. G. Tacker and V. Rampton, *Hypersound in Solid State Physics*, edited by I. G. Mikhaylov and V. A. Shutilov (Mir, Moscow, 1975) (English translation).
10. O. V. Rudenko and S. I. Soluyan, *Teoreticheskie osnovy nelineynoy akustiki* [*Theoretical Foundations of Nonlinear Acoustics*] (Nauka, Moscow, 1975).
11. L. K. Zarembo and V. A. Krasil'nikov, *Vvedenie v nelineynuyu akustiku* [*An Introduction to Nonlinear Acoustics*] (Nauka, Moscow, 1966).

12. W. Crill and O. Weis, Phys. Rev. Lett. **35**, 588 (1975).
13. W. E. Bron, M. Rossinelli, Y. H. Bai, and F. Keilmann, Phys. Rev. B **27**, 1370 (1983).
14. V. A. Krasil'nikov and V. V. Krylov, *Vvedenie v fizicheskuyu akustiku* [*Introduction to Physical Acoustics*] (Nauka, Moscow, 1984).
15. W. B. Gauster and D. H. Habing, Phys. Rev. Lett. **18**, 1058 (1967).
16. Yu. V. Pogorel'skiy, Fiz. Tverd. Tela **24**, 2361 (1982) [Sov. Phys. Solid State **24**, 1340 (1982)].
17. S. M. Avanesyan, V. E. Gusev, and N. I. Zheludev, Appl. Phys. A **40**, 163 (1986).
18. G. L. Bir and G. E. Pikus, *Simmetriya i deformatsionnye effekty v poluprovodnikakh* [*Symmetry and Deformation Effects in Semiconductors*] (Nauka, Moscow, 1972).
19. A. Kozlov and V. P. Plesskiy, Akust. Zh. **34**, 663 (1988) [Sov. Phys. Acoust. **34**, 381 (1988)].
20. P. M. Karageorgiy-Alkalaev, Fiz. Tekh. Poluprovodn. **2**, 216 (1968) [Sov. Phys. Semicond., **2**, 181 (1968)].
21. V. N. Deev and P. A. Pyatakov, Pis'ma Zh. Tekh. Fiz. **12**, 928 (1986) [Sov. Tech. Phys. Lett. **12**, 384 (1986)].
22. V. N. Deev and P. A. Pyatakov, Zh. Tekh. Fiz. **56**, 1909 (1986) [Sov. Phys. Tech. Phys. **31**, 1142 (1986)].
23. V. L. Gurevich, *Kinetika fononnykh sistem* [*The Kinetics of Phonon Systems*] (Nauka, Moscow, 1980).
24. L. E. Gurevich, *Osnovy fizicheskoy kinetiki* [*The Principles of Physical Kinetics*] (Gostekhizdat, Moscow, 1950).
25. A. A. Galeev and V. I. Karpman, Zh. Eksp. Teor. Fiz. **44**, 592 (1963) [Sov. Phys. JETP, **17**, 403 (1963)].
26. R. Gutfeld, *Fizicheskaya akustika* [*Physical Acoustics*], edited by W. Mason (Mir, Moscow, 1973), Vol. 5.
27. E. M. Lifshitz and L. P. Pitaevskiy, *Fizicheskaya kinetika* [*Physical Kinetics*] (Nauka, Moscow, 1979).
28. V. E. Gusev, Kvantovaya Elektron. **11**, 2197 (1984) [Sov. J. Quantum Electron. **14**, 1464 (1984)].
29. S. M. Avanesyan and V. E. Gusev, Solid State Commun. **54**, 1065 (1985).
30. S. M. Avanesyan, M. Bonch-Osmolovskii, T. I. Galkina, V. E. Gusev *et al.*, Izvestia Akademii Nauk SSSR, seria fizicheskaya **50**, 2258 (1986) [Bulletin of the Academy of Sciences of the USSR Physical Series, **50**, 175 (1986)].
31. A. I. Akhieser, J. Phys. USSR **1**, 277 (1939).

Chapter 4

Optical Excitation of Sound in Semiconductors

Thermoelastic sound generation from interband absorption of light in semiconductors may be accompanied by excitation of acoustic waves by nonthermal mechanisms (Sec. 3.4). In this case, the optoacoustic (OA) signal will be essentially dependent on the nature of the space-time evolution of the photoexcited charge carriers and the physical parameters of the semiconductors responsible for this process.

4.1. Laser Generation of Longitudinal Acoustic Waves by the Concentration-Deformation Mechanism

Excitation of sound by the concentration-deformation mechanism was first achieved in silicon.[1] The nonthermal nature of sound generation was easily identified in this experiment, since photoexcitation of the electron-hole (EH) pairs resulted in silicon compression ($d > 0$), while heating at room temperature resulted in its expansion. An optoacoustic signal excited from interband absorption in Si of radiation $\lambda_1 \approx 1.06$ μm begins with the rarefactional phase, which fundamentally differentiates it from thermoelastic pulses in the case $\beta > 0$.[1] Interest in optoacoustic diagnostics[2-4] of semiconductor materials has grown with the development of microelectronics and acoustoelectronics. Pulsed optoacoustic measurements are actively used to investigate laser annealing and melting of semiconductors.[5-7]

It is necessary to account for thermoelastic (3.6) and deformation (3.68) acoustic sources simultaneously to describe the optoacoustic effect in nonpiezoelectric semiconductors. The wave equation (2.30) for the potential φ of vibrational velocity v defining longitudinal wave generation takes the form[4,8]

$$\frac{\partial^2 \varphi}{\partial t^2} - c_L^2 \frac{\partial^2 \varphi}{\partial z^2} = -\frac{1}{\rho_0} \frac{\partial}{\partial t} (K\beta T - dn) . \tag{4.1}$$

Here n is the concentration of photoexcited EH plasma, i.e., the increment of plasma concentration relative to its equilibrium value in the absence of optical irradiation. Assuming that charge carrier motion and phonon heat conduction have a diffusive character at high temperatures, we utilize the following simple model[8,9] to describe fields $n(t,z)$ and $T(t,z)$:

$$\frac{\partial n}{\partial t} = D \frac{\partial^2 n}{\partial z^2} - \frac{I_0}{\hbar \omega_L} g(z)\, f(t) - \frac{n}{\tau_R} ,$$

$$\frac{\partial T}{\partial t} = \chi \frac{\partial^2 T}{\partial z^2} - \frac{(\hbar\omega_L - E_g)}{\hbar\omega_L \rho_0 c_p} I_0 g(z)\, f(t) + \frac{E_g}{\rho_0 c_p} \frac{n}{\tau_{RT}}, \quad (4.2)$$

$$D\frac{\partial n}{\partial z}\bigg|_{z=0} = sn|_{z=0}, \; \chi\frac{\partial T}{\partial z}\bigg|_{z=0} = -\frac{E_g}{\rho_0 c_p} s_T n|_{z=0},$$

where D is the ambipolar diffusion coefficient of the EH plasma;[10] τ_R and τ_{RT} are the recombination time and the nonradiative recombination time of the EH pairs; s and s_T are their surface recombination and nonradiative surface recombination rates. The function $g(z)$ describes the spatial distribution of the optical radiation.

Simplified description (4.2) can be obtained from a more general description[11–13] if we linearize and uncouple the carrier diffusion equation, the lattice heat conduction equation, and the equation describing the propagation of optical radiation. Model (4.2) neglects the dependence of the physical parameters of the system on temperature and carrier concentration, the reverse effect of heating and photogeneration of EH plasma on the spatial distribution of optical radiation,[13] heat-induced carrier drift,[12] surface bending of the semiconductor energy bands,[14] the reverse effect of excited deformation waves on carrier motion,[15] etc. At the same time, model (4.2) takes into account the fundamental physical processes occurring from interband light absorption by the semiconductors.

The first terms on the right-hand side of Eqs. (4.2) describe the diffusion of nonequilibrium EH plasma and phonon heat conduction, while the second terms describe photogeneration of nonequilibrium carriers and instantaneous (on a picosecond time scale[11]) heating from their relaxation at the edge of the energy bands. The last terms describe nonequilibrium carrier recombination and crystal heating from the energy released in nonradiative recombination. The boundary conditions in model (4.2) account for the possibility of crystal temperature elevation from nonradiative surface recombination of EH pairs. Constructing a linear theory of the optoacoustic effect in semiconductors based on Eqs. (4.1) and (4.2) (Ref. 8) is a necessary step in order to analyze more complex models and nonlinear regimes (Chaps. 5 and 6).

The optical sound generation process is conveniently described by means of the spectral approach outlined in Chaps. 2 and 3. The fact that Eqs. (4.2) can be solved successively (initially the equation for n and then the equation for T) simplifies the solution. Consistent with Eq. (2.35) it is sufficient to represent the solution of these equations in spectral form:*

$$\hat{n}(\omega,p) = \frac{I_0}{\hbar\omega_L}\tilde{f}(\omega)\frac{1}{D(p^2-p_2^2)}\left[\frac{s+Dp}{s+Dp_2}\hat{g}(p_2) - \hat{g}(p)\right],$$

$$\hat{T}(\omega,p) = \frac{E_g}{\rho_0 c_p}\frac{1}{\chi(p^2-p_3^2)}\left\{\frac{I_0}{\hbar\omega_L}\tilde{f}(\omega)\left[\frac{\hbar\omega_L - E_g}{E_g}\left(\frac{p}{p_3}\hat{g}(p_3) - \hat{g}(p)\right)\right.\right.$$
$$\left.\left. + \frac{s_T}{s+Dp_2}\left(\frac{p}{p_3} - 1\right)\hat{g}(p_2)\right] + \frac{1}{\tau_{RT}}\left[\frac{p}{p_3}\hat{n}(\omega,p_3) - \hat{n}(\omega,p)\right]\right\},$$

*p is the Laplace variable of the transformed z coordinate.

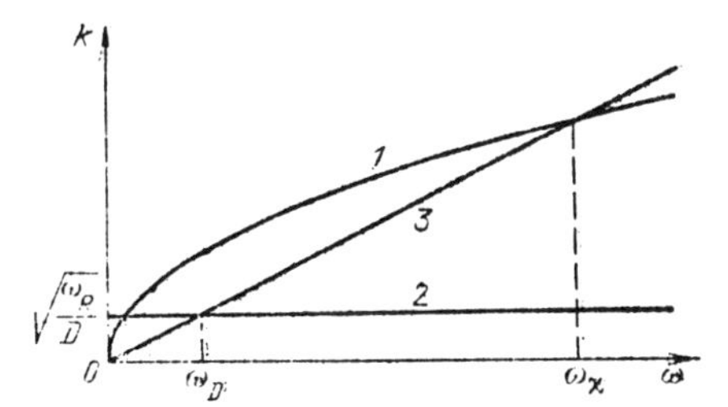

FIG. 4.1. Dispersion curves of the temperature (1), concentration (2), and acoustic (3) waves.

$$p_2 = \left(\frac{1}{D\tau_R} - i\frac{\omega}{D}\right)^{1/2}, \quad p_3 = \left(-i\frac{\omega}{\chi}\right)^{1/2}, \quad \mathrm{Re}\, p_2, p_3 > 0. \tag{4.3}$$

Relations (4.3) together with Eq. (2.35) provide a complete description of the acoustic signal spectrum far from the photoexcited region with an arbitrary ratio between the parameters of the problem. It is clear that in each specific case they can be significantly simplified.

Physically in solutions (4.3) $p_3 = -ik_T$, while $p_2 = -ik_n$, where k_T and k_n are the wave numbers of the temperature and concentration waves. Figure 4.1 shows the frequency dependences of the moduli of the wave vectors of these waves: $|k_T| = (\omega/\chi)^{1/2}$ and

$$|k_n| = [(\omega^2 + \omega_R^2)/D^2]^{1/4}. \tag{4.4}$$

Here we introduce the designation ω_R for the inverse carrier recombination time ($\omega_R \equiv \tau_R^{-1}$). The relative configuration of the curves in Fig. 4.1 takes into account that $\chi \sim 10^{-2} D$ in typical semiconductors.[11,13] Hence, in the frequency range where we can assume free diffusion of EH plasma ($\omega \gg \omega_R$): $|k_n| \ll |k_T|$. At the same time, in the low-frequency range ($\omega \ll \omega_R$) diffuse carrier motion is limited by their recombination ($|k_n| \approx (\omega_R/D))^{1/2}$ and $|k_n| \gg |k_T|$ as $\omega \to 0$. The equality $|k_T| = |k_n|$ holds at frequency ω_τ:

$$\omega_\tau = \chi\omega_R/(D^2 - \chi^2)^{1/2} \approx \chi\omega_R/D \ll \omega_R. \tag{4.5}$$

It is convenient to introduce the characteristic frequency ω_D at which the moduli of the wave numbers of the acoustic $|k_a| = \omega/c_L$ and concentration $|k_n|$ waves are equal. This characteristic frequency ω_D is analogous to the characteristic frequency $\omega_\chi = c_L^2/\chi$ (Sec. 3.2):

$$\omega_D = \omega_{D0}[1/2 + (1/4 + \omega_R^2/\omega_{D0}^2)^{1/2}]^{1/2}, \tag{4.6}$$

where $\omega_{D0} = c_L^2/D$ is the corresponding frequency in the absence of carrier recombination. According to Eq. (4.6) $\omega_D \approx \omega_{D0} \ll \omega_\chi$ for $\omega_R \ll \omega_{D0}$ and this quantity grows with diminishing of bulk recombination time. Figure 4.1 shows the case where $\omega_\tau < \omega_D < \omega_\chi$.

When the inequality $D \gg \chi$ holds, the characteristic frequencies are comparable, if $\omega_R \approx D\omega_\chi/\chi \gg \omega_\chi$. Since $\omega_\chi \sim 10^{11}$–10^{13} s^{-1} consistent with Sec. 3.2, this is possible with subpicosecond recombination times τ_R. Such fast recombination in model (4.2) should be treated as instantaneous recombination, causing thermalization without inertia of the absorbed light energy and dominance of thermoelastic sound generation. Hence, the deformation sound excitation mechanism can be expected only in the case $\omega_\tau \ll \omega_D \ll \omega_\chi$ which is similar to the situation shown in Fig. 4.1. We also note that diffusion equation (4.2) cannot generally be used to describe the dynamics of EH plasma on a subpicosecond time scale.

In typical semiconductors at high temperatures ($T_0 \gtrsim 300$ K) $D \lesssim 10$–50 cm^2/s,[11,16] while $c_L \sim (5$–$10) \times 10^5$ cm/s. Consequently, $\omega_D \gtrsim 10^{10}$–10^{11} s^{-1} and this quantity grows with ω_R. Hence, description (4.3), (2.35) of the acoustic wave spectra can be simplified in analyzing the optoacoustic effect within the frequency band $\omega \lesssim 10^9$–10^{10} s^{-1} by using the inequality

$$\omega \ll \omega_D < \omega_\chi. \tag{4.7}$$

The expressions for the vibrational velocity spectra of an acoustic wave (acoustic pulses) excited at the free semiconductor surface by deformation $\tilde{v}_n(\omega)$ and thermoelastic $\tilde{v}_T(\omega)$ mechanisms $[\tilde{v}_f(\omega) \equiv \tilde{v}_n(\omega) + \tilde{v}_T(\omega)]$ can be reduced to

$$\tilde{v}_n(\omega) = \frac{I_0}{\hbar\omega_L} \frac{d}{\rho_0 c_L^2} \tilde{f}(\omega) \frac{i\omega}{Dp_2^2} \left[\frac{-Di\omega/c_L}{s + Dp_2} \hat{g}(p_2) + \frac{\hat{g}(i\omega/c_L) - \hat{g}(-i\omega/c_L)}{2} \right],$$

$$\begin{aligned}
\tilde{v}_T(\omega) = {} & \frac{I_0}{\hbar\omega_L} \frac{K\beta E_g}{\rho_0^2 c_L^2 c_p} \tilde{f}(\omega) \frac{i\omega}{\chi p_3^2} \left\{ \frac{\hbar\omega_L - E_g}{E_g} \left[\frac{i\omega/c_L}{p_3} \hat{g}(p_3) \right.\right. \\
& \left. - \frac{\hat{g}(i\omega/c_L) - \hat{g}(-i\omega/c_L)}{2} \right] + \frac{1}{\tau_{RT}} \left[\frac{i\omega/c_L}{D(p_3^2 - p_2^2)p_3} \left(\frac{s + Dp_3}{s + Dp_2} \hat{g}(p_2) \right.\right. \\
& \left.\left. - \hat{g}(p_3) \right) - \frac{1}{Dp_2^2} \left(\frac{-Di\omega/c_L}{s + Dp_2} \hat{g}(p_2) + \frac{\hat{g}(i\omega/c_L) - \hat{g}(-i\omega/c_L)}{2} \right) \right] \\
& \left. + \frac{s_T}{s + Dp_2^2} \frac{i\omega/c_L}{p_3} \hat{g}(p_2) \right\}.
\end{aligned} \tag{4.8}$$

To clarify the physical nature of the thermoelastic component of the acoustic wave we have explicitly isolated in representation (4.8) the parts of the signal attributable to lattice heating from electron relaxation to the bottom of the conduction band and hole relaxation in the valence band (the first square brackets) from nonradiative bulk recombination of the EH pairs (the second square brackets) and from nonradiative surface carrier recombination. If there are no nonradiative recombination channels ($\tau_{RT} \gg \tau_R$, $s_T \ll s$), then $\tilde{v}_T(\omega)$ will not contain information on the dynamics of the EH plasma. The thermoelastic sound generation process itself in this case will be identical to the case examined previously in Chap. 3. Hence, since we are interested primarily in the specific nature of optical sound generation in semiconductors, for definiteness we will treat the nonradiative recombination channels as the dominant channels ($\tau_{RT} \approx \tau_R$, $s_T \approx s$). Such a situation is characteristic of indirect-band-gap semiconductors.[10]

As in the case of thermoelastic generation of sound (Sec. 2.2), the spectral transfer function $K_{n,T}(\omega) \equiv \tilde{v}_{n,T}(\omega)/f(\omega)$ can also be introduced due to relation (4.8) for acoustic wave excitation in a semiconductor. The characteristic frequencies where it is possible to expect a significant variation of $K_{n,T}(\omega)$ and their relation to the physical parameters of the semiconductor can be determined qualitatively by analyzing the poles of the denominators of solution (4.8).

For definiteness, we therefore set $\hat{g}(p) = \alpha(\alpha + p)^{-1}$ corresponding to linear absorption of light by a homogeneous medium (see Sec. 2.2). Then the multiplier $(\alpha + p_3)^{-1}$ determines the characteristic frequency $\omega_T = \alpha^2\chi$ which is related to the cooling time of the region heated from instantaneous thermalization of absorbed energy (Sec. 3.2). By analogy, the multiplier $(\alpha + p_2)^{-1}$ determines the frequency $|\omega| \sim [(\alpha^2 D)^2 + \omega_R^2]^{1/2}$ which characterizes the reduction time of carrier concentration in the carrier generation region due to the joint effect of diffusion and recombination processes.

Surface recombination may also be manifested in the frequency range $|\omega| \sim [(s^2/D)^2 + \omega_R^2]^{1/2}$ [the multiplier $(s + Dp_2)^{-1}$ in Eq. (4.8)]. As in the case of thermoelastic generation (Sec. 3.2) the characteristic frequency $\omega_a = \alpha c_L$ determined by equality between the wave number of sound and the light absorptivity [the multiplier $(\omega^2 + \omega_a^2)^{-1}$] is conserved. Finally, bulk recombination is manifested in pure form at frequencies $\omega \sim \omega_R$ (the multiplier p_2^{-2}). Variations of $K_T(\omega)$ over the frequency range $\omega \sim \chi\omega_R/(D-\chi) \approx \chi\omega_R/D \approx \omega_\tau$ (4.5) [the multiplier $(p_3^2 - p_2^2)^{-1}$] are related to the cooling of the region heated from recombination of the diffusing carriers.

If the OA signal is to contain information on any semiconductor parameter, at the very least it is necessary to excite sound at a characteristic frequency dependent on this parameter [i.e., this frequency must fall within the band $\omega \lesssim \tau_L^{-1}$ determined by $\tilde{f}(\omega)$]. Moreover, this frequency must satisfy inequality (4.7) for description (4.8) to be correct. It therefore follows from these estimates of the characteristic frequencies that bulk recombination inhibits optoacoustic determination of the frequencies $\omega_n = \alpha^2 D$ and $\omega_s = s^2/D$. In order to assure that information on these parameters is not lost, the bulk recombination time τ_R must at least exceed the laser irradiation time τ_L.

It is useful to recast solution (4.8) using the designations given here for the characteristic frequencies. We take advantage of the fact that the relation $\omega_n = \omega_a^2/\omega_{D0}$ exists between the frequencies ω_n, ω_{D0}, and ω_a (analogous to the relation $\omega_T = \omega_a^2/\omega_\chi$). It is also convenient to introduce the parameter $m_{D0} = \omega_a/\omega_{D0}$ analogous to the parameter $m_\chi = \omega_a/\omega_\chi$ (3.23) which characterizes the spatial dimensions of the light absorption region relative to the diffusion of nonrecombining EH pairs. It is important that $m_{D0} \gg m_\chi$ due to the condition $D \gg \chi$. Then

$$\omega_a = m_{D0}\omega_{D0},\quad \omega_n = m_{D0}^2\omega_{D0},\quad \omega_\chi = (m_{D0}/m_\chi)\omega_{D0},$$

$$\omega_T = m_{D0}m_\chi\omega_{D0} \ll \omega_n,\quad \omega_\tau = (m_\chi/m_{D0})\omega_R,$$

and Eq. (4.8) is given as

$$\tilde{v}_n(\omega) = \tilde{f}(\omega)(-1)\frac{-i\omega}{\omega_R - i\omega}\left[m_{D0}\frac{(-i\omega)^{1/2}}{\omega_S^{1/2} + (\omega_R - i\omega)^{1/2}}\frac{(-i\omega)^{1/2}}{m_{D0}\omega_{D0}^{1/2} + (\omega_R - i\omega)^{1/2}}\right.$$

$$\left. + \frac{-i\omega m_{D0}\omega_{D0}}{\omega^2 + (m_{D0}\omega_{D0})^2}\right], \tag{4.9}$$

$$\tilde{v}_T(\omega) = \tilde{f}(\omega) B\left(\frac{\hbar\omega_L - E_g}{E_g}\right)\Bigg[m_\chi \frac{(-i\omega)^{1/2}}{(m_{D0}m_\chi\omega_{D0})^{1/2} + (-i\omega)^{1/2}}$$

$$+ \frac{-i\omega m_{D0}\omega_{D0}}{\omega^2 + (m_{D0}\omega_{D0})^2}\Bigg] - B\frac{\omega_R}{(-i\omega)}\tilde{v}_n(\omega) + \tilde{f}(\omega)B$$

$$\times\Bigg\{\frac{(m_{D0}m_\chi)^{1/2}(-i\omega)^{1/2}}{[(m_{D0}m_\chi\omega_{D0})^{1/2} + (-i\omega)^{1/2}]}\frac{\omega_S^{1/2}}{[\omega_S^{1/2} + (\omega_R - i\omega)^{1/2}]}$$

$$- m_\chi\frac{\omega_\tau}{\omega_\tau + i\omega}\Bigg[\left(\frac{m_{D0}}{m_\chi}\right)^{1/2}\frac{[\omega_S^{1/2} - (m_{D0}/m_\chi)^{1/2}(-i\omega)^{1/2}]}{[\omega_S^{1/2} + (\omega_R - i\omega)^{1/2}]}$$

$$\times\frac{(-i\omega)^{1/2}}{[m_{D0}\omega_{D0}^{1/2} + (\omega_R - i\omega)^{1/2}]} - \frac{(-i\omega)^{1/2}}{(m_{D0}m_\chi\omega_{D0})^{1/2} + (-i\omega)^{1/2}}\Bigg]\Bigg\}.$$

Here the vibrational velocity in the acoustic wave is normalized to the characteristic value $v_0 = I_0 d/\hbar\omega_L \rho_0 c_L^2$, while $B = K\beta E_g/d\rho_0 c_p$ is the dimensionless parameter ($|B| \sim 0.1$) [see (I.10)].

According to Eq. (4.9) transfer functions $K_{n,T}(\omega)$ are dependent on the frequencies ω_R, ω_S, ω_{D0}, and the two parameters m_{D0} and m_χ. It is clearly evident that it is the parameters m_{D0} *and* m_χ that primarily determine the relative value of the different terms in Eq. (4.9). This substantially simplifies classification of the different sound generation conditions as a function of m_{D0} and m_χ. However, before carrying out such a classification we consider the concentration-deformation acoustic wave generation mechanism in isolation from the thermoelastic mechanism.

In the absence of carrier recombination $\omega_R, \omega_S \ll \omega$ the spectral transfer function K_n takes the form

$$K_n \approx (-1)(-i\omega)m_{D0}[(-i\omega)^{-1/2}/(m_{D0}\omega_{D0}^{1/2} + (i\omega)^{1/2})$$

$$+ \omega_{D0}/(\omega^2 + (m_{D0}\omega_{D0})^2)]. \qquad (4.10)$$

For definiteness we focus on the excitation of sound in the frequency range

$$\omega \ll \omega_{D0} \ll \omega_\chi. \qquad (4.11)$$

Then the terms in the square brackets of (4.10) are comparable when $\omega \sim m_{D0}^2\omega_{D0}$ if $m_{D0} \ll 1$. If $m_{D0} \gtrsim 1$, the first term corresponding to diffusion of the EH pairs will be predominant in frequency range (4.11).

Therefore, diffusion has no effect on the sound excitation process in the frequency range $\omega \gg m_{D0}^2\omega_{D0}$. When carrier recombination is accounted for this statement remains valid, since recombination only reduces the relative value of the first term in Eq. (4.10). This is because the recombination processes suppress diffusion. The spectral transfer function K_n in the frequency range $\omega \gg m_{D0}^2\omega_{D0}$ takes the form

$$K_n(\omega) \approx (-1)\frac{-i\omega}{\omega_R - i\omega}\frac{-i\omega\omega_a}{\omega^2 + \omega_a^2}. \qquad (4.12)$$

In the frequency range

$$\omega \leqslant m_{D0}^{2}\omega_{D0}\,, \tag{4.13}$$

it is generally necessary to use representation (4.9). The role of diffusion processes grows with diminishing frequency and hence they must be accounted for only if the first term is equal to or greater than the second term at least as $\omega \to 0$. A corresponding comparison yields the estimate

$$\omega_R, \omega_S \lesssim m_{D0}^{2}\omega_{D0}. \tag{4.14}$$

Otherwise EH pair motion is suppressed by pair recombination across all of frequency band (4.13). Inequality (4.14) has a clearly evident physical meaning:

$$l_D \equiv \sqrt{D\tau_R} \gtrsim \alpha^{-1}, \quad l_S = D/s \gg \alpha^{-1}. \tag{4.15}$$

The first inequality means that the distance over which the EH pairs diffuse during their lifetime will exceed their generation length. The left-hand side of the second inequality of (4.15) defines the characteristic extent of the effect of surface recombination on carrier motion. When the second inequality of (4.15) does not hold, the transit time of the carriers from the bulk of the absorption range to the boundary at speed s will be far less than their lifetime: $\alpha^{-1}/s \ll \alpha^{-2}/D = (\alpha^{-1}/l_D)^2\tau_R \leqslant \tau_R$ and surface recombination suppresses the diffusion of EH pairs from the crystal boundary.

Therefore, optoacoustic diagnostics of EH plasma motion processes can be implemented if the bulk and surface recombination frequencies themselves fall within the test range of (4.13). The expression for spectral transfer function K_n (4.9) can be substantially simplified only when the characteristic frequencies are quite different. We identify the most interesting physical cases.

If ω_R, $\omega_S \ll m_{D0}^2\omega_{D0} \sim \omega$, the transfer function is dependent only on the characteristic frequency ω_n: the inverse of the carrier diffusion time from the light absorption region,

$$K_n \approx (-1)m_{D0}\frac{-i\omega}{\omega_n}\left[\frac{1}{(-i\omega/\omega_n)^{1/2}[1+(-i\omega/\omega_n)^{1/2}]}+1\right]. \tag{4.16}$$

In the frequency range $\omega \ll m_{D0}^2\omega_{D0}$ the transfer function

$$K_n \approx (-1)\frac{(-i\omega)^2}{\omega_R - i\omega}\frac{1}{m_{D0}\omega_{D0}}\left[\frac{m_{D0}^2\omega_{D0}}{[\omega_S^{1/2}+(\omega_R-i\omega)^{1/2}][m_{D0}\omega_{D0}^{1/2}+\omega_R^{1/2}]}+1\right]$$

is also dependent on the bulk and surface recombination frequencies. In this situation the following limiting cases are possible.

(a) $\omega_R \ll \omega_S \sim \omega \ll m_{D0}^2\omega_{D0}$; here surface recombination is manifested to a complete extent:

$$K_n \approx (-1)\left(\frac{\omega_S}{\omega_{D0}}\right)^{1/2}\frac{-i\omega/\omega_S}{1+(-i\omega/\omega_S)^{1/2}}; \tag{4.17}$$

and the transfer function is independent of the bulk recombination time.

(b) $\omega \sim \omega_R \ll \omega_S$, $m_{D0}^2\omega_{D0}$; surface recombination only has an effect on the modulus of the transfer function:

$$K_n \approx (-1)\frac{\omega_R}{m_{D0}\omega_{D0}}\left[\left(\frac{m_{D0}^2\omega_{D0}}{\omega_S}\right)^{1/2}+1\right]\frac{(-i\omega/\omega_R)^2}{1-i\omega/\omega_R}; \tag{4.18}$$

(c) $\omega_S \ll \omega_R \sim \omega \ll m_{D0}^2\omega_{D0}$; information on surface recombination is lost:

$$K_n \approx (-1)\left(\frac{\omega_R}{\omega_{D0}}\right)^{1/2}\frac{(-i\omega/\omega_R)^2}{(1-i\omega/\omega_R)^{3/2}}. \tag{4.19}$$

We note the relation $\omega_S = \omega$ corresponds to equality of the surface recombination rate s and the phase velocity of the concentration wave $\sqrt{D\omega}$. Therefore the effect of surface recombination will primarily be manifested for $s \gg \sqrt{D\omega}$.

Thermoelastic generation of acoustic waves by interband light absorption in semiconductors results from instantaneous and recombination heating: $\tilde{v}_T$ $\tilde{v}_{TI} + \tilde{v}_{TR}$. Here we have denoted the thermoelastic signal component related to nonradiative EH pair recombination by $\tilde{v}_{TR}$: $\tilde{v}_{TR} \equiv \tilde{v}\mathrm{T}(\hbar\omega_\mathrm{L} = \mathrm{E_g})$. The component $\tilde{v}_{TI}$ characterizes sound generation from instantaneous heating and is independent of the parameters responsible for EH plasma motion. The corresponding component of the spectral transfer function is conveniently represented as

$$K_{TI} \approx B\frac{\hbar\omega_L - E_g}{E_g} m_\chi(-i\omega)\left[\frac{1}{(-i\omega)^{1/2}[m_\chi\omega_\chi^{1/2} + (-i\omega)^{1/2}]} + \frac{\omega_\chi}{\omega^2 + (m_\chi\omega_\chi)^2}\right]. \tag{4.20}$$

The frequency relation $K_{TI}(\omega)$ has been discussed in detail in Sec. 3.2. It is important for the subsequent analysis that if the frequency lies in range (4.11), then for $\omega \gg m_\chi^2\omega_\chi = m_\chi m_{D0}\omega_{D0}$ phonon heat conduction [the first term in square brackets in (4.20)] can be neglected, while for $\omega \ll m_\chi m_{D0}\omega_{D0}$ it plays a dominant role. Transfer function (4.20) coincides with (4.16) in the frequency range $\omega \sim m_\chi^2\omega_\chi \equiv \omega_T$ accurate to the substitutions $B(\hbar\omega_L - E_g)E_g^{-1}m_\chi \rightleftarrows (-m_{D0})$ and $\omega_T \rightleftarrows \omega_n$. Consequently, analysis of K_n in the range $\omega \sim \omega_n = m_{D0}^2\omega_{D0}$ is analogous to the analysis of K_T in the range $\omega \sim \omega_T$ carried out in Sec. 3.2. Therefore in this section we limit the analysis to excitation of sound simultaneously accounting for the two generation mechanisms in frequency ranges bounded by strong inequalities:

$$\omega_3 \ll m_\chi m_{D0}\omega_{D0} \ll \omega_2 \ll m_{D0}^2\omega_{D0} \ll \omega_1. \tag{4.21}$$

Here indices 1–3 label the spectral ranges.

The analysis carried out above has demonstrated that the absorption range is diffusively and thermally thick at frequencies $\omega \sim \omega_1$, while at frequencies $\omega \sim \omega_2$ it becomes thin for freely diffusing EH pairs, and at frequencies $\omega \sim \omega_3$ the absorption range becomes even thermally thin. We note that in the second and third frequency ranges the absorption range may remain diffusively thick if the recombination processes suppress EH pair motion [for example, when inequality (4.15) does not hold].

We classify the regimes of pulsed laser sound generation as a function of the parameters m_{D0} and m_χ. Consistent with Eq. (4.21) an increase of the parameter m_{D0} [and a simultaneous increase of the parameter $m_\chi \sim 10^{-2}m_{D0}$] will cause a high-frequency shift of the boundaries of these spectral ranges. Therefore the band

$\omega \sim \omega_1$ falls within spectral range (4.11) analyzed here only for $m_{D0} \ll 1$, while the band $\omega \sim \omega_2$ will only fall in this range for $m_\chi m_{D0} \ll 1$ [i.e., in fact for $m_{D0} \lesssim 1$].

Two facts are fundamental in the proposed classification. First, sound generation efficiency drops as we go from the first range ($\omega \sim \omega_1$) to the second and then drops further in the crossover to the third range. This is a direct consequence of Eqs. (4.12) and (4.17)–(4.19) for the spectral transfer functions of the deformation mechanism and will also hold when $\tilde{v}_n$ and $\tilde{v}_T$ are simultaneously accounted for. Second, we are considering pulsed sound excitation and assume that the acoustic waves are generated across the entire frequency band $\omega \lesssim \tau_L^{-1}$. Consequently, the spectral components with the maximum frequencies $\omega \approx \tau_L^{-1}$ will make the primary contribution to the signal. Hence, since we are interested in the generation of sound of the highest possible frequency we consider the maximum frequencies of those permitted by inequality (4.11): $\omega \sim (10^{-1}–10^{-2})\omega_{D0}$. This means that for $m_{D0} \ll 1$ we can account for sound excitation in the range $\omega \sim \omega_1$ only, while for $m_{D0} \sim 1$ we only consider sound excitation in the range $\omega \sim \omega_2$, and for $m_{D0} \gg 1$ we consider sound excitation only in the range $\omega \sim \omega_3$.

A preliminary analysis of the different sound excitation regimes from interband light absorption in nonpiezoelectric semiconductors makes it possible to identify a common feature that is useful to note before classifying the regimes. According to solution (4.9) the acoustic pulse v_{TR} excited thermoelastically by recombination heating of the crystal contains the contribution of $v_{TR}^{(1)}$ which is rather easily related to the deformation component v_n.

$$\tilde{v}_{TR}^{(1)} = -B \frac{\omega_R}{(-i\omega)} \tilde{v}_n. \tag{4.22}$$

Hence $|K_{TR}^{(1)}| \geqslant |K_n|$ in the frequency range $\omega \lesssim B\omega_R \sim 0.1\omega_R$. It turns out that when the other components of the thermoelastic wave are accounted for, this only enhances the role of thermoelastic generation, which is always dominant in the frequency range $\omega \ll B\omega_R$.

Another statement is also nearly always valid: sound excitation by the deformation mechanism is predominant in the spectral range $\omega \gtrsim \omega_R$. This conclusion can break down only under strong surface recombination with a thermally thin generation region. When we consider deformation and thermoelastic generation simultaneously it is necessary to analyze the transfer functions near the frequency $\omega_\tau \approx \chi\omega_R/D \ll B\omega_R$ as well (in addition to the frequency ranges $\omega \sim \omega_S$ and $\omega \sim \omega_R$).

We classify three primary regimes: $m_{D0} \ll 1$, $m_{D0} \sim 1$, $m_{D0} \gg 1$.

1. Regime $m_{D0} \ll 1$. Here the spectral range $\omega \sim \omega_1$ (4.21) plays the predominant role, while the thermal diffusivity and EH plasma diffusion have no effect on the sound generation process and the spectra of the acoustic pulses take the form

$$\tilde{v}_n(\omega) = \tilde{f}(\omega) \frac{i\omega/\omega_a}{R - i\omega/\omega_a} \frac{-i\omega/\omega_a}{1 + (\omega/\omega_a)^2},$$

$$\tilde{v}_T(\omega) = \tilde{f}(\omega) B\left(\frac{\hbar\omega_L - E_g}{E_g} + \frac{R}{R - i\omega/\omega_a}\right) \frac{-i\omega/\omega_a}{1 + (\omega/\omega_a)^2}. \tag{4.23}$$

According to Eq. (4.23) the form of the spectral transfer functions $K_{n,T}(\omega)$ will be determined by the parameter $R=\omega_R/\omega_a$. Physically, the parameter R represents the ratio of the sound transit time through the light absorption region to the EH pair recombination time.

In the case $R \ll 1$ carrier recombination has no effect on sound generation in the frequency range $\omega \sim \omega_a$:

$$\tilde{v}_n \approx \tilde{f}(\omega)(-1)\frac{-i\omega/\omega_a}{1+(\omega/\omega_a)^2},$$

$$\tilde{v}_T \approx \tilde{f}(\omega)\frac{B(\hbar\omega_L - E_g)}{E_g}\frac{-i\omega/\omega_a}{1+(\omega/\omega_a)^2}. \tag{4.24}$$

The frequency dependence of the transfer functions coincides with Eq. (2.25) here. The relative contribution of the thermoelastic and deformation mechanisms to the OA effect is given by the parameter

$$|K_T|/|K_n| = |B|(\hbar\omega_L - E_g)/E_g.$$

The deformation sound generation mechanism predominates for $\hbar\omega_L \lesssim 2E_g$ ($R \ll 1$) since $|B| \sim 0.1$. The phase shift between K_T and K_n is determined by the sign of B: $\Delta\varphi(R \ll 1) = \varphi_T - \varphi_n = \pi(1+\text{sgn}\,B)/2$. The acoustic waves are excited by these mechanisms in antiphase if the signs of d and β coincide and in phase if these signs are opposed. Therefore the deformation and thermoelastic sound mechanisms are in competition for $\text{sgn}(d\beta) > 0$.

In the case $R \gg 1$ acoustic pulse spectra (4.3) take the form

$$\tilde{v}_n(\omega) \approx \tilde{f}(\omega)(-1)\frac{-i\omega/\omega_a}{R}\frac{-i\omega/\omega_a}{1+(\omega/\omega_a)^2},$$

$$\tilde{v}_T(\omega) \approx \tilde{f}(\omega)B\frac{\hbar\omega_L}{E_g}\frac{-i\omega/\omega_a}{1+(\omega/\omega_a)^2}. \tag{4.25}$$

A comparison of Eqs. (4.24) and (4.25) reveals that going from $R \ll 1$ to $R \gg 1$ leads to differentiation of the profile of the deformation component of the acoustic wave and reduces its generation efficiency by a factor of R. The physical cause of this differentiation lies in the fact that for $R \gg 1$ the sound excitation efficiencies from carrier photogeneration and recombination are identical, and these processes excite acoustic waves of different polarities (since the first process generates $\partial n/\partial t > 0$, while the second yields $\partial n/\partial t < 0$). The asynchronous-phase combination of waves excited from carrier generation and recombination will also lead to differentiation of the profile $v_n(\tau)$. The reduction in wave amplitude is caused by the saturation of carrier concentration growth for $R \gg 1$ in the frequency band of interest $\omega \sim \omega_a$. Indeed, at these frequencies the concentration will not rise over a time ω_a^{-1} but rather over a time $\tau_R = \omega_a^{-1}/R \ll \omega_a^{-1}$.

The relative role of the deformation and thermoelastic generation mechanisms is dependent on frequency in the range $R \gg 1$:

$$|K_T|/|K_n| \approx |B|(\hbar\omega_L/E_g)(\omega_R/\omega).$$

Thermoelastic acoustic wave excitation predominates at low frequencies $\omega \lesssim |B|(\hbar\omega_L/E_g)\omega_R$, while concentration-deformation excitation predominates at high frequencies $\omega \gtrsim |B|(\hbar\omega_L/E_g)\omega_R$.

The role of recombination is clearly evident from a comparison of $\tilde{v}_n$ and the component $\tilde{v}_{TR}$ of the thermoelastic wave excited by recombination heating. From Eq. (4.23) $\tilde{v}_{TR} = -BR(-i\omega/\omega_a)^{-1}\tilde{v}_n$. Hence, there will be a fixed phase shift $\Delta\varphi = \Delta\varphi(R \ll 1) + \pi/2$ between K_{TR} and K_n; the profile of v_{TR} can be obtained by integrating the profile of v_n. We obtain by means of Eqs. (4.12) or (4.23) for the spectral transfer function K_n,

$$\varphi_n = \arctan\frac{\omega}{\omega_R}, \quad |K_n| = \frac{\omega/\omega_a}{[R^2 + (\omega/\omega_a)^2]^{1/2}}\frac{\omega/\omega_a}{[1 + (\omega/\omega_a)^2]},$$

$$|K_T| = |B|\frac{R}{(\omega/\omega_a)}|K_n|.$$

Therefore, the reduction of the recombination time (growth of R) will serve to reduce the efficiency of the deformation mechanism and will enhance the efficiency of the thermoelastic mechanism at all frequencies, yet primarily at low frequencies. Physically, this is attributed to acceleration of energy transfer from the EH plasma to the thermal lattice vibrations. The frequency range $\omega \lesssim |B|R\omega_a = |B|\omega_R$ in which thermoelastic sound generation due to recombination heating predominates will expand with increasing R.

Section 5.2 will provide analytic representations for the profiles of acoustic pulses of spectral composition (4.23) for the case of a Gaussian laser pulse envelope in analyzing a sound generation experiment in Si. Here, we consider the limiting cases of acoustically thick ($\omega_1 \gg \omega_a$) and acoustically thin ($\omega_1 \ll \omega_a$) absorption regions. We note from the outset that sound generation in these cases with a free surface is inefficient in the same manner as in subsequent general regimes 2 and 3. However, all these regimes can be implemented for spectroscopic purposes.

When the wave number of the acoustic wave far exceeds the absorption coefficient of light ($\omega_1 \gg \omega_a = m_{D0}\omega_{D0}$), Eq. (4.23) can be further simplified:

$$\tilde{v}_n^{(0)} \approx \tilde{f}(\omega)\omega_a/(\omega_R - i\omega),$$
$$\tilde{v}_T^{(0)} \approx \tilde{f}(\omega)(-B)\omega_a[(\hbar\omega_L/E_g)(-i\omega)^{-1} - (\omega_R - i\omega)^{-1}]. \tag{4.26}$$

The profiles of acoustic pulses excited in this case will depend on the ratio of the laser pulse duration τ_L and EH recombination time $\tau_R = \omega_R^{-1}$. In order to achieve a compact representation of the acoustic wave profiles in the case of a Gaussian laser pulse envelope ($f(t) = \exp[-(2t/\tau_L)^2]$) we introduce a special designation for the following integral:

$$U(c,\tau) \equiv (2\pi)^{-1}\int_{-\infty}^{\infty}[\exp(-i\omega\tau)/\tau_L]\tilde{f}(\omega)/(c - i\omega)d\omega$$
$$= (\sqrt{\pi}/4)\exp[-(c\tau_L/4)^2 - c\tau]\operatorname{erfc}(c\tau_L/4 - 2\tau/\tau_L).$$

If dimensionless time $\tau = \tau/\tau_L$ and the parameter $c = c\tau_L$ are introduced, then

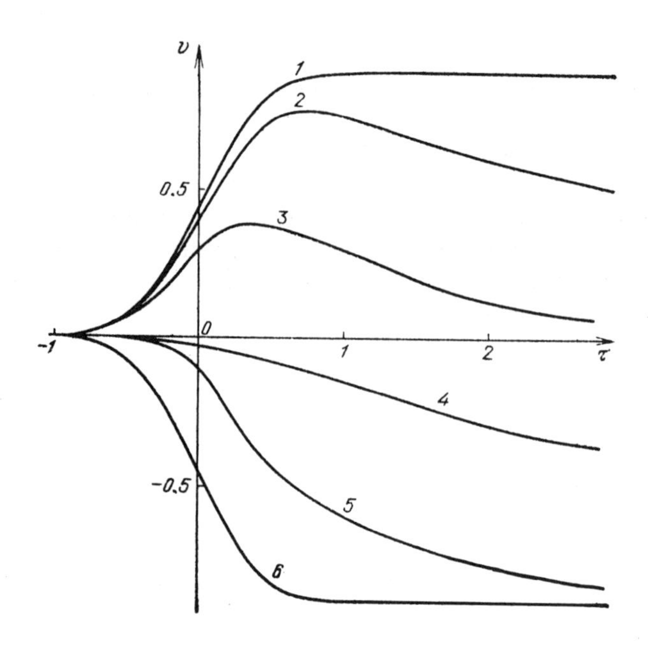

FIG. 4.2. Profiles of the deformation and thermoelastic acoustic wave components for $\omega_a \ll \omega_R$, τ_L^{-1}; $v = v_n/\omega_a\tau_L - c = 0$ (1); 0.2 (2); 1 (3), $v = v_{TR}/B\omega_a\tau_L - c = 0.2$ (4); 1 (5); ∞ (6).

$$U(c,\tau) = (\sqrt{\pi}/4)\exp[-(c/4)^2 - c\tau]\operatorname{erfc}(c/4 - 2\tau),$$

while the following relations are valid for pulses excited by the deformation and thermoelastic mechanisms:

$$\begin{aligned} v_n^{(0)} &= \omega_a\tau_L U(\tau_L/\tau_R,\tau), \\ v_T^{(0)} &= -B\omega_a\tau_L[\hbar\omega_L E_g^{-1} U(0,\tau) - U(\tau_L/\tau_R,\tau)]. \end{aligned} \tag{4.27}$$

Profiles (4.27) are shown in Fig. 4.2 for different ratios between the optical excitation duration and the carrier recombination time ($c = \tau_L/\tau_R$). For definiteness, we shall henceforth assume $B > 0$, which corresponds to generation of sound by the two mechanisms considered here (in the absence of recombination) in antiphase.

Consistent with representation (4.27) and Fig. 4.2 the duration of the leading edge of the pulse $v_n^{(0)}(\tau)$ is of the order of the light irradiation duration τ_L; the duration of the trailing edge is determined by the carrier recombination time for $\tau_R \gtrsim \tau_L$, while it is determined by the light pulse duration for $\tau_R \lesssim \tau_L$. From Eqs. (4.26) and (4.27) for $\tau_R \gg \tau_L$ the profile $v_n^{(0)}$ is the integral of the light pulse envelope. For $\tau_R \ll \tau_L$ the profile $v_n^{(0)}$ mimics the light pulse envelope. The duration of crystal thermal expansion is determined by the maximum of times τ_L and τ_R (Fig. 4.2). Therefore, the characteristic time of acoustic signal variation will not exceed $\max\{\tau_L,\tau_R\}$ and the boundaries on the applicability of solutions (4.26) and (4.27) can be given as

$$\omega_a \ll \min\{\omega_R, \tau_L^{-1}\}.$$

In terms of physical meaning there will be no sound generation in the crystal bulk in this subregime. The calculated acoustic pulse profiles reflect the temporal behavior of the deformation of the crystal surface [see solution (3.100) of the wave equation which explicitly identifies the signal excited at the boundary]. We recall that the vibrational velocity and longitudinal deformation are linearly related in a traveling acoustic wave:

$$v = \frac{\partial u}{\partial t} = -c_L \frac{\partial u}{\partial z} = -c_L \mathscr{D}.$$

Consequently, $|v_n^{(0)}|$ and $|v_T^{(0)}|$ in fact determine the time dependence of plasma concentration and crystal temperature near the surface ($z \ll \alpha^{-1}$).

When the light absorption region is acoustically thin ($\omega \ll \omega_a$) a further simplification of spectra (4.23) makes it possible to obtain the relation $\tilde{v}_{n,T}^{(1)} \approx -(-i\omega/\omega_a)^2 \tilde{v}_{nT}^{(0)}$. The acoustic pulse profiles are correspondingly proportional to the second derivatives of profiles (4.27):

$$v_{n,T}^{(1)} \approx -(\omega_a \tau_L)^{-2} \frac{\partial}{\partial \tau^2} (v_{n,T}^{(0)}). \tag{4.28}$$

Figure 4.3 shows the profiles of the two components of an acoustic signal (v_n and v_T) for different values of the parameter $c = \tau_L/\tau_R$. We note that $\hbar\omega_L = E_g$ was assumed in plotting these profiles for definiteness and, consequently, the figures actually show the v_{TR} profiles. The $v_n^{(1)}$ profile undergoes a transformation from the first to the second derivative of the light pulse envelope with increasing recombination rate [Fig. 4.3(a)].

Analysis reveals that the characteristic acoustic pulse duration is always of the order of τ_L in this subregime. The situation is that the contribution of the carrier recombination processes to $v_n^{(1)}$ and the thermoelastic component $v_T^{(1)}$ for $\tau_R \gg \tau_L$ (where the duration of these processes are of the order of τ_R) are small compared to the contribution to the resulting signal of acoustic waves excited from photogeneration of EH pairs. All components have a duration of the order of τ_L for $\tau_R \lesssim \tau_L$. Consequently, solution (4.28) is valid for $\tau_L^{-1} \ll \omega_a \ll \omega_{D0}$ independent of the recombination time.

2. Regime $m_{D0} \sim 1$. Beginning with this regime spatial redistribution of photogenerated EH pairs begins to have an effect on the acoustic signal spectrum. The spectral range $\omega \sim \omega_2$ (4.21) has a dominant effect on the generation process. The case of greatest interest from the viewpoint of determining semiconductor parameters is when one of the characteristic frequencies falls within this spectral range (characteristic frequencies ω_S, ω_R, or ω_τ). Several subregimes can therefore be classified on this basis.

(a) $\omega_R \ll \omega_S \sim \omega_2$. The deformation mechanism is dominant since $\omega_2 \gg \omega_R$. Representation (4.17) holds for $K \sim K_n$ which makes it possible to obtain the following expressions for the modulus of the transfer function and the frequency-dependent phase shift:

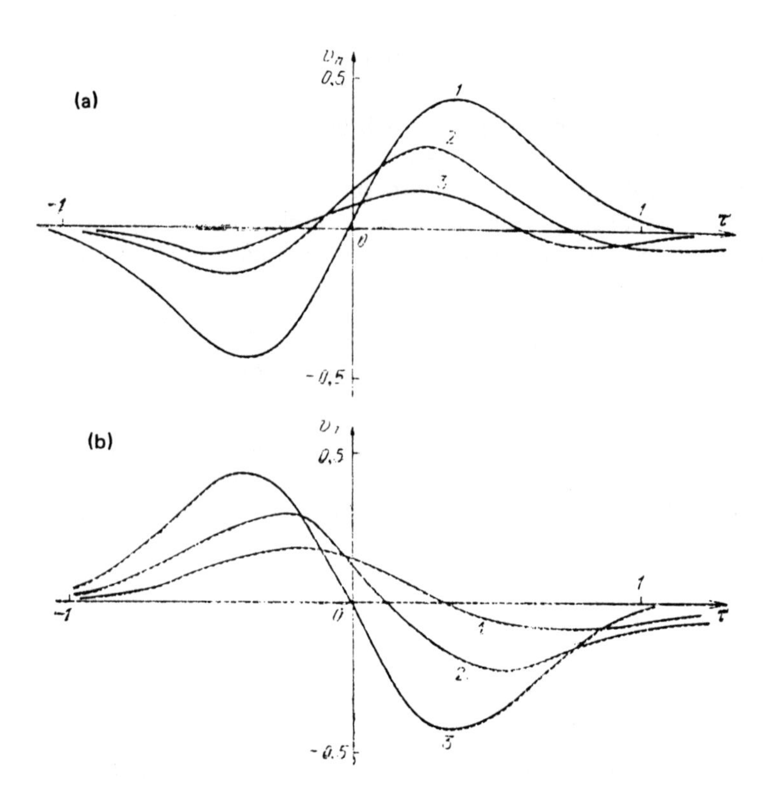

FIG. 4.3. Acoustic pulse profiles in regime $\tau_L^{-1} \ll \omega_a \ll \omega_{D0}$ for values of c equal to (a) 0 (1), 2 (2), 3 (3); (b) 1 (1), 2 (2), ∞ (3).

$$|K| = \left(\frac{\omega_S}{\omega_{D0}}\right)^{1/2} \frac{\omega/\omega_S}{[1 + (2\omega/\omega_S)^{1/2} + \omega/\omega_S]^{1/2}},$$
$$\Delta\varphi = -\arctan[1 + (2\omega_S/\omega)^{1/2}]. \tag{4.29}$$

Consequently, the phase of the transfer function shifts by $-\pi/4$ in the frequency range $\omega \sim \omega_S$ (as $\omega \ll \omega_S$ goes to $\omega \gg \omega_S$).

The simplest analytical representation for the acoustic pulse profiles whose spectrum contains anomalies of the type $(c + \sqrt{-i\omega})^{-1}$ can be obtained by modeling the laser pulse envelope by the function $f(t) \equiv \exp(-2|t|/\tau_L)$. Then, introducing the designation

$$W \equiv (2\pi)^{-1} \int_{-\infty}^{\infty} \tau_L^{-1/2} \exp(-i\omega\tau) \tilde{f}(\omega)(c + \sqrt{-i\omega})^{-1} d\omega$$

and the dimensionless variables $\tau = \tau/\tau_L$, $c = c(\tau_L)^{1/2}$ we obtain

$$W(c,\tau < 0) = \frac{\sqrt{2}}{c + \sqrt{2}} \exp(2\tau),$$

$W(c,\tau > 0)$

$$= \frac{2}{c^2+2}\left[\frac{c}{\sqrt{2}}\exp^{(-2\tau)} + \exp^{(-2\tau)}\text{erfi}\sqrt{2\tau} - \frac{c^2+2}{c^2-2}\left|e^{2\tau}\right|\text{erfc}\sqrt{2\tau}\right.$$
$$\left. + \frac{2\sqrt{2c}}{c^2-2}\text{erfc}(c\sqrt{\tau})\right]. \quad (4.30)$$

The profiles of the derivative of the function $W(c,\tau)$ for different values of the parameter c are shown in Fig. 4.4. With use of Eq. (4.30) the acoustic wave profile with spectrum (4.17) can be given as

$$v_n \approx -(\omega_{D0}\tau_L)^{-1/2}\frac{\partial W[(\omega_S\tau_L)^{1/2},\tau]}{\partial\tau}. \quad (4.31)$$

In this case $c=(\omega_S\tau_L)^{1/2}$. The acoustic wave profile (specifically, the ratio of the amplitudes and durations of peaks of different polarity) is essentially dependent on the carrier surface recombination rate and the ambipolar diffusion coefficient. The pulse profile v_n goes from a bipolar ($\omega_S \ll \tau_L^{-1}$, Fig. 4.4, curve 1) to a tripolar profile ($\omega_S=\tau_L^{-1}$, Fig. 4.4, curve 2) with increasing characteristic frequency $\omega_S = s^2/D$.

(b) $\omega_2 \sim \omega_R \ll \omega_S$.

(c) $\omega_S \ll \omega_R \sim \omega_2$. The deformation sound generation mechanism also predominates in these subregimes and spectral transfer functions (4.18) and (4.19) can be used here.

(d) $\omega_2 \lesssim \omega_\tau$. Here the thermoelastic sound generation mechanism dominates and the transfer function takes the form

$$K \approx K_T \approx B(\hbar\omega_L/E_g)(-i\omega/\omega_a).$$

The OA signal in this subregime contains no information on the electronic subsystem of the semiconductor.

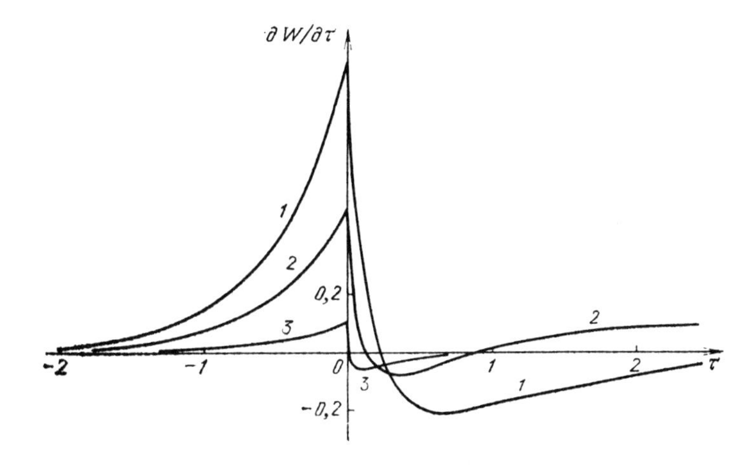

FIG. 4.4. The ancillary function $\partial W/\partial\tau$ for the values of parameter c equal to 0(1), 1(2), 10(3).

3. The regime $m_{D0} \gg 1$. Phonon heat conduction begins to manifest an effect on the acoustic signal spectrum in certain subregimes beginning with this regime.

(a) $\omega_R \ll \omega_S \sim \omega_3$. This does not occur in this subregime. There are virtually no changes compared to 2(a). There is only greater basis for considering the limiting process $\omega \ll \omega_S$ since the frequency band $\omega \sim \omega_3$ unlike $\omega \sim \omega_2$ has no lower limit. The modulus of transfer function (4.29) in the range $\omega \ll \omega_S$ diminishes with increasing surface recombination rate (as $\sim s^{-1}$). This is due to the limitation of carrier concentration growth by surface recombination.

(b) $\omega_3 \sim \omega_R \ll \omega_S$. Here thermoelastic generation may compete with generation by deformation mechanism (4.18) due to the fact that the transfer function

$$K_T \approx B(\hbar\omega_L/E_g)(-i\omega/\omega_\chi)^{1/2} \tag{4.32}$$

and the efficiency of thermoelastic generation diminish more slowly with diminishing frequency. A comparison of Eqs. (4.18) and (4.32) demonstrates that the thermoelastic mechanism is dominant in the range $\omega \sim \omega_R$ if

$$\omega_R \ll B^2 \min\{m_\chi m_{D0}\omega_{D0}, m_\chi \omega_S/m_{D0}\}.$$

This means that Eq. (4.32) can be used for the transfer function in the frequency range $\omega_4 \sim \omega_R \ll B^2 m_\chi\, m_{D0}\omega_{D0}$ in subregime 3(b), if very fast surface recombination exists: $\omega_S \gtrsim m_{D0}^2\omega_{D0} \gg \omega_{D0}$. In this case since $B^2 \sim 10^{-2}$ frequencies $\omega \sim \omega_4$ fall within the high-frequency portion of the range (4.11) only when $1 \ll m_\chi \ll m_{D0}$. With moderately strong surface recombination ($\omega_S \lesssim m_{D0}^2\omega_{D0}$) thermoelastic generation is dominant for

$$\omega \sim \omega_R \ll B^2 m_\chi \omega_S/m_{D0} \sim 10^{-4}\omega_S.$$

(c) $\omega_S \ll \omega_R \sim \omega_3$. Nothing changes compared to case 2(c) in this subregime. Consequently, the predominance of thermoelastic generation in the frequency range $\omega \sim \omega_R$ can be achieved only in the case of strong surface recombination $\omega_S \gg 10^4\omega_R$.

(d) $\omega_3 \lesssim \omega_\tau$. As is always the case, thermoelastic generation predominates in this subregime. However, unlike 2(d), the transfer function may be dependent on ω_τ:

$$K_T \approx B\left(\frac{m_\chi}{m_{D0}}\right)\left(-i\frac{\omega}{\omega_\tau}\frac{\omega_R}{\omega_{D0}}\right)^{1/2}\left[\frac{\hbar\omega_L}{E_g} + \frac{-i\omega/\omega_\tau}{1+(-i\omega/\omega_\tau)^{1/2}}\frac{\omega_R^{1/2}}{\omega_S^{1/2}+\omega_R^{1/2}}\right]. \tag{4.33}$$

Frequency ω_τ has an evident physical meaning. It is equal to the inverse of the cooling time of the region heated by recombination of diffusing carriers ($\omega_\tau^{-1} = D\tau_R/\chi = l_D^2/\chi$).

Transfer function (4.33) is different from Eq. (4.32) only when bulk recombination predominates over surface recombination ($\omega_R \gtrsim \omega_S$). In the opposite case ($\omega_S \gg \omega_R$) this anomaly does not appear in the acoustic wave spectrum. If we set $\hbar\omega_L = 2E_g$, $\omega_S \ll \omega_R$ in Eq. (4.33) for convenience, then we obtain for the frequency-dependent component of the phase of the transfer function

$$\Delta\varphi = -\arctan\frac{(\omega/\omega_\tau)^{3/2} + \sqrt{2}(\omega/\omega_\tau)}{(\omega/\omega_\tau)^{3/2} + 2\sqrt{2}(\omega/\omega_\tau) + 4(\omega/\omega_\tau)^{1/2} + 2\sqrt{2}}.$$

Consequently, the phase of the transfer function shifts by $-\pi/4$ in the frequency range $\omega \sim \omega_\tau$ (going from $\omega \ll \omega_\tau$ to $\omega \gg \omega_\tau$). The thermoelastic component profile in the case $f(t) = \exp(-2|t|/\tau_L)$ can be represented by means of function $W(c,\tau)$ (4.30) introduced earlier as

$$v_T \approx \frac{B}{(\omega_\chi \tau_L)^{1/2}} \frac{\partial}{\partial \tau} \left\{ \frac{\hbar\omega_L}{E_g} W(0,\tau) + \frac{\omega_R^{1/2}}{\omega_S^{1/2} + \omega_R^{1/2}} [f(\tau) - (\omega_\tau \tau_L)^{1/2} W((\omega_\tau \tau_L)^{1/2},\tau)] \right\}.$$

The acoustic pulse profile in this subregime will depend on the ratio of the laser pulse duration and the cooling time of the region heated by nonradiative carrier recombination $[c = (\omega_\tau \tau_L)^{1/2}]$.

All these longitudinal acoustic pulse excitation regimes can be implemented experimentally using the same semiconductor material and optical radiation of different frequencies ω_L. This is directly related to the strong dependence of the interband light absorption coefficient on the difference between the energies of optical quantum $\hbar\omega_L$ and band gap E_g. For example, as we go from radiation at wavelength $\lambda_1 \approx 1.06$ μm to its third harmonic $\lambda_3 \approx 0.35$ μm, the interband light absorptivity in crystalline silicon will vary by five orders of magnitude (from $\alpha \approx 10$ cm^{-1} to $\alpha \approx 10^6$ cm^{-1}). On the other hand, using optical radiation with fixed parameters it is possible to induce thermoelastic generation of sound in certain types of semiconductor materials and sound generation by the concentration-deformation mechanism in others.

We consider as an example excitation of longitudinal acoustic pulses in Si and Ge by pulsed optical radiation with parameters $\lambda_1 = 1.06$ μm, $\tau_L \simeq 20$ ns, $I_s \simeq 1$ MW/cm^2. The theory outlined above demonstrates that the predominance of one or the other sound generation mechanism is primarily determined by the ratio of the laser irradiation time τ_L to the recombination time τ_R of the EH pairs. This ratio must be estimated. In this case of intense laser irradiation, the lifetime of the EH pairs is determined by their nonlinear recombination. In silicon and germanium the primary nonlinear recombination channel is nonradiative Auger recombination, with time scale

$$\tau_R = (\gamma n^2)^{-1}. \tag{4.34}$$

Here γ is the Auger-recombination constant. The values of the parameters for Si and Ge that we henceforth utilize for the estimates are listed in Table 4.1. We note that expression (4.34) holds for the recombination time if the photoexcited carrier concentration far exceeds the equilibrium free-carrier concentration.

In order to identify the sound generation mechanism it is necessary to estimate the characteristic carrier concentrations that are achieved in interband light absorption. For purposes of simplicity we neglect surface recombination. Then if the EH plasma concentration grows over time τ_{ex}, while plasma penetrates the crystal to a depth l_{ex}, obviously, the following estimate will be valid:

$$n \sim (I_0/\hbar\omega_L)\tau_{ex}/l_{ex}. \tag{4.35}$$

Here $I_0/\hbar\omega_L$ is the number of EH pairs generated by optical radiation per unit of time and unit of irradiated area. The structure of Eq. (4.2) for the carrier concentration indicates that

TABLE 4.1. Physical parameters of germanium and silicon.

Material	γ cm^6/s	D cm^2/s	c_L 10^6 cm/s	α, cm^{-1} $\lambda_1 = 1.06\ \mu$m	α, cm^{-1} $\lambda_2 = 0.53\ \mu$m	E_g eV (300 K)	I_0/I_S λ_1	I_0/I_S λ_2
Ge	2×10^{-31}	65	0.5	1.4×10^4	5×10^5	0.66	0.61	0.53
Si	4×10^{-31}	35	1	10	10^4	1.12	0.7	0.62

Material	χ cm^2/s	ρ_0 g/cm^3	C_p 10^6 erg/$(g\times K)$	K 10^{11} dyne/cm^2	β $10^{-5}K^{-1}$	d, eV
Ge	0.35	5.3	3.4	7.7	1.8	−7
Si	0.2	2.3	9.5	9.8	1.2	8

$$l_{\text{ex}} \sim \max\{\alpha^{-1}, l_D = (D\tau_{\text{ex}})^{1/2}\}, \quad \tau_{\text{ex}} \sim \min\{\tau_L, \tau_R\}. \tag{4.36}$$

If we can neglect the effect of rising temperature on the values of the physical parameters, relations (4.34)–(4.36) form a closed system to estimate the characteristic values of n, τ_R, and l_D.

The following are achieved in Si with the optical irradiation parameters listed above: $\omega_{D0} \simeq 3\times10^{10}$ s^{-1}, $\omega_a \simeq 10^7$ s^{-1}, $m_{D0} \simeq 3\times10^{-4}$. The characteristic frequency $\omega \sim \tau_L^{-1}$ falls within the spectral band range $\omega \sim \omega_1$. In this regime (the first regime according to the classification considered above) the absorption range is diffusively thick. This means that $l_{\text{ex}} \sim \alpha^{-1}$. Further estimates by means of Eqs. (4.34)–(4.36) yield $\tau_R \sim 5\times10^{-6}$ s $\gg \tau_L$. Therefore, the concentration grows linearly during the entire laser irradiation process $\tau_{\text{ex}} \sim \tau_L$. The characteristic frequency $\tau_L^{-1} \sim 5\times10^7$ s^{-1} substantially exceeds the recombination frequency $\omega_R \sim 2\times10^5$ s^{-1} and, consequently, the deformation sound generation mechanism dominates. Since the parameter $R = \omega_R/\omega_a = 2\times10^{-2} \ll 1$, recombination will have no effect on the sound excitation process. The spectrum of pulse $\tilde{v}_n$ is described by relation (4.24). The pulse profile is a bipolar profile with the first phase being a rarefactional phase (since $d > 0$ in Si). The duration of both phases is determined by the sound transit time through the EH pair generation region [$\omega_a^{-1} \sim (\alpha c_L)^{-1} \sim 100$ ns], while the duration of the transition region is of the order of the laser irradiation time τ_L. All these conclusions have been confirmed by experiment[4] (Fig. 4.5, profile 1).

Under the same conditions $\omega_{D0} \simeq 4\times10^9$ s^{-1}, $\omega_a \simeq 7\times10^9$ s^{-1}, $m_{D0} \simeq 2$, $m_\chi \simeq 10^{-2}$ are achieved in Ge. Therefore, the characteristic frequency $\tau_L^{-1} \sim 5\times10^7$ s^{-1} falls within the spectral range $\omega \lesssim \omega_\chi = m_\chi m_{D0}\omega_{D0}$ (corresponding to transition from regime 2 to regime 3 in the classification). Estimates from Eqs. (4.34)–(4.36) reveal that in this case the depth of the surface layer occupied by nonequilibrium carriers in Ge is determined by carrier diffusion $l_{\text{ex}} \sim (D\tau_R)^{1/2}$. Their recombination effectively limits the growth of EH plasma concentration: $\tau_{\text{ex}} \sim \tau_R \sim 5\times10^{-9}$ s. Since $\omega_R \sim 2\times10^8$ s^{-1} will not substantially exceed τ_L^{-1} in the

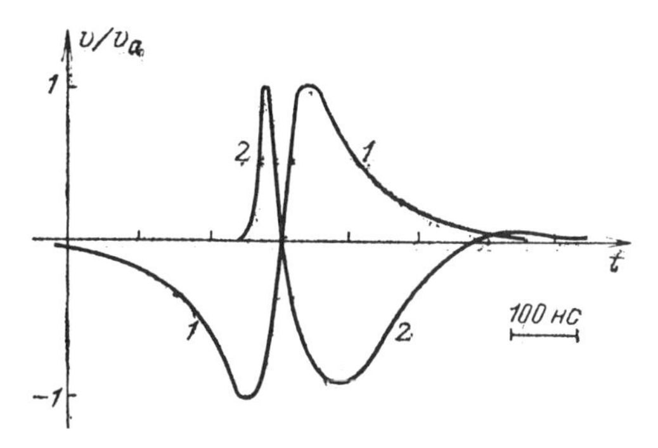

FIG. 4.5. Longitudinal acoustic pulse profiles in Si (1) and Ge (2) excited by laser radiation with parameters $\lambda_1 = 1.06$ μm, $\tau_L = 20$ ns, $I_0 = 1$ MW/cm^2. Vertical scales are not identical.

frequency range $\omega \sim \tau_L^{-1}$ both generation mechanisms ($|K_T| \sim 0.4|K_n|$ for $|B| \sim 0.1$) will make a comparable contribution to the OA effect.

An increase in the optical irradiation intensity in regime $l_{ex} = (D\tau_R)^{1/2}$ will serve to reduce the carrier recombination time $\tau_R \sim I_0^{-1}$. For example, for $I_S \sim 25$ MW/cm^2 we have $\omega_R \sim 5 \times 10^9$ s$^{-1} \gg \tau_L^{-1}$: thermoelastic sound generation dominates. It is interesting that in this case the frequency $\omega_\tau \simeq \chi\omega_R/D \simeq 3 \times 10^7$ s^{-1} falls within the same spectral range as frequencies τ_L^{-1} and ω_χ. We can therefore expect a simultaneous dependence of the acoustic wave profile on the cooling time of the photoexcitation region $[\sim (\alpha^2\chi)^{-1} \sim \omega_\chi^{-1}]$ and the cooling time of the recombination heating region $(\sim \omega_\tau^{-1})$. Figure 4.5 shows the longitudinal pulse profile in Ge recorded experimentally at such irradiation intensities.[17] The duration of the compressional phase is of the order of the optical irradiation time τ_L. The latter fact is characteristic of all regimes in which the thermoelastic generation mechanism is dominant and in which heat conduction plays a significant role (see Chap. 3). The structure of function $W(c,\tau)$ (4.30) is a formal substantiation; this function can be used to describe the acoustic wave profile [both the spectral composition (4.33) and (4.20)].

A more general statement can also be formulated: the first phase of the acoustic pulse in regimes with $m_{D0} \gtrsim 1$ has a duration of the order of τ_L. This is due to the fact that in these regimes the signal rises over a time of the order of τ_L [see, for example, Eq. (4.30)] and the change in polarity due to the free character of the boundary will occur over an identical time. The rarefactional pulse profile (Fig. 4.5, curve 2) will contain information on the characteristic times ω_χ^{-1} and ω_τ^{-1}.

In conclusion, we recall that the classification of acoustic pulse generation conditions carried out here made extensive use of inequality (4.7) and stronger inequality (4.11). The left-hand sides of these inequalities will no longer hold in the frequency range $\omega \gtrsim 10^9$ s^{-1}. Certain anomalies of acoustic pulse excitation by absorption of subnanosecond light pulses have been investigated in Ref. 9. However, inequalities (4.7) and (4.11) may no longer hold in the range $\omega \lesssim 10^9$ s^{-1} as well due to acceleration of the diffusion processes (increased carrier mobility) from the transition to lower temperatures.

4.2. Sound Excitation From Fast Expansion of Photoexcited Electron-Hole Plasma

Section 4.1 analyzed the OA effect in the frequency range $\omega \ll \omega_D < \omega_\chi$ (4.7) in which the wave vectors of the concentration and thermal waves far exceed the wave vector of the acoustic wave (Fig. 4.1). In other words, the phase velocities of the concentration and thermal waves are essentially subsonic:

$$\omega/|k_n| \ll \omega/k_a = c_L, \ \omega/|k_T| \ll \omega/k_a = c_L. \tag{4.37}$$

The velocity of the concentration wave c_D is essentially subsonic in the frequency range $\omega \ll \omega_{D0} \ll \omega_\chi$ (4.11) and in the case of diffusion that is not limited by recombination:

$$c_D \equiv \omega/(\omega/D)^{1/2} \ll (D\omega_{D0})^{1/2} \equiv c_L. \tag{4.38}$$

From the physical viewpoint, this means that the results of Sec. 4.1 cannot be used to analyze the OA effect if phonon thermal conduction or motion of the photoexcited carriers occur at transonic or supersonic velocities. Indeed, over the optical irradiation period the energy transported by diffusive phonon heat conduction travels over distances $l_\chi \sim (\chi\tau_L)^{1/2}$. During this time, the EH pairs diffuse over a distance of the order $l_D = (D\tau_L)^{1/2}$. We naturally have the following definition of the characteristic phonon heat conduction V_χ and EH diffusion V_D velocities:

$$V_\chi \equiv l_\chi/\tau_L = (\chi/\tau_L)^{1/2}, \ V_D \equiv l_D/\tau_L = (D/\tau_L)^{1/2}. \tag{4.39}$$

According to Eqs. (4.37)–(4.39) in spectral range (4.11) the following inequalities must hold:

$$V_\chi \ll c_L, \ V_D \ll c_L. \tag{4.40}$$

Since the phonon propagation velocity does not exceed the longitudinal sound velocity, the relation $V_\chi \lesssim c_L$ will always hold. The equality $V_\chi \approx c_L$ will hold only at very low (helium) temperatures, when the phonon heat conduction process manifests a ballistic character (Sec. 3.5). Of course in this case diffusion-type equation (4.2) cannot be used to describe this process.

At the same time such fundamental physical constraints on the velocity V_D of directional carrier motion at a level of $V_D \sim c_L$ do not exist. Chaotic carrier motion occurs at velocities exceeding the Fermi velocity $v_F = \hbar(3\pi^2 n)^{1/3}/m^*$ (m^* is the effective mass of the EH pair). Hence, even at moderate concentrations of photoexcited EH plasma the chaotic velocity may far exceed c_L. For example,[18] $m^* \approx 0.4 \times 10^{-27}$ g in Ge and $v_F \gtrsim 10^7$ cm/s for $n \gtrsim 2\times 10^{17}$ cm^{-3}. Here the velocity of directional motion may be transonic: $V_D \gtrsim c_L$. It is important that the motion of EH plasma manifests a diffusional character and may be described by Eq. (4.2). This requires the internal pressure gradient of the EH plasma at each point in space to be compensated with the density of retarding forces resulting from carrier scattering by defects, impurities, thermal lattice vibrations, etc.[19,20] Otherwise, we speak of hydrodynamic expansion of the EH plasma[15,19] (Sec. 6.2).

We analyze the sound generation process for the case of rapid (transonic and supersonic) expansion of EH plasma from a photoexcited surface. In this case the diffusion velocity V_D defined in accordance with Eq. (4.39) coincides with the drift (or hydrodynamic) velocity of the nonequilibrium carriers v_d. The latter is determined by equality of the diffusion and hydrodynamic fluxes of EH pairs:

$$-D\frac{\partial n}{\partial z} \approx n v_d. \tag{4.41}$$

The characteristic spatial scale of the region occupied by the EH plasma prior to the end of optical irradiation is determined (in the absence of recombination) either by the light absorption length α^{-1} or the diffusion length $l_D = (D\tau_L)^{1/2}$. Hence, according to Eq. (4.41) it is necessary to use strongly absorbed radiation $\alpha^{-1} \lesssim (D\tau_L)^{1/2}$ in order to initiate the highest velocity of EH plasma drift. However, in this case the characteristic velocity of hydrodynamic expansion of photoexcited plasma $v_d \sim D(\partial n/\partial z)/n \sim D/l_D = (D/\tau_L)^{1/2}$ in fact coincides with free diffusion velocity (4.39).

A comparison of (4.41) to the boundary condition for photoexcited carrier concentration (4.2) demonstrates that the surface recombination velocity s represents the velocity of directional carrier motion towards the surface. It can therefore be expected that an OA process in the case of fast (near sonic) surface recombination velocity cannot be analyzed in spectral range (4.11). The expression $\omega_S = s^2/D = (s/c_L)^2 \omega_{D0} \equiv M_S^2 \omega_{D0}$ is valid for the surface recombination frequency ω_S. Consequently, the characteristic frequency ω_S in fact falls within range (4.11) only for subsonic surface recombination rates s ($M_S < 1$).

The results of many experiments[18,20–24] have indicated supersonic expansion of EH plasma photogenerated near the surface of a semiconductor at low temperatures. Traditionally, the light scattering and luminescence spectra are recorded in order to analyze the space-time evolution of EH plasma. The spectral line shifts in these experiments are attributable to fast plasma drift from the semiconductor surface at velocities $v_d \gtrsim 10^7$ cm/s $\sim v_F$ (Refs. 20–22).

Optical recording of nonequilibrium EH pair spatial motion over time [18,23,24] represents convincing evidence of fast ($v_d \gtrsim c_L$) plasma expansion. However, the results of such research indicate that the velocities of EH plasma expansion are substantially below Fermi velocities ($v_d \lesssim c_L \ll v_F$). It is therefore important to analyze alternative (not strictly optical) diagnostic methods for analyzing the spatial evolution of nonequilibrium carriers. In this respect pulsed OA spectroscopy may yield certain advantages. Acoustic wave generation is particularly efficient in this case when the sources travel at transonic velocities (Sec. 1.3). We can therefore expect that when a source crosses the sound barrier, the wave forms of the acoustic pulses will change qualitatively.

General solution (2.35), (4.3) from Ref. 25 contains a description of the OA effect by fast diffusive expansion of EH plasma. If assumptions (4.7) of subsonic diffusion are not used to simplify (4.3), the spectra $\tilde{v}_n$ of the acoustic waves excited by the concentration-deformation mechanism will differ from Eqs. (4.8) and (4.9) only in the substitution of the multiplier in front of the square brackets:

$$p_2^{-2} \rightarrow [(\omega/c_L)^2 + p_2^2]^{-1},$$

$$(\omega_R - i\omega)^{-1} \rightarrow [(\omega^2/\omega_{D0}) + (\omega_R - i\omega)]^{-1}.$$

We consider sound excitation in the spectral range

$$\omega_S, \omega_R \ll \omega \ll \omega_n, \omega_a. \quad (4.42)$$

The right-hand side of this inequality shows that we assume a diffusely and acoustically thin light absorption region. This is consistent with the need to use strongly absorbed radiation to initiate fast plasma expansion. The right-hand side of Eq. (4.42) holds for frequencies $\omega \lesssim 10^9\ s^{-1}$ if $\alpha > 2\times 10^4\ cm^{-1}$. It is sufficient to use radiation of energy $\hbar\omega_L \gtrsim 2E_g$ for this purpose for typical semiconductors.

The left-hand side of inequality (4.42) indicates that we can assume free diffusion of the EH pairs. The violation of this condition in the frequency range 10^7–$10^9\ s^{-1}$ may be due primarily to nonlinear recombination of the EH pairs. The fact that acceleration of diffusion processes (an increase in D) serves to decelerate the growth of plasma concentration n (4.35) and inhibit a sharp drop in the characteristic lifetimes of the EH pairs (4.34) with increasing optical irradiation intensity is important. The maximum EH plasma concentrations and the maximum attainable optical radiation intensities for $\tau_R \gg \tau_L$ can be estimated from Eqs. (4.34)–(4.36). It is also necessary to account for the dependence of the diffusion coefficient on the EH plasma concentration in the case of a degenerate EH plasma in Eqs. (4.35) and (4.36): $D = D(n)$ (see Sec. 5.3).[11,19]

The deformation sound excitation mechanism in spectral band (4.42) dominates the thermoelastic generation mechanism associated with recombination heating of the semiconductor. Thermoelastic generation by instantaneous heating is independent of the character of EH plasma motion. A description of this process in the case of the diffusive phonon heat conduction regime is given in Sec. 3.2 while the ballistic regime is discussed in Sec. 3.6.

The vibrational velocity spectrum $\tilde{v}_n$ of the acoustic wave can be represented in frequency band (4.42) as

$$\tilde{v}_n \approx \tilde{f}(\omega)(-1)\frac{(-i\omega/\omega_{D0})^{1/2}}{1-(-i\omega/\omega_{D0})}. \quad (4.43)$$

Transfer function K_n is dependent on the single characteristic frequency ω_{D0}. We obtain for the modulus of the transfer function and the variable component of the phase shift

$$|K_n| = \left(\frac{\omega/\omega_{D0}}{1+(\omega/\omega_{D0})^2}\right)^{1/2}, \quad \Delta\varphi = -\arctan\left(\frac{\omega}{\omega_{D0}}\right). \quad (4.44)$$

According to Eq. (4.44) sound generation will be most efficient in the frequency band $(2-\sqrt{3})\omega_{D0} \leqslant \omega \leqslant (2+\sqrt{3})\omega_{D0}$. The maximum of the transfer function corresponds to $\omega = \omega_{D0}$. Clearly, the characteristic frequency τ_L^{-1} of the acoustic waves excited by absorption of pulsed laser radiation falls within the efficient generation band in the case of EH plasma expansion at transonic velocities. Indeed, for $\tau_L^{-1} \sim \omega_{D0}$ we have $V_D = (D/\tau_L)^{1/2} \sim c_L$ from Eq. (4.39).

In analyzing the acoustic pulse profiles it is convenient to introduce into Eq. (4.43) the dimensionless frequency $\omega = \omega\tau_L$ and the Mach number of directional diffusive EH pair motion $M_D = V_D/c_L$:

$$\tilde{v}_n \approx M_D^{-1}\tilde{f}(\Omega)(-i\Omega)^{1/2}/(-i\Omega - M_D^{-2}). \quad (4.45)$$

Equation (4.45) defines the acoustic signal spectrum as a function of the characteristic diffusion velocity of the EH pairs during optical irradiation. This expression describes the essential spectral transformation that occurs in the transition from slow (subsonic, $M_D \ll 1$) diffusion:

$$\tilde{v}_n \approx M_D(-i\Omega)^{1/2}\tilde{f}(\Omega), \qquad (4.46)$$

to fast (supersonic, $M_D \gg 1$) diffusion:

$$\tilde{v}_n \approx M_D^{-1}(-i\Omega)^{-1/2}\tilde{f}(\Omega). \qquad (4.47)$$

The acoustic pulse profile undergoes a transformation from a bipolar to a unipolar profile with increasing velocity of expansion of the photoexcited carriers according to Eqs. (4.46) and (4.47). The acoustic wave amplitude initially grows in proportion to M_D (for $M_D \ll 1$) [see Eq. (4.46)], and then decays (for $M_D \gg 1$) in proportion to M_D^{-1} [see Eq. (4.47)]. Consequently, at certain intermediate diffusion velocities ($M_D \sim 1$) optimum conditions may be established for sound generation by the concentration-deformation mechanism.

The profiles of acoustic pulses with spectral composition (4.45) for model $f(t) = \exp(-2|t|/\tau_L)$ may be represented by known mathematical functions [$\tau = (t - z/c_L)/\tau_L$]:

$$v_n(\tau<0) = \frac{M_D}{M_D^2-1}\left[\frac{2M_D}{M_D^2+1}\exp\left(\frac{2\tau}{M_D^2}\right) - \exp(2\tau)\right],$$

$$v_n(\tau>0) = \frac{2M_D^2}{M_D^4-1}\exp\left(\frac{2\tau}{M_D^2}\right)\operatorname{erfc}\left(\frac{\sqrt{2\tau}}{M_D}\right) - \frac{M_D}{M_D^2-1}\exp(2\tau)\operatorname{erfc}\sqrt{2\tau}$$

$$-\frac{M_D}{M_D^2+1}\exp(-2\tau)\operatorname{erfi}\sqrt{2\tau}. \qquad (4.48)$$

Figure 4.6 shows the transformation of the acoustic pulse profile with increasing diffusion velocity ($d<0$). The quantity $M_D^{-1}v_n$ is plotted on the y axis (curve 1) for the case of slow ($M_D \ll 1$) diffusion. This profile is similar to profile 1 shown in Fig. 4.4. In the case of transonic ($M_D = 1$) diffusion $2v_n$ (curve 2) is plotted on the y axis, while in the case of supersonic ($M_D \gg 1$) diffusion $M_D v_n$ is plotted on this axis (curve 3).

Using Eq. (4.48), we can find the analytic representation for the amplitude $v_a^{(1)}$ of the first phase of the acoustic signal (this is the compressional phase for $d<0$). Acoustic waves that immediately separate from the boundary after excitation (i.e., travel in the same direction as the diffusing carriers) make the predominant contribution to this part of the signal:

$$v_a^{(1)} = |v^{(1)}(\tau_{\max})| = M_D\left[\frac{2}{M_D(M_D^2+1)}\right]^{M_D^2/(M_D^2-1)},$$

$$\tau_{\max} = \frac{M_D^2}{2(M_D^2-1)}\ln\frac{2}{M_D(M_D^2+1)}. \qquad (4.49)$$

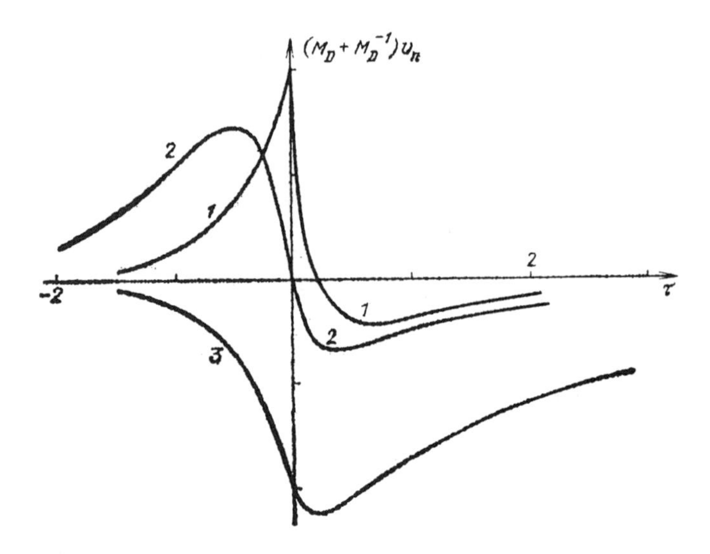

FIG. 4.6. Acoustical pulse profiles for different EH plasma diffusion velocities.

The amplitude of the first phase of the acoustic wave is plotted as a function of the EH pair diffusion velocity in Fig. 4.7. The optimum generation conditions for an acoustic wave traveling without reflection from the boundary are therefore achieved at transonic carrier diffusion velocities ($M_D \sim 1$). This can be attributed to the quasisynchronous nature of wave generation for $V_D \approx c_L$.

In order to avoid confusion we note that the quasisynchronous generation of a coherent acoustic wave is not generally attributed to the fact that the carriers travel at transonic velocities, but rather that under spatially inhomogeneous photoexcitation, the regions with maximum carrier concentration gradients may also travel at transonic velocities. In the case of diffusive EH plasma expansion, the carrier drift velocity V_D is dependent on both time and the spatial coordinate (4.41). Actually, our value of V_D (4.39) is only an order-of-magnitude estimate of the characteristic values of V_D within spectral range (4.42). The rise in amplitude of the acoustic wave identified for $V_D \approx c_L$ indicates that the definition of V_D used here may also be employed for the characteristic velocity of nonstationary, spatially distributed acoustic sources G_{tz} (2.29) resulting from the diffusive expansion of the EH plasma.

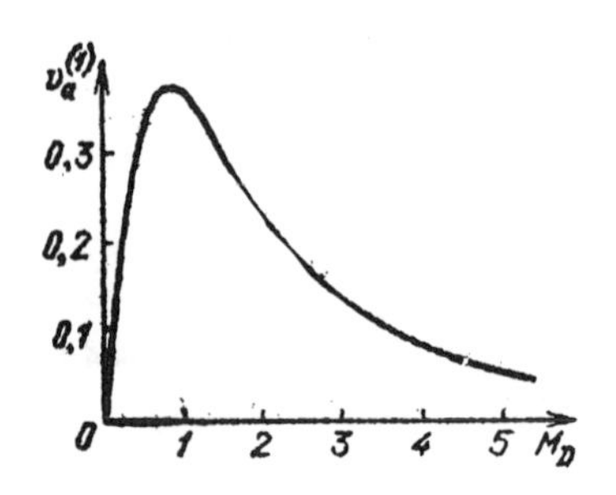

FIG. 4.7. Amplitude of the first phase of the acoustic wave plotted as a function of the characteristic Mach number of diffuse carrier motion.

In accordance with Eq. (4.49) the amplitude of the first signal phase $v_a^{(1)} \sim M_D^{-2}$ for $M_D \gg 1$ at the same time that the amplitude of the wave overall will decay more slowly according to Eq. (4.47): $v_a \sim M_D^{1}$. Analysis of solution (4.48) reveals that if the amplitude of the second phase for $M_D \sim 1$ is approximately one-half the amplitude of the first phase (Fig. 4.6, curve 2), then for $M_D \gg 1$ the second phase will make the primary contribution to the acoustic signal (Fig. 4.6, curve 3), where $v_a^{(2)} \sim M_D^{-1}$.

This analysis demonstrates that a unipolar acoustic pulse (4.47) may be excited from supersonic diffusion of nonequilibrium EH plasma near a free semiconductor surface (Fig. 4.6, curve 3). The very position of this pulse on the time scale (near $\tau \approx 0$) indicates that waves excited near the surface ($z \approx 0$) make the primary contribution here. For comparison, we note that the peak of the first phase of the acoustic signal enters the detector at successively earlier instants in time with increasing plasma expansion velocity. According to Eq. (4.49) for $M_D \gg 1$, $\tau_{\max} \approx -(\frac{3}{2})\ln M_D$, $|\tau_{\max}| > 1$.

A secondary analysis reveals that the very same expression (4.47) can be used in approximation (4.42) and with $M_D \gg 1$ to describe the vibrational velocity spectrum of the crystal boundary $\tilde{v}_n(\omega, z=0)$ while for the surface carrier concentration $\tilde{n}\ (\omega, z=0) \propto \tilde{v}_n$. This makes it possible to assume that fast carrier diffusion smooths out the spatial inhomogeneities in carrier concentration to such an extent to make sound generation in the crystal bulk inefficient. In this case the acoustic wave emitted in the crystal will be determined by surface motion where this motion in turn will be determined by a local action (the plasma surface concentration). The situation is analogous to a semiconductor crystal under weakly inhomogeneous illumination [see Eq. (4.26)] and sound generation by the surface piezoeffect (Sec. 3.4). On a qualitative level it is clearly evident that in the limiting case of a homogeneous distribution of the photoexcited carriers throughout the sample bulk, surface atoms will be the only atoms under asymmetrical conditions along the z axis and will be driven by laser-induced stresses. The remaining atoms are displaced only in an acoustic wave field.

We estimate the change in acoustic signal amplitude from a transition from room temperatures to helium temperatures. We consider the irradiation of germanium by pulsed optical radiation at $\lambda_2 \simeq 0.53\ \mu$m, $\tau_L \simeq 20$ ns. Using the data in Table 4.1 at high temperatures, we obtain $m_{D0} = 65$, $\omega_n \gg \omega_a = 2.5 \times 10^{11}\ \text{s}^{-1}$. Consequently, regime 3 of the general classification is implemented. In this case, the right-hand side of inequality (4.42) will hold, even at room temperatures $T_r \gtrsim 300$ K. The inequality will therefore also be valid at low temperatures $T_{\text{He}} \sim 2$–4 K since carrier mobility and therefore the diffusion coefficient grow.[10] We determine using Eqs. (4.34)–(4.36) for an optical radiation intensity $I_S(T_k) \sim 10\ \text{MW/cm}^2$, $\tau_R \simeq 10^{-9}$ s. Therefore, $\tau_L^{-1} \sim 10\omega_\tau \sim 5 \times 10^{-2}\omega_R$ and we can obtain, using Eq. (4.33), the following for the estimate:

$$\tilde{v} \sim \tilde{v}_T \sim B(\hbar\omega_L/E_g)M_T(-i\Omega)^{1/2}. \tag{4.50}$$

Here $M_T \equiv V_T/c_L \sim 8 \times 10^{-3}$ at room temperatures in Ge.

In the crossover to helium temperatures, the supersonic EH plasma expansion velocities are achieved in Ge if the diffusion coefficient rises to $D \gtrsim c_L^2\ \tau_L \sim 5 \times 10^3$ cm^2/s. If the laser radiation intensity is reduced by an order of magnitude [to

$I_S(T_{He}) \sim 1$ MW/cm^2], for all $M_D \gtrsim 1$ the left-hand side of inequality (4.42) will hold ($\omega_R \ll \tau_L^{-1}$) according to Eqs. (4.34)–(4.46). Therefore the results of the present section are valid. Using Eqs. (4.47) and (4.50), we find the relation

$$\frac{|v_n(T_{He}, M_D \gg 1)|}{|v_T(T_r)|} \sim \frac{I_S(T_{He})}{I_S(T_r)} \frac{E_g}{\hbar\omega_L |B| M_T(T_r) M_D(T_{He})}.$$

Therefore, in spite of the reduced irradiation intensity $I_S(T_{He}) \sim 10^{-1} I_S(T_k)$, the amplitude of the acoustic wave rises in the crossover to helium temperatures if $M_D \lesssim 30$. This is due to the change in the sound generation mechanism and its increased efficiency. An even stronger increase in the acoustic signal may be expected from plasma expansion at transonic velocities $M_D \approx 1$.

In order to identify possible anomalies of acoustic wave generation from hydrodynamic EH plasma expansion, we use a simple model for describing such motion.[18] This model assumes that photogeneration of EH pairs, pair interaction, and motion produce near the irradiated crystal surface ($z=0$) an EH liquid layer of constant (equilibrium) density n^* corresponding to minimum plasma free energy:[18,26,27]

$$n^* z_e(t) = \frac{I_0}{\hbar\omega_L} \int_{-\infty}^{t} f(t')dt'. \tag{4.51}$$

Here $z_e(t)$ is the spatial coordinate of the EH plasma front edge whose formation is the characteristic differentiating feature of hydrodynamic expansion conditions. Relation (4.51) assumes inequality (4.42) holds for $\omega \sim \tau_L^{-1}$.

Differentiating this expression with respect to time we find the velocity of the plasma front

$$v_e(t) = \frac{I_0}{n^* \hbar\omega_L} f(t). \tag{4.52}$$

This relation describes the acceleration and subsequent stopping of the plasma from pulsed optical irradiation. This relation is in qualitative agreement with the experimental observations that indicate a rapid (supersonic) plasma drift solely during light pulse irradiation.[18] Traditionally, the model (4.51) and (4.52) is modified in minor ways for a qualitative analysis of optical experiments.[18,26]

In this model the space-time evolution of the EH plasma is given by

$$n = n^* \{\theta(z) - \theta[z - z_e(t)]\}. \tag{4.53}$$

We use expression (2.36) for the sound velocity profile of an acoustic wave excited near a free surface $z=0$ ($G = -dn$).[27] We substitute Eq. (4.53) and carry out time integration

$$v = -\frac{1}{2}\left\{ \int_{\tau}^{\infty} v_e(t)\delta[z_e(t) - c_L t - c_L \tau]\, dt - \int_{\infty}^{\tau} v_e(t)\delta[z_e(t) + c_L t - c_L \tau] dt \right\}. \tag{4.54}$$

Here we carry out normalization: $v = v\rho_0 c_L / dn^*$. Utilizing the properties of the Dirac δ function, we finally obtain

$$v = \frac{1}{2}\left[\sum_i \frac{v_e(t_i)}{|v_e(t_i) - c_L|} - \sum_j \frac{v_e(t_j)}{|v_e(t_j) + c_L|} \right], \tag{4.55}$$

where t_i and t_j are the roots of equations

$$z_e(t) = \pm c_L(t - \tau), \tag{4.56}$$

that satisfy the inequalities

$$\tau \leqslant t_i \leqslant \infty \quad \text{and} \quad -\infty \leqslant t_j \leqslant \tau,$$

respectively. The first sum in Eq. (4.55) defines the contribution to the acoustic signal from waves that travel along the z axis in a positive direction immediately after generation. The second sum corresponds to waves that have been reflected by the free boundary of the solid.

Descriptions (4.55) and (4.56) make it possible to derive the acoustic pulse profile for an arbitrary given law of EH plasma front motion where the acoustic sources are δ localized. However, even without specifying the form of the relation $v_f(t)$, we can derive certain conclusions regarding the wave form of the excited longitudinal acoustic wave.

Formally, according to solution (4.55), the vibrational velocity of the acoustic wave will grow without limit at times $\tau_{i,j} = \tau(t_{i,j})$ corresponding to the times $t_{i,j}$ where the plasma edge crosses the sound barrier $[v_e(t_i) = c_L, v_e(t_j) = -c_L]$ during its motion away from or towards the crystal boundary. Therefore, the appearance of peaks in the acoustic pulse profile with changing external irradiation that drives the wave sources may serve to indicate that these sources have achieved supersonic velocities. In case (4.52) the EH plasma front travels only away from the crystal boundary by virtue of $f(t) \geqslant 0$: $v_e \geqslant 0$. Consequently, acoustic waves reflected by the crystal boundary contain no resonance anomalies. Based on our assumed pulsed irradiation, the acoustic sources are initially accelerated and must then be decelerated. Therefore, when the sources cross the sound barrier, the anomalies in the acoustic wave profile will appear in pairs: one will appear upon the crossover from $v_e < c_L$ to $v_e > c_L$ while the other appears in the reverse crossover. The unlimited growth of the acoustic signal for $v_e = c_L$ is due to the synchronous sound generation in this regime and the δ localization of sources.

To limit the signal, we need only account for the finite width of the expanding plasma front. Indeed, we interpret the δ function in, for example, the first integral of (4.54):

$$\delta|z_e - c_L(t - \tau)] = \lim_{\Delta z_e \to 0} \frac{\theta[z_e - c_L(t - \tau)] - \theta[z_e - \Delta z_e - c_L(t - \tau)]}{\Delta z_e},$$

which physically corresponds to linear diminishing of plasma concentration near the edge at a distance Δz_e. We expand the front velocity near phase synchronism into a series:

$$v_e(t) = v_e(t_i) + \sum_{k=1}^{\infty} v_e^{(k)}(t_i)(t - t_i)^k (k!)^{-1},$$

where $v_e^{(k)}$ is the kth time derivative. We obtain for the amplitude of vibrational velocity $|v_i|$ of an acoustic wave at instants τ_i corresponding to the sound barrier crossover $v_e(t_i) = c_L$:

$$|v_i| = \lim_{\Delta z_e \to 0} \frac{c_L}{2(\Delta z_e)} \left(\frac{(m+1)!\Delta z_e}{|v_e^{(m)}(t_i)|} \right)^{1/(m+1)}. \tag{4.57}$$

Here $v_e^{(m)}(t_i)$ is the first nonzero (at the crossover point) velocity derivative with respect to time. The sound barrier crossover occurs with odd m. Each time the source velocity reaches sound speed yet does not exceed this velocity, an additional (unpaired) anomaly appears in the acoustic pulse profile. It can also be identified by means of Eq. (4.57) although in this case with an even m. Therefore, solutions (4.55) and (4.56) in the case of synchronous generation of acoustic waves must be supplemented by relation (4.57) in which Δz_e is a small, yet finite quantity.

We note that if the quasisynchronous sound generation process continues for a sufficient time it may be necessary to account for the limitation on acoustic wave amplitudes (4.57) imposed by an acoustic nonlinearity (Chap. 6).

The arrival of the signal at the acoustic detector before the arrival of acoustic waves excited at the boundary $(z=0)$ at the initial instant of generation $(t=0)$ represents even more convincing evidence of supersonic motion of the EH plasma edge. We henceforth assume $v_e(t<0) = 0$ for definiteness. In order to observe the leading arrival of the signal, the source position sufficient only be ahead of the distance travelled by the boundary-excited acoustic waves at only one of the crossover times from supersonic to subsonic velocities:

$$z_e(t_i) > c_L t_i, \quad v_e(t_i) = c_L. \tag{4.58}$$

We recall that the leading arrival of the peak of the first acoustic pulse phase excited by diffusive expansion of EH plasma is described by relations (4.49).

For illustrative purposes we consider a model of a parabolic light pulse envelope:

$$f(t) \equiv [1 - (1 - t/\tau_L)^2][\theta(t) - \theta(t - 2\tau_L)].$$

The time dependence of the plasma front velocity (4.52) is conveniently represented as $M_e(t) = M_0 f(t)$. Here, the hydrodynamic Mach number of the plasma front $(M_e = v_e/c_L)$ and dimensionless time $t = t/\tau_L$ are introduced. The maximum attainable velocities are determined by the optical irradiation parameters $M_0 = I_0/\hbar\omega_L n^* c_L$. In this model the condition $M_0 < 1$ corresponds to totally subsonic motion of the sources, $M_0 = 1$ corresponds to attainment of the sound barrier, and $M_0 > 1$ corresponds to sound-barrier crossover. The coordinate of the plasma front grows as

$$z_e \equiv \frac{z_e}{c_L \tau_L} = \begin{cases} M_0(t^2 - t^3/3), & 0 \leqslant t \leqslant 2, \\ 4M_0/3, & t \geqslant 2. \end{cases} \tag{4.59}$$

Figure 4.8 provides the time dependences of the coordinate of the plasma front for different values of the parameter M_0. Oblique lines 1′–4′ correspond to functions $z = t - \tau$. Therefore, the horizontal coordinates of the intersection points of these lines with curves $z = z_e(t)$ define the roots of the first equation in (4.56). A unique root exists for $M_0 < 1$ (Fig. 4.8, the intersection of 1 and 1′). For $M_0 = 1$ Eq. (4.56)

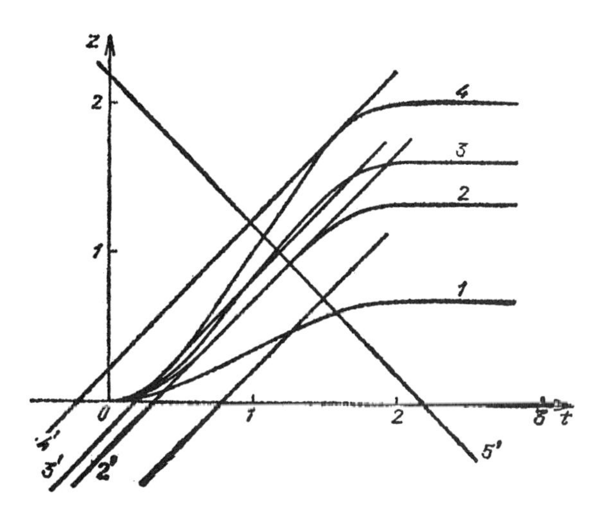

FIG. 4.8. Graphic solution of Eqs. (4.56). The relations $z = z_e(t)$ (4.59) are plotted for values of M_0 of 0.5 (1); 1 (2); 1.2 (3); and 1.5 (4).

may have a set of solutions determined by the nature of the tangency of $z = t - \tau$ and $z = z_e(t)$ with a certain τ_i (the intersection of 2 and 2′). Finally, a characteristic anomaly of the supersonic condition $M_0 > 1$ is the existence of a range of values of τ where this equation has three different roots (Fig. 4.8, intersection of 3 and 3′). Physically, the latter corresponds to the case where acoustic waves excited at three different times by the plasma front arrive at the detector at any of these times τ.

Inequality (4.58) which determines the possibility for ahead signal arrival, corresponds in Fig. 4.8 to the case where the line $z = t - \tau$ tangential to $z = z_e(t)$, intersects the axis of abscissae at $t < 0$ (line 4′). Using Eqs. (4.58) and (4.59), we find that in this case satisfaction of the condition $M_0 > 4/3$ is sufficient. Note that $z = z_e(t)$ may have only a single common point with the oblique line $z = -(t - \tau)$ (Fig. 4.8, line 5′). This corresponds to unique root of the second equation of (4.56).

Profiles of acoustic waves excited at different maximum EH plasma front velocities are plotted in Fig. 4.9 by means of Eqs. (4.55), (4.56), and (4.59). These plots provide clear evidence of the significant change in the acoustic signal profile in the transition from subsonic EH plasma expansion ($M_0 < 1$, curve 1) to attainment of the sound barrier ($M_0 = 1$, curve 2), and above this barrier ($M_0 > 1$, curve 3).

In conclusion, we note one additional possibility for efficient generation of coherent deformation waves by fast expansion of EH plasma. At supersonic charge carrier velocities the process whereby the carriers emit phonons due to scattering by lattice vibrations may become a stimulated process.[28] A new effective crystal heating channel for EH plasma kinetic energy transfer to the phonon subsystem is in fact set up for $v_d > c_L$. In this case, the plasma is decelerated. It is interesting that phonons that travel quasicollinearly to the z axis are emitted for $|v_d - c_L|/c_L \ll 1$. This results in quasisynchronous amplification of the thermal pulse comoving with the fast carriers. A coherent acoustic pulse will in turn be synchronously excited from nonlinear interaction of random acoustic waves (Sec. 3.6).

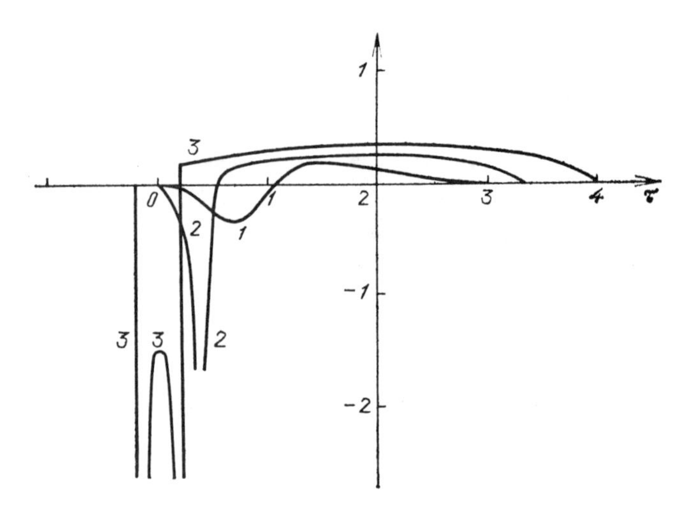

FIG. 4.9. Profiles of acoustic waves excited at different maximum EH plasma front velocities. The parameter M_0 is equal to 0.5 (1); 1 (2); 1.5 (3).

4.3. Optical Excitation of Rayleigh Waves by Interband Light Absorption in Semiconductors

The first experiments on optical generation of surface acoustic waves (SAW's) in semiconductors[29–31] acted as an impetus for the development of theoretical research.[32–34] The possibility for effective excitation of Rayleigh SAW's by the concentration-deformation mechanism was first noted in Ref. 32. A detailed theoretical model describing the spectra of SAW pulses induced by interband light absorption in nonpiezoelectric semiconductors was developed in Ref. 34.

We utilize the results of a general theory of Rayleigh SAW generation by distributed volumetric sources (Sec. 3.3). According to Eq. (3.42), in order to determine the spectrum of the SAW vibrational velocity component normal to the surface ($z=0$) outside the generation region it is necessary to find the spectral representation of the space-time evolution of the sources.

For simplicity, we consider the excitation of sound in a two-dimensional geometry (radiation focused along a strip of width a perpendicular to the x axis). In the case of interband light absorption in Eq. (3.42) $T \to T-(d/K\beta)n$. Equations (4.2) can be used to describe the concentration fields $n(t,z,x)$ of photogenerated EH pairs and the crystal temperature $T(t,z,x)$ by carrying out the substitutions $g(z) \to g(z)H(x)$ and $\partial^2/\partial z^2 \to \partial^2/\partial z^2+\partial^2/\partial x^2$. The function $H(x)$ describes the distribution of the radiation intensity at the crystal surface (Sec. 3.3). Then spectral representations $\underset{\sim}{\hat{n}}(\omega,p,k_\perp)$ and $\underset{\sim}{\hat{T}}(\omega,p,k_\perp)$ differ from Eqs. (4.3) only in the additional multiplier $\underset{\sim}{H}(k_\perp)$ $[\underset{\sim}{f}(\omega) \to \underset{\sim}{f}(\omega)\, \underset{\sim}{H}(k_\perp)]$ and the alteration of p_2 and p_3:

$$p_2=\left(k_\perp^2+\frac{1}{D\tau_R}-i\frac{\omega}{D}\right)^{1/2},\quad p_3=\left(k_\perp^2-i\frac{\omega}{\chi}\right)^{1/2}. \tag{4.60}$$

Relations (3.42), (4.3), and (4.60) provide a complete description of the SAW spectrum.

We can analyze the relative position of the dispersion curves of the surface, concentration, and temperature waves completely analogously to Sec. 4.1 (Fig. 4.1). It is only necessary to substitute the longitudinal sound velocity c_L by the Rayleigh wave velocity c_R. The expressions for the spectra of SAW's excited by the concentration-deformation ($\widetilde{w}_n$) and thermoelastic ($\widetilde{w}_T$) mechanisms ($\widetilde{w} = \widetilde{w}_n + \widetilde{w}_T$) are simplified in frequency range (4.7):

$$\widetilde{w}_n(\omega) = \frac{I_0}{\hbar\omega_L}\frac{d}{\rho_0 c_L^2}\mathscr{F}\widetilde{f}(\omega)\widetilde{H}\left(\frac{\omega}{c_R}\right)\frac{\omega^2}{c_R D p_2^2}\left[\frac{s + D|\omega|/c_R^*}{s + Dp_2}\widehat{g}(p_2) - \widehat{g}\left(\frac{|\omega|}{c_R^*}\right)\right],$$

$$\widetilde{w}_T(\omega) = \frac{I_0}{\hbar\omega_L}\mathscr{F}\frac{K\beta E_g}{\rho_0 c_L^2 c_p}\mathscr{F}\widetilde{f}(\omega)\widetilde{H}\left(\frac{\omega}{c_R}\right)\frac{(-i\omega)^2}{c_R \chi p_3^2}\left\{\frac{\hbar\omega_L - E_g}{E_g}\left[\frac{|\omega|/c_R^*}{p_3}\widehat{g}(p_3)\right.\right.$$
$$\left. - \widehat{g}\left(\frac{|\omega|}{c_R^*}\right)\right] + \frac{1}{\tau_{RT}}\left[\frac{|\omega|/c_R^*}{D(p_3^2 - p_2^2)p_3}\left(\frac{s + Dp_3}{s + Dp_2}\widehat{g}(p_2) - \widehat{g}(p_3)\right)\right.$$
$$\left.\left. + \frac{1}{Dp_2^2}\left(\frac{s + D|\omega|/c_R^*}{s + Dp_2}\widehat{g}(p_2) - \widehat{g}\left(\frac{|\omega|}{c_R^*}\right)\right)\right] - \frac{s_T}{s_T + Dp_2}\widehat{g}(p_2)\right\}. \tag{4.61}$$

Here, Eqs. (4.60) for p_2 and p_3 adopt their previous form (4.3), as $k_\perp = \omega/c_R$ represents the wave number of the Rayleigh wave, while inequality (4.7) correlates with $k_\perp \ll k_n, k_T$. We note that the difference between c_R and $c_R^* \equiv c_R/[1 - (c_R/c_L)^2]^{1/2}$ is not significant in estimates of different expressions using type (4.7) strong inequalities.

It is useful to compare solution (4.61) to the corresponding representation for longitudinal wave spectra (4.8). At first glance, insignificant changes lead to a significant reduction in the information content of the OA signal. This largely applies to thermoelastic generation. In the case of a function $\widehat{g}(p)$ that diminishes monotonically with increasing p [in, for example, an optically homogeneous medium: $\widehat{g}(p) = \alpha(\alpha + p)^{-1}$] strong inequality $(|\omega|/c_R^* p_3)\widehat{g}(p_3) \ll \widehat{g}(|\omega|/c_R^*)$ takes place in spectral range (4.7). This corresponds to heat conduction having no effect on thermoelastic generation of sound by instantaneous heating. The situation is analogous to the generation of longitudinal acoustic waves near a rigid boundary. Moreover, heat conduction will, generally speaking, not influence the SAW spectra. The term in $\widetilde{w}_T(\omega)$ proportional to $(p_3^2 - p_2^2)^{-1}$, describing the effect of the cooling of the recombination heating region on the signal, will be significant in accordance with Sec. 4.2, only in the case of strong manifestation of heat conduction [see Eq. (4.33)]. It does not exceed $(|\omega|/c_R^* p_3)\widehat{g}(p_3)$ within an order of magnitude and is also small compared to $\widehat{g}(|\omega|/c_R^*)$.

The remaining contributions to the spectrum of thermoelastic component $\widetilde{w}_T$ can, by regrouping, be represented as

$$\widetilde{w}_T(\omega) \approx B\widetilde{w}_n(\omega) - \widetilde{f}(\omega)\widetilde{H}\left(\frac{\omega}{c_R}\right)B\frac{\hbar\omega_L}{E_g}\frac{(-i\omega)}{c_R}\widehat{g}\left(\frac{|\omega|}{c_R^*}\right). \tag{4.62}$$

The first term in Eq. (4.62) only yields a correction to the total acoustic signal $\widetilde{w} = \widetilde{w}_n + \widetilde{w}_T$ for $|B| \sim 0.1$ and can be dropped. The vibrational velocity is normalized to the characteristic quantity $v_0 = I_0\, d\mathscr{F} / \hbar\omega_L \rho_0 c_L^2$ in Eq. (4.62) and further on.

We can also neglect the contribution of

$$\hat{g}(p_2)(D|\omega|/c_R)/(s + Dp_2) \ll \hat{g}(|\omega|/c_R^*)$$

to the spectrum (4.61) of the OA-signal deformation component. However, EH pair diffusion continues to affect sound excitation.

For the subsequent analysis it is convenient to rewrite $\widetilde{w}_n(\omega)$ and $\widetilde{w}_T(\omega)$ by introducing the characteristic frequencies $\omega_R = \tau_R^{-1}$, $\omega_n = \alpha^2 D$, $\omega_a = \alpha c_R^*$, $\omega_{D0} = (c_R^*)^2/D$, $\omega_S = s^2/D$ and the parameter $m_{D0} \equiv \omega_a/\omega_{D0}$ in the case $\hat{g}(p) = \alpha(\alpha + p)^{-1}$, $\tau_{RT} = \tau_R$, $s_T = s$. Then $\omega_a = m_{D0}\omega_{D0}$, $\omega_n = m_{D0}^2\omega_{D0}$, and the SAW spectral components take the form

$$\widetilde{w}_n = \widetilde{f}(\omega)\widetilde{H}\left(\frac{\omega}{c_R}\right)\frac{-1}{c_R}\frac{(-i\omega)^2}{\omega_R - i\omega} \times \left[\frac{\omega_S^{1/2}}{\omega_S^{1/2} + (\omega_R - i\omega)^{1/2}}\frac{m_{D0}\omega_{D0}^{1/2}}{m_{D0}\omega_{D0}^{1/2} + (\omega - i\omega)^{1/2}} - \frac{m_{D0}\omega_{D0}}{m_{D0}\omega_{D0} + |\omega|}\right], \tag{4.63}$$

$$\widetilde{w}_T = \widetilde{f}(\omega)\widetilde{H}\left(\frac{\omega}{c_R}\right)(-B)\frac{(-i\omega)}{c_R}\frac{m_{D0}\omega_{D0}}{m_{D0}\omega_{D0} + |\omega|}. \tag{4.64}$$

As in the case of analyzing excitation of longitudinal waves (Sec. 4.1) we shall henceforth be interested in the generation of SAW's in the spectral range $\omega \ll \omega_{D0}$ [compare to Eq. (4.11)]. Then the contribution to the spectrum $\widetilde{w}_n$ bearing information on carrier diffusion [the first term in square brackets in (4.63)] will never exceed the component independent of EH plasma motion. This contribution is negligible in the range $\omega \gg m_{D0}^2\omega_{D0}$:

$$\widetilde{w}_n = \widetilde{f}(\omega)\widetilde{H}\left(\frac{\omega}{c_R}\right)\frac{(-i\omega)^2}{c_R(\omega_R - i\omega)}\frac{m_{D0}\omega_{D0}}{m_{D0}\omega_{D0} + |\omega|}. \tag{4.65}$$

The effect of the first term in curly brackets in (4.63) in the spectral range $\omega \lesssim m_{D0}^2\omega_{D0}$ will be significant only when the following conditions hold

$$\omega_R \ll m_{D0}^2\omega_{D0};\quad \omega_R, \omega \ll \omega_S.$$

The first of these inequalities, as previously (4.14), denotes that recombination limits carrier diffusion, at least not within the whole test frequency band ($\omega \lesssim m_{D0}^2\,\omega_{D0}$). The second inequality demonstrates that carrier diffusion is manifested in the spectrum of the SAW deformation component only in the case where surface recombination is predominant over bulk recombination ($\omega_S \gtrsim \omega_R$) and only at frequencies where diffusion is limited by surface recombination ($\omega \lesssim \omega_S$).

On the qualitative level the lack of an effect of heat conduction and the possibility for diffusion influence on the OA effect are related to the following: strong surface recombination of EH pairs shifts the maximum of the EH plasma concentration from the surface into the crystal, although it does not shift the temperature maximum. The distribution of EH pairs becomes nonmonotonic along the z axis, while the force exerted on the crystalline lattice from carrier photogeneration ($\sim n_{tz}$) changes the sign at a certain distance from the surface ($z=0$).

For simplicity, as in Sec. 4.1, we will be interested in cases in which the characteristic frequencies ω_R, ω_S, ω_n are substantially different. Then in the frequency range $\omega \leqslant m_{D0}^2 \omega_{D0}$ it is possible to identify the following interesting physical cases.

(a) $\omega_R \ll m_{D0}^2 \omega_{D0} \sim \omega \ll \omega_S$. The spectral transfer function is dependent on the carrier diffusion time ω_n^{-1} out of the photogeneration region:

$$K_n \simeq (-1)\frac{\omega_n}{c_R}\left(-i\frac{\omega}{\omega_n}\right)\left[\frac{1}{1+(-i\omega/\omega_n)^{1/2}}-1\right]\widetilde{H}\left(\frac{\omega}{c_R}\right). \tag{4.66}$$

(b) $\omega_R \ll \omega_S \sim \omega \ll m_{D0}^2 \omega_{D0}$. The modulus and phase of the transfer function are dependent on the characteristic surface recombination frequency:

$$K_n \approx (-1)\frac{\omega_S}{c_R}\left(-i\frac{\omega}{\omega_S}\right)\left[\frac{1}{1+(-i\omega/\omega_S)^{1/2}}-1\right]\widetilde{H}\left(\frac{\omega}{c_R}\right). \tag{4.67}$$

(c) $\omega \sim \omega_R \ll \omega_S$, $m_{D0}^2 \omega_{D0}$. Carrier motion and surface recombination have an effect solely on the modulus of the transfer function:

$$K_n \approx \frac{\omega_R}{c_R}\left(\frac{\omega_R^{1/2}}{\omega_S^{1/2}}+\frac{\omega_R^{1/2}}{m_{D0}\omega_{D0}^{1/2}}\right)\frac{(-i\omega/\omega_R)^2}{(1-i\omega/\omega_R)^{1/2}}\widetilde{H}\left(\frac{\omega}{c_R}\right). \tag{4.68}$$

Analysis of the spectra (4.66)–(4.68) of SAW's excited by the concentration-deformation mechanism makes possible a rather simple classification of the generation regimes of high-frequency $[\omega \sim (10^{-2}\text{–}10^{-1})\omega_{D0}]$ Rayleigh waves in semiconductors.

(1) $m_{D0} \ll 1$. In this regime the light absorption region is a diffusively thick region. We employ Eqs. (4.64) and (4.65) to obtain the following representation for the transfer function:

$$K = \frac{-i\omega}{c_R}\left[\frac{(-i\omega)}{\omega_R - i\omega} - B\frac{\hbar\omega_L}{E_g}\right]\frac{\omega_a}{\omega_a + |\omega|}\widetilde{H}\left(\frac{\omega}{c_R}\right). \tag{4.69}$$

As in the corresponding volume wave generation regime (4.23), the form of the transfer function is dependent on the single parameter $R \equiv \omega_R/\omega_a$. If, as before, we assume for definiteness $|B| \sim 0.1$, $h\omega_L/E_g \lesssim 2$, then according to Eq. (4.69) the efficiencies of the two sound generation mechanisms will be comparable at frequencies $\omega \sim |B|\ (\hbar\omega_L/E_g)\omega_R \ll \omega_R$. The deformation sound generation mechanism is dominant in the high-frequency range $\omega \gtrsim \omega_R$.

We note that unlike the OA effect of bulk waves (Sec. 4.1) the dynamics of EH plasma have no effect on a thermoelastic SAW signal which may compete with $\widetilde{w}_n$. Therefore when the role of $\widetilde{w}_T$ is significant, the thermoelastic generation efficiency will be independent, for example, of the rate of nonradiative recombination processes (the quantity $\omega_R = \tau_R^{-1}$). The possible predominance of the thermoelastic mechanism is related only to the diminishing efficiency of the deformation mechanism with increasing frequency ω_R.

(2) $m_{D0}\sim 0.3$–0.1. OA diagnostics of ω_n^{-1} (4.66) are possible in this intermediate regime. The deformation mechanism of SAW generation predominates (since $\omega\sim\omega_n \gg \omega_R$).

(3) $m_{D0}\gtrsim 1$. The deformation sound excitation mechanism predominates in diagnostics of the surface recombination velocity (4.67) (since $\omega\sim\omega_S \gg \omega_R$). In OA determination of bulk recombination time (4.68) in the case $(\omega_R/\omega_S)^{1/2} + (\omega_R/\omega_n)^{1/2} \lesssim B(\hbar\omega_L/E_g)$ it is necessary to account for the thermoelastic contribution

$$K_T \approx -B\frac{\hbar\omega_L}{E_g}\left(\frac{\omega_R}{c_R}\right)\left(-i\frac{\omega_R}{c_R}\right)\tilde{H}\left(\frac{\omega}{c_R}\right),$$

even at frequencies $\omega\sim\omega_R$.

As an illustration of the capabilities of this theory we employ the theory to analyze experimental results.[30] This study recorded SAW pulses excited in crystalline silicon by optical radiation at $\lambda_1 \simeq 1.06$ μm ($\tau_L^{(1)} \approx 20$ ns) and at the second harmonic $\lambda_2 \simeq 0.53$ μm ($\tau_L^{(2)} \approx 15$ ns). In order to generate weakly diffracted SAW's, the laser radiation was focused by a cylindrical lens onto a strip of length b and width a ($b \gg a$). Since $\lambda_R \sim a$ holds in the experiment for the characteristic wavelength of the Rayleigh wave, the results of two-dimensional theory considered in this section remain valid in the range $a \ll x \ll b^2/a$.

The fundamental and most representative result of this experiment is the change in polarity of the registered SAW from doubling of the laser radiation frequency absorbed by the Si. Analysis reveals that this effect is related to the change of the dominant SAW excitation mechanism. A value of $s \lesssim 3\times 10^3$ cm/s (Ref. 16) is achieved for the recombination rate on a polished silicon surface. Representation (4.69) is therefore valid for the transfer function over the spectral range $\omega_S = s^2/D \lesssim 3\times 10^5$ s$^{-1} \ll \omega \ll \omega_{D0} = (c_R^*)^2/D = 8\times 10^9$ s^{-1}, which contains the characteristic frequencies of the Rayleigh waves excited in experiment.[30] In this case the characteristic frequency of the generated waves is determined by the product of spectral multipliers $\tilde{f}(\omega)$, $\tilde{H}(\omega/c_R)$, and $|\hat{g}(|\omega|/c_R^*)| \equiv \omega_a/(\omega_a + |\omega|)$:

$$\omega \sim \min\{\tau_L^{-1}; c_R/a, \alpha c_R^* \equiv \omega_a\}. \tag{4.70}$$

In Si, $c_R \simeq 4.6\times 10^5$ cm/s, $c_R^* \approx 1.1 c_R$. The predominance of the deformation or the temperature mechanism of sound excitation will depend on the ratio ω/ω_R. Therefore, in order to determine the generation mechanism it is necessary to estimate the characteristic frequency of the SAW's and the bulk carrier recombination time ($\omega_R = \tau_R^{-1}$).

Figure 4.10(a) shows the profile of a Rayleigh wave excited with the following parameters of the fundamental ($\lambda = \lambda_1$) radiation: $a_1 = 0.43$ mm [at full width at half maximum (FWHM) of the intensity distribution], $\mathscr{E}_0^{(1)} = I_0^{(1)}\tau_L^{(1)} \approx 30$ mJ/cm^2. In this case according to the data listed in Table 4.1 we have $(\tau_L^{(1)})^{-1} \approx 5\times 10^7$ s$^{-1} \gg c_R/a_1 \approx 10^7$ s$^{-1} \gtrsim \alpha_1 c_R^* \approx 5\times 10^6$ s^{-1}. Hence the SAW profile is dependent on the acoustic radiation intensity distribution both in the cross section of the light beam and in the sample bulk. The characteristic SAW frequency is $\omega_1 \sim \alpha_1 c_R^* \sim 5\times 10^6$ s^{-1}.

Estimates recovered from Eqs. (4.34)–(4.36) reveal that neither diffusion nor recombination have an effect on the increase in carrier concentration during laser

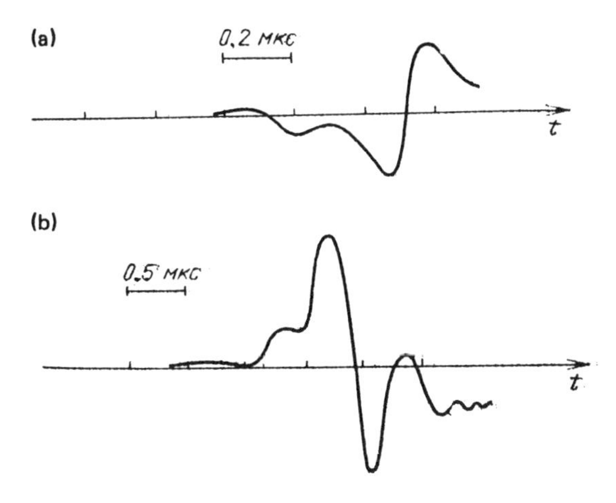

FIG. 4.10. Experimentally recorded profiles of SAW's excited in Si by laser radiation at a wavelength of 1.06 (a) and 0.53 μm (b).

irradiation. Characteristic EH plasma concentrations $n_1 \approx 10^{18}$ cm^{-3} ($\tau_R^{(1)} \approx 2$ μs) are achieved by the end of laser action. Hence $\omega \gg \omega_R^{(1)} \approx 5\times 10^5$ s^{-1} and the deformation SAW generation mechanism predominates. This conclusion remains valid when additional crystal heating from free-carrier absorption of light is taken into account.[30]

Figure 4.10(b) shows the profile of a SAW recorded with the following second harmonic radiation parameters ($\lambda = \lambda_2$): $a_2 \simeq 1.1$ mm, $\mathscr{E}_0^{(2)} \approx 36$ mJ/cm^2. Therefore the optical radiation energy density remained virtually unchanged, although the SAW pulse was inverted. In this case we have $\alpha_2 c_R^* \approx 5\times 10^9$ s^{-1} $\gg (\tau_L^{(2)})^{-1} \approx 7\times 10^7$ s$^{-1} \gg c_R/a_2 \approx 4\times 10^6$ s^{-1}. The characteristic frequency of the excited SAW's (4.70) is determined by the transverse dimensions of the light beam: $\omega_2 \sim c_R/a_2 \sim 4\times 10^6$ s^{-1}.

In this regime the increase in the nonequilibrium carrier concentration is limited by carrier recombination and diffusive motion from the photogeneration region according to Eqs. (4.34)–(4.36). Estimates yield the following values of the EH plasma parameters: $n_2 \approx 3\times 10^{19}$ cm^{-3}, $\tau_R^{(2)} \approx 3$ ns. Hence $\omega_2 \ll B(\hbar\omega_L^{(2)}/E_g)\times\omega_R^{(2)}$ $\sim 7\times 10^7$ s^{-1} and, consequently, the thermoelastic SAW generation mechanism is dominant. In this case the profile of the Rayleigh wave is the first derivative of the light intensity distribution in the beam cross section. The change in SAW polarity in the transition from the deformation generation mechanism to the thermoelastic generation mechanism is due to the fact that $\text{sgn}(\beta d) > 0$ in Si at room temperatures.

In conclusion, we note that the comparative simplicity of SAW spectra (4.63), (4.64) makes it possible, without substantially complicating the equations, to account for the possible appearance of radiative recombination channels.[34]

4.4. Excitation of Longitudinal Sound by Laser Irradiation of Piezoelectric Semiconductors

Interband light absorption in piezoelectric semiconductors produces, in addition to thermoelastic and concentration-deformation mechanisms, an additional acoustic wave generation mechanism associated with the piezoeffect.

An inhomogeneous electrical field that induces acoustic wave excitation in accordance with Sec. 3.4 arises from the loss of neutrality of the photogenerated EH plasma. The latter will necessarily occur in the case of spatial inhomogeneous photoexcitation of a semiconductor due to the difference in the mobilities μ of charge carriers of different sign ($\mu_e \neq \mu_h$).

The ambipolar EH plasma diffusion process (4.2) is itself the result of an electrical field (the Dember field[10]) that inhibits the spatial separation of electrons and holes. Indeed, the electron and hole fluxes

$$j_e = n_e(v_d)_e = -\mu_e n_e E - D_e \frac{\partial n_e}{\partial z},$$

$$j_h = n_h(v_d)_h = \mu_h n_h E - D_h \frac{\partial n_h}{\partial z},$$

even under conditions of quasineutrality ($n_e \approx n_h \equiv n$) become identical ($j_e = j_n$) only under exposure to an electrical field

$$E = -\frac{D_e - D_h}{\mu_e + \mu_h} n^{-1} \frac{\partial n}{\partial z} \tag{4.71}$$

associated with local charge separation. Field (4.71) is also called a Dember field[10] $E \equiv E_D$. Since the diffusion coefficients of the electrons and holes $D_{e,h}$ are proportional to the corresponding mobilities $\mu_{e,h}$, according to Eq. (4.71) the electric fields induced by carrier photogeneration are high in the case of significant difference in electron and hole mobilities. If we assume for definiteness that $\mu_e \gg \mu_h$, then $E_D \simeq -(D_e/\mu_e) n^{-1} \partial n/\partial z$.

The ratio D_e/μ_e is dependent only on the electron energy distribution function and the structure of the conduction band. The Einstein relation (q_0 is the electron charge modulus) is valid in a parabolic band for a nondegenerate electron gas.[10] For degenerate electrons $q_0 D_e/\mu_e = (2/3) E_F(n) [E_F(n) = (\hbar^2/2m_e^*)(3\pi_n^2)^{2/3}$ is the Fermi energy, m_e^* is the effective mass of the electrons]. In both cases electrical field E_D grows with increasing concentration of photogenerated plasma at a sublinear rate. This is due to screening of the electrical fields at high free-carrier concentrations.[35] Hence, the piezogeneration of acoustic waves by a Dember field against a thermoelastic and concentration-deformation background is to be anticipated primarily at low photoexcited carrier concentrations n':

$$n \simeq n_0 + n', \quad n' \ll n_0.$$

Using the previous designation $n' = n$ (Sec. 4.1), we obtain

$$E_D \simeq -\frac{D_e - D_h}{\mu_e + \mu_h} n_0^{-1} \frac{\partial n}{\partial z}. \tag{4.72}$$

In such an approximation (a linear approximation) field (4.72) defines the piezoelectric sources of acoustic waves that grow with increasing optical irradiation intensity in the same manner as concentration-deformation sources.

The piezogeneration of sound by optical excitation of free charge carriers from the impurity levels was investigated theoretically in Refs. 36 and 37. The authors incorporated the possibility of a breakdown of quasineutrality of the charge distribution in the semiconductor. The deviation of charge-carrier motion from ambipolar diffusion must be accounted for beginning at the frequencies where the wave vector of the acoustic wave begins to exceed the inverse charge screening length l_s^{-1}. In a nondegenerate (Boltzmann) EH plasma

$$l_s \equiv (4\pi n_0 q_0^2/\varepsilon k_B T)^{-1/2}, \tag{4.73}$$

Screening length (4.73) is called the Debye-Hückel screening length (or simply Debye length).[35] The permittivity of the crystal lattice at the acoustic frequency which is actually a statical quantity figures in here. The Thomas-Fermi screening length plays an analogous role in a degenerate plasma; the expression for this length can be obtained by substitution of $\frac{3}{2}k_B T$ by $E_F(n_0)$ in Eq. (4.73).

We estimate the screening length in the case of nondegenerate electron statistics. This is a typical case for weakly doped semiconductors at room temperature ($T_r \sim 300$ K). For definiteness we shall henceforth utilize the parameters of GaAs in the calculations. Assuming $\varepsilon \sim 10\varepsilon_0$, we obtain

$$l_s \approx 10^2 n_0^{-1/2}, \tag{4.74}$$

where l_s is measured in centimeters, n_0 is measured in cm^{-3}.

References 36 and 37 neglected the effect of crystal boundaries on the sound generation process, which makes it impossible to utilize the results of these studies in the case of light absorption near the semiconductor surface. Piezogeneration of transverse acoustic waves by Dember field (4.72) near the free surface of a semiconductor was investigated in a theoretical study.[38] Here, we employ the mathematical apparatus developed in Chap. 2 to compare the efficiencies of piezogeneration and the concentration-deformation mechanism of longitudinal acoustic waves excitation in the EH plasma ambipolar diffusion process.

In the physical case of interest to us $G(t,z) = -dn - eE_D \approx -dn + e(D_e/\mu_e) \times n_0^{-1}\, \partial n/\partial z$ in inhomogeneous wave equation (2.29) where e is effective piezomodulus of the crystal. Consequently, $G = -d\{\hat{n}(\omega,p) - L[-\tilde{n}(\omega,0) + p\hat{n}(\omega,p)]\}$. Then, using Eq. (2.35) and a spectral representation of the EH plasma concentration field (4.3) we obtain the following representation for the transfer function of the OA conversion:

$$\begin{aligned} K &= K_n + K_E \\ &= (-1)\frac{\omega_{D0}}{\omega_{D0} + i\omega}\left\{\left[\frac{m_{D0}(-i\omega)^{1/2}}{\omega_S^{1/2} + (-i\omega)^{1/2}}\,\frac{(-i\omega)^{1/2}}{m_{D0}\omega_{D0}^{1/2} + (-i\omega)^{1/2}}\right.\right. \\ &\quad \left. + \frac{-i\omega m_{D0}\omega_{D0}}{\omega^2 + (m_{D0}\omega_{D0})^2}\right] \end{aligned}$$

$$- L\alpha\left[\left(\frac{\omega_S}{\omega_{D0}}\right)^{1/2}\frac{(-i\omega)^{1/2}}{\omega_S^{1/2}+(-i\omega)^{1/2}}\frac{(-i\omega)^{1/2}}{m_{D0}\omega_{D0}^{1/2}+(-i\omega)^{1/2}}\right.$$

$$\left.\left.-\frac{-i\omega m_{D0}\omega_{D0}}{\omega^2+(m_{D0}\omega_{D0})^2}\right]\right\}. \qquad (4.75)$$

The designations and normalizations in Sec. 4.4 are conserved in solution (4.75). The inequality $\omega_R \ll \omega$ was used to find Eq. (4.75). On the one hand, this assumption is consistent with neglecting thermoelastic sound generation and, on the other, is typical for the frequency range $\omega > 10^6\ \mathrm{s}^{-1}$ in the case of weak doping of semiconductors and insignificant deviation of the EH plasma concentration from equilibrium ($n \ll n_0$). For example, in GaAs even the nonlinear bimolecular recombination frequency $\omega_R \sim \mathscr{B} n_0$ exceeds $10^6\ \mathrm{s}^{-1}$ only for $n_0 \gtrsim 3\times 10^{17}\ \mathrm{cm}^{-3}$ ($\mathscr{B} \approx 3\times 10^{-11}\ \mathrm{cm}^3\,\mathrm{s}^{-1}$ [Ref. 39]).

The characteristic parameter L introduced in Eq. (4.75) has dimension of length: $L \equiv e(D_e - D_h)/d(\mu_e + \mu_h)n_0 \approx eD_e/d\mu_e n_0$. Using $|e| \approx 1.6\times 10^{-5}\ \mathrm{C/cm^2}$, $d \approx 8.6$ eV, for estimates for GaAs, we find

$$|L| \approx 3\times 10^{11} n_0^{-1}, \qquad (4.76)$$

where L is measured in centimeters.

Analysis of spectral transfer functions (4.75) identifies a fundamentally different effect of surface recombination on the sound piezogeneration process and the concentration-deformation sound excitation process. An increase in the surface recombination rate (frequency ω_S) will always inhibit the efficiency of the deformation mechanism. At the same time, the efficiency of acoustic wave piezoexcitation diminishes at frequencies for which the absorption range is acoustically thin ($\omega \lesssim \omega_a = m_{D0}\omega_{D0}$). An increase in ω_S enhances the piezogeneration at frequencies $\omega \gtrsim m_{D0}\omega_{D0}$.

The primary mechanisms of Dember field-induced sound generation in semiconducting piezoelectrics and the possibility for the predominance of piezogeneration over the concentration-deformation mechanism are clearly revealed in an analysis of the limiting OA-conversion regimes.

1. The weak interband light absorption regime ($m_{D0} \ll 1$, $\alpha \ll c_L/D \sim 2\times 10^4\ \mathrm{cm}^{-1}$).

At frequencies for which the absorption range is diffusive and acoustically thin ($\omega \ll m_{D0}^2\omega_{D0} \ll m_{D0}\omega_{D0}$), the efficiencies of both generation mechanisms will grow with increasing frequency. The relative efficiency of these two sound generation mechanisms will grow with frequency independent of the surface recombination rate:

$$|K_E|/|K_n| \approx |L|(c_L/D)(\omega/\omega_{D0})^{1/2}.$$

Surface recombination has virtually no effect on spectral transfer function (4.75) in the frequency band $m_{D0}^2\omega_{D0} \ll \omega \ll m_{D0}^{2/3}\omega_{D0}$:

$$K_E \approx L\alpha K_n \approx (-1)L\alpha\frac{(-i\omega)m_{D0}\omega_{D0}}{\omega^2+(m_{D0}\omega_{D0})^2}. \qquad (4.77)$$

According to Eq. (4.77), the relative efficiency of these two sound excitation mechanisms for $\omega \sim \omega_a$ is independent of frequency:

$$|K_E|/|K_n| = |L|\alpha. \tag{4.78}$$

The phase shift is determined by the sign of the parameter $L\alpha$, i.e., effectively the sign of the ratio e/d. The efficiency of both generation mechanisms begins to diminish according to Eq. (4.77) at frequencies $\omega \gtrsim m_{D0}\omega_{D0} \equiv \omega_a$ for which the absorption range is diffusive and acoustically thick.

However, unlike the concentration-deformation mechanism, there exists one additional effective acoustic wave generation region for the piezomechanism near $\omega \sim \omega_{D0}$. The piezogeneration efficiencies in the frequency ranges $\omega \sim \omega_{D0}$ and $\omega \sim \omega_a$ are comparable under sufficiently strong surface recombination ($\omega_S \gg \omega$). In this case in the spectral band $m_{D0}^{2/3}\omega_{D0} \ll \omega \ll \omega_S$

$$K_E \approx (L\alpha)\frac{\omega_{D0}}{\omega_{D0} + i\omega}\left(\frac{-i\omega}{\omega_{D0}}\right)^{1/2}.$$

The highest OA-conversion efficiency is achieved at frequency $\omega \sim \omega_{D0} \ll \omega_S$ ($|K_E| \sim \omega^{1/2}$ for $\omega \ll \omega_{D0}$, and $|K_E| \sim \omega^{-1/2}$ for $\omega \gg \omega_{D0}$). At this frequency

$$|K_E|/|K_n| \sim |L| c_L/D. \tag{4.79}$$

Therefore the relative efficiency $|K_E|/|K_n|$ grew by a factor of $m_{D0}^{-1} \gg 1$ compared to Eq. (4.78) due to the diminishing efficiency of the deformation mechanism. We note that the relative efficiency of piezogeneration will also grow with a further increase of frequency (for example, for $\omega \sim \omega_S \gg \omega_{D0}$):

$$|K_E|/|K_n| \approx |L|(c_L/D)(\omega_S/\omega_{D0})^{1/2} \gg |L| c_L/D,$$

while the efficiency itself diminishes. We also recall (Sec. 4.2) that inequality $\omega_S \gg \omega_{D0}$ holds at supersonic surface recombination velocities ($M_S^2 = s^2/c_L^2 \gg 1$).

2. Strong interband light absorption regime ($m_{D0} \gg 1$, $\alpha \gg c_L/D \sim 2 \times 10^4$ cm^{-1}).

The efficiency of both mechanisms grows with frequency in the frequency range $\omega \lesssim \omega_{D0}$ where the relative efficiency of piezogeneration by Dember field also increases

$$|K_E|/|K_n| \simeq |L|(c_L/D)(\omega/\omega_{D0})^{1/2} = |L|(\omega/D)^{1/2},$$

through frequencies $\omega \sim m_{D0}^{2/3}\omega_{D0}$ independent of ω_S. The behavior of the spectral transfer functions in the higher-frequency range is essentially dependent on the surface recombination frequency.

In the case of weak surface recombination ($\omega_S \ll \omega$), representation (4.43), as expected, is valid for K_n in the range $\omega \ll m_{D0}\omega_{D0}$. The concentration-deformation mechanism is most efficient at frequencies $\omega \sim \omega_{D0}$. For $\omega \gtrsim \omega_{D0}$ the quantity $|K_n|$ begins to diminish ($|K_n| \sim (\omega_{D0}/\omega)^{1/2}$) where this decay continues for $\omega \gtrsim m_{D0}\omega_{D0}$ as well. The following representation holds for K_E in the region $\omega \ll m_{D0}^{2/3}\omega_{D0}$:

$$K_E \approx L(c_L/D)i\omega/(\omega_{D0} + i\omega).$$

Therefore, for $\omega \gtrsim \omega_{D0}$ the efficiency of the piezomechanism reaches saturation at the level $|K_E| \sim |L| c_L/D$. Analysis reveals that the decay in piezogeneration efficiency will begin only at frequencies $\omega \gtrsim m_{D0}\omega_{D0}$, for which the absorption range is acoustically thick. Consequently, the relative efficiency of piezogeneration is maximized at frequencies $\omega_S \ll \omega \sim m_{D0}\omega_{D0} \equiv \omega_a$:

$$|K_E|/|K_n| \approx |L|(c_L/D)m_{D0}^{1/2} \gg |L|(c_L/D). \qquad (4.80)$$

The increase in the surface recombination rate will serve to increase $|K_E|$ in the range $\omega \gtrsim m_{D0}\omega_{D0}$ and hence it is interesting to consider the case of strong surface recombination ($\omega_S \gg \omega$). For definiteness we assume that inequality $M_S \gg m_{D0} \gg 1$ holds. In this case ω_S is the highest characteristic frequency of the problem (specifically, $\omega_S \gg m_{D0}^2\omega_{D0}$). In this situation the following relation is valid for K_n in the range $\omega \ll m_{D0}\omega_{D0}$,

$$K_n \approx m_{D0}^{-1} i\omega/(\omega_{D0} + i\omega).$$

Therefore the deformation mechanism efficiency goes to saturation for $\omega \gtrsim \omega_{D0}$. Decay is initiated at $\omega \gtrsim m_{D0}\omega_{D0}$. Consequently, an increase of ω_S expands the range of efficient acoustic wave excitation by the concentration-deformation mechanism. However, the maximum values of $|K_n|$ diminish here by a factor of $m_{D0} \gg 1$ times.

At the same time, an increase in the surface recombination frequency will have a significantly different effect on K_E. The spectral band of efficient piezogeneration shifts towards higher frequencies without altering the maximum values of $|K_E|$. At frequencies $m_{D0}^{2/3}\omega_{D0} \ll \omega \ll m_{D0}^2\omega_{D0}$ we have

$$K_E \approx -L\frac{c_L}{D}\frac{\omega^2}{\omega^2 + (m_{D0}\omega_{D0})^2}.$$

Therefore the OA-conversion piezomechanism will only go to saturation in the range $\omega \gtrsim m_{D0}\omega_{D0}$. The efficiency will diminish only at frequencies $\omega \gtrsim m_{D0}^2\omega_{D0}$, for which the absorption range becomes diffusively thick. The relative efficiency of piezogeneration in this case is maximized at frequencies $\omega \sim m_{D0}^2\omega_{D0} \sim \omega_n \ll \omega_S$:

$$|K_E|/|K_n| \approx |L|(c_L/D)M_S \gg |L|(c_L/D). \qquad (4.81)$$

Using Eqs. (4.78)–(4.81), we find conditions where the piezomechanism of sound generation will predominate over the concentration-deformation mechanism ($|K_E| \geqslant |K_n|$). The latter inequality imposes an upper limit on the equilibrium EH pair concentration n_0 according to Eq. (4.76). At the same time, the ambipolar diffusion condition $k_a \lesssim l_s^{-1}$ establishes a lower limit on n_0 according to Eq. (4.74).

Piezogeneration may be dominant ($|K_E| \gg |K_n|$) in the weak interband absorption regime ($m_{D0} \ll 1$). For example, for $\alpha \sim 2\times10^3$ cm$^{-1} \gg c_L/D \sim 2\times10^4$ cm^{-1} ($c_L \simeq 5\times10^5$ cm s^{-1}, $D \simeq 20$ cm^2 s^{-1} in GaAs), the piezomechanism of sound excitation is dominant at frequencies $\omega \sim \omega_a \sim 10^9$ s^{-1} for 5×10^{10} cm$^{-3} \ll n_0 \ll 5\times10^{14}$ cm^{-3} and at frequencies $\omega \sim \omega_{D0} \sim 10^{10}$ s^{-1} for 5×10^{12} cm$^{-3} \ll n_0 \ll 5\times10^{15}$ cm^{-3}.

The results obtained here cannot be used when the piezogeneration mechanism is dominant in the strong interband light absorption regime ($m_{D0} \gg 1$). For example, for $\alpha \sim 2\times10^6$ cm^{-1}($m_{D0} \simeq 10^2$) Dember field-induced piezoexcitation of sound at frequencies $\omega \sim \omega_a \sim 10^{12}$ cm^{-1} begins to dominate at $n_0 \lesssim 5\times10^{16}$ cm^{-3}. However, the EH plasma begins to deviate from quasineutrality at such frequencies for $n_0 \lesssim 5\times10^{16}$ cm^{-3}. It is therefore necessary to account for the deviation of EH plasma motion from ambipolar diffusion in order to investigate the possibility of achieving $|K_E| \gg |K_n|$ in the high-frequency portion of the spectrum.[40,41] We note that in all ranges of ω, n_0 examined above, the Maxwellian charge relaxation frequencies $\omega_M \equiv \tau_M^{-1} \equiv (D_e + D_h)/l_s^2$ far exceed the frequencies of the excited acoustic waves ($\omega_M \gg \omega$).

In conclusion we note that SAW excitation by interband absorption of light in piezoelectric semiconductors was first observed experimentally in Refs. 39 and 40, the theoretical description of this process was proposed in Refs. 43 and 44.

This chapter has considered optical excitation of sound in semiconductors. An analysis of the transfer functions was limited to a linear approximation when the temperature field and the EH plasma concentration field have no effect on light absorption or the dynamics of the nonequilibrium carriers. This effect must be taken into account at higher radiation intensities. This will be carried out in Chap. 5.

REFERENCES

1. W. B. Gauster and D. H. Habing, Phys. Rev. Lett. **18**, 1058 (1967).
2. M. A. Olmstead, N. M. Amer, and S. Kohn, Appl. Phys. A **32**, 141 (1983).
3. I. Suemune, H. Yamamoto, and M. Yamanishi, J. Appl. Phys. **58**, 615 (1985).
4. S. M. Avanesyan, V. E. Gusev, and N. I. Zheludev, Appl. Phys. A **40**, 163 (1986).
5. R. Tsu, J. E. Baglin *et al.*, Proceedings of the Symposium Laser and Electron Beam Processing of Electronic Materials, edited by C. L. Anderson, G. K. Celler, and G. A. Rozgonyi (PUBLISHER, New York, 1980), Vol. 80-1, p. 382.
6. N. Baltzer, M. von Allmen, and M. W. Sigrist, Appl. Phys. Lett. **43**, 826 (1983).
7. I. A. Veselovskiy, B. M. Zhiryakov *et al.*, Kvantovaya Elektron. **12**, 381 (1985) [Sov. J. Quantum Electron. **9** (1985)].
8. V. E. Gusev and E. G. Petrosyan, Akust. Zh. **33**, 223 (1987) [Sov. Phys. Acoust. **33**, 135 (1987)].
9. S. M. Advanesyan and V. E. Gusev, Kvantovaya Elektron. **13**, 1241 (1986) [Sov. J. Quantum Electron. **10**, 812 (1986)].
10. R. Smith, *Poluprovodniki* [*Semiconductors*], edited by A. N. Penin (Mir, Moscow, 1982) (English translation).
11. M. I. Gallant and H. M. van Driel, Phys. Rev. B **26**, 2133 (1982).
12. E. Y. Yoffa, Phys. Rev. B **21**, 2415 (1980).
13. M. G. Grimaldi, P. Baeri, and E. Rimini, Appl. Phys. A **33**, 107 (1984).
14. V. A. Sablikov and V. B. Sandomirskiy, Fiz. Tekh. Poluprovodn. **17**, 81 (1983) [Sov. Phys. Semicon. **17**, 50 (1983)].
15. V. E. Gusev, Pis'ma Zh. Eksp. Teor. Fiz. **45**, 288 (1987) [JETP Lett. **45**, 362 (1987)].
16. V. A. Zavaritskaya, A. V. Kudinov *et al.*, Fiz. Tekh. Poluprovodn. **18**, 2160 (1984) [Sov. Phys. Semicon. **18**, 1347 (1984)].
17. V. E. Gusev , B. V. Zhdanov *et al.*, Fiz. Tekh. Poluprovodn. **23**, 366 (1989) [Sov. Phys. Semicon. **23**, 226 (1989)].
18. M. A. Tamor, M. Greenstein, and J. P. Wolfe, Phys. Rev. B **27**, 7353 (1983).
19. M. Combescot and J. Bok, J. Lumin. **30**, 1 (1985).
20. A. Forchel, H. Schweizer, and G. Mahler, Phys. Rev. Lett. **51**, 50 (1983).
21. C. L. Collins and P. Y. Yu, Solid State Commun. **51**, 123 (1984).
22. H. Nather and L. Quagliano, J. Lumin. **30**, 50 (1985).
23. A. Forchel, B. Laurich *et al.*, J. Lumin. **30**, 67 (1985).
24. J. P. Wolfe, J. Lumin. **30**, 82 (1985).
25. V. E. Gusev, B. V. Zhdanov *et al.*, Akust. Zh. **35**, 454 (1989) [Sov. Phys. Acoust. **35**, 266 (1989)].
26. M. Combescot, Solid State Commun. **30**, 81 (1979).
27. V. E. Gusev, Akust. Zh. **33**, 863 (1987) [Sov. Phys. Acoust. **33**, 501 (1987)].
28. V. E. Gusev, Fiz. Tverd. Tela, **31**, 97 (1989) [Sov. Phys. Solid. State, **31**, 774 (1989)].

29. M. Schmidt and K. Dransfeld, Appl. Phys. A **28**, 211 (1982).
30. S. M. Avanesyan, V. E. Gusev *et al.*, Akust. Zh. **32**, 562 (1986) [Sov. Phys. Acoust. **32**, 356 (1986)].
31. S. M. Avanesyan, V. E. Gusev *et al.*, Pis'ma Zh. Tekh. Fiz. **12**, 1067 (1986) [Sov. Tech. Phys. Lett. **12**, 442 (1986)].
32. Yu. V. Pogorel'skiy, Fiz. Tverd. Tela **24**, 2361 (1982) [Sov. Phys. Solid State **24**, 1340 (1982)].
33. V. I. Emel'yanov and I. F. Uvarova, Akust. Zh. **31**, 481 (1985) [Sov. Phys. Acoust. **31**, 284 (1985)].
34. V. E. Gusev and A. A. Karabutov, Fiz. Tekh. Poluprovodn. **20**, 1070 (1986) [Sov. Phys. Semicon. **20**, 674 (1986)].
35. V. F. Gantmakher and I. B. Levinson, *Rasseyanie nositeley toka v metallakh i poluprovodnikakh* [*Charge Carrier Scattering in Metals and Semiconductors*] (Nauka, Moscow, 1984).
36. P. M. Karageorgiy-Alkalaev, Fiz. Tekh. Poluprovodn. **2**, 216 (1968) [Sov. Phys. Semicond. **2**, 181 (1968)].
37. Yu. V. Gulyaev, G. N. Shekrdin, and B. B. Elenkrig, Pis'ma Zh. Tekh. Fiz. **6**, 924 (1980) [Sov. Tech. Phys. Lett. **6**, 400 (1980)].
38. O. I. Kozlov and V. P. Plesskiy, Akust. Zh. **34**, 663 (1988) [Sov. Phys. Acoust. **34**, 381 (1988)].
39. D. G. McLean, M. G. Roe, A. I. D'Souza, and P. E. Wigen, Appl. Phys. Lett. **48**, 992 (1986).
40. V. N. Deev and P. A. Pyatakov, Pis'ma Zh. Tekh. Fiz. **12**, 928 (1986) [Sov. Tech. Phys. Lett. **12**, 384 (1986)].

Chapter 5

Nonlinear Regimes of Laser Sound Generation

Pulsed optical irradiation may substantially alter the physical parameters of a medium dependent on temperature and the concentration of photoexcited EH pairs. The light absorptivity may be essentially dependent on irradiation intensity. In such cases optoacoustic phenomena are essentially dependent on the incident laser radiation intensity, although nonlinear acoustic effects in the excitation region continue to be weakly manifested. Certain of such nonlinear regimes of acoustic wave generation are analyzed in this chapter.

5.1. Effect of Changes in Thermophysical Parameters on Optoacoustic Excitation of Sound

The local temperature rise generated by high-power thermooptical converters may run into the single or even tens of degree range. Under such conditions the thermophysical parameters of the medium can, generally speaking, no longer be treated as constant. As demonstrated by estimates (Table 5.1), the temperature dependence of the thermal coefficient of volume expansion is most strongly manifested. This effect has come to be called a "thermal" nonlinearity (to differentiate it from an acoustic nonlinearity).

Equations (2.11) and (2.12) can be used to describe optoacoustic excitation of sound under conditions of a thermal nonlinearity. These equations are easily recast for the temperature increment (rather than entropy):

$$\frac{\partial^2 \varphi}{\partial t^2} - c_0^2 \, \Delta\varphi = -c_0^2 \frac{\partial(\beta T')}{\partial t}, \tag{5.1}$$

$$\rho_0 c_p \frac{\partial T'}{\partial t} = \varkappa \, \Delta T' - \operatorname{div}\langle \mathbf{S} \rangle, \tag{5.2}$$

since the parameters are ordinarily functions of temperature itself. Since the thermal coefficient of volume expansion changes under laser irradiation, it cannot be factored out from under the derivative in Eq. (5.1) [as was done in deriving Eq. (2.11)]. We limit the analysis to a linear relation $\beta(T)$:[1, 2]

$$\beta(T) = \beta(T_0) + \left(\frac{d\beta}{dT}\right)_{T_0} T'. \tag{5.3}$$

TABLE 5.1. Temperature coefficients for different materials

Thermophysical parameters	Water	Ethanol	Al	Hg
$c^{-1}(\partial c/\partial T)$, K^{-1}	1.7×10^{-3}	-2.9×10^{-3}	-1.5×10^{-1}	2.5×10^{-4}
$c_p^{-1}(\partial c_p/\partial T)$, K^{-1}	2×10^{-4}	4.2×10^{-3}	5×10^{-4}	10^{-5}
$\rho^{-1}(\partial\rho/\partial T)$, K^{-1}	2×10^{-4}	1.1×10^{-3}	6.9×10^{-5}	1.8×10^{-4}
$\beta^{-1}(\partial\beta/\partial T)$, K^{-1}	5×10^{-2}	3×10^{-3}	7×10^{-4}	10^{-4}

The degree of manifestation of the nonlinearity can be characterized by the dimensionless parameter

$$q=\left(\frac{\alpha\mathscr{E}_0}{\rho_0 c_p}\right)\beta^{-1}(T_0)\left(\frac{d\beta}{dT}\right)_{T_0} \tag{5.4}$$

($\alpha\mathscr{E}_0$ is the absorbed energy per unit volume, $\mathscr{E}_0=I_0\tau_L$). For $q\ll1$ the coefficient of volume expansion in sound excitation remains approximately unchanged: $\beta(T)=\text{const}$. For $q\gtrsim1$ the thermal nonlinearity becomes fundamental. As follows from Eq. (5.4) the thermal nonlinearity becomes dominant if the temperature of the medium T_0 corresponds to the density extremum of the material $[\beta(T_0)\to0]$ as occurs in, for example, water at $T_0=4.0\,^\circ\text{C}$. With use of Eq. (5.3), Eq. (5.1) is reduced to (we consider only the plane geometry of the problem)

$$\frac{\partial^2\varphi}{\partial t^2}-c_0^2\frac{\partial^2\varphi}{\partial z^2}=-c_0^2\beta\,(T_0)\frac{\partial T'}{\partial t}-c_0^2\left(\frac{d\beta}{dT}\right)_{T_0}\frac{\partial}{\partial t}(T')^2. \tag{5.5}$$

It is clear from Eq. (5.5) that the nonlinearity on the right-hand side causes the spectrum of the optoacoustic signal to no longer be proportional to the light intensity spectrum.

Equation (5.5) has the same structure as Eq. (2.30), with the source function

$$G(t,z)=\rho_0c_0^2\left[\beta(T_0)+\left(\frac{d\beta}{dT}\right)_{T_0}T'\right]T'.$$

Consequently the expression for vibrational velocity (2.36) takes the form

$$v_{r,f}(\tau)=\frac{1}{2}\frac{\partial}{\partial\tau}\int_0^\infty dz'\left\{\beta_0\left[T'\left(\tau+\frac{z'}{c_0},z'\right)\pm T'\left(\tau-\frac{z'}{c_0},z'\right)\right]\right.$$
$$\left.+\left(\frac{d\beta}{dT}\right)_{T_0}\left[T'^2\left(\tau+\frac{z'}{c_0},z'\right)\pm T'^2\left(\tau-\frac{z'}{c_0},z'\right)\right]\right\} \tag{5.6}$$

[here β_0 denotes $\beta_0=\beta(T_0)$]. Therefore, the thermal nonlinearity will produce additional terms in the solution: the first square bracket corresponds to linear thermo-optical sources; the wave forms of acoustic pulses excited by such sources were discussed above. The sources in the second square bracket in Eq. (5.6) are attributable to the thermal nonlinearity.

The behavior of temperature field T is essentially dependent on the ratio of the light absorptivity α and the thermal diffusivity (see Sec. 3.2). It is therefore necessary to consider the case of nonheat-conducting ($\alpha\chi/c_L \ll 1$) and strongly absorbing ($\alpha\chi/c_L \sim 1$) media separately. Since the thermal coefficient of volume expansion $\beta^{-1}(d\beta/dT)$ is comparable to the temperature coefficients of the other parameters for metals, it is not advisable to consider the thermal nonlinearity separately for heat-conducting media. We therefore will limit the analysis to the thermal nonlinearity of dielectrics. In this case we can neglect the diffusion term in Eq. (5.2) and directly obtain

$$\frac{\partial T'}{\partial t} = \frac{\alpha I_0}{\rho_0 c_p} e^{-\alpha z} f(t) \tag{5.7}$$

[expression (2.14) is used for the averaged Umov-Poynting vector]. Then for a short laser pulse $f(t) = \tau_L \delta\ (t)$ expression (5.6) subject to Eq. (5.7) will take the form

$$v_{r,f} = \frac{\alpha \mathscr{E}_0 c_0}{2\rho_0 c_p} \beta_0 [\exp(-\alpha c_0 |\tau|) + \beta_0^{-1} q \exp(-2\alpha c_0 |\tau|)] \left\{ \begin{matrix} 1 \\ \operatorname{sgn} \tau \end{matrix} \right., \tag{5.8}$$

where unity corresponds to the case of a rigid boundary, and sgn τ corresponds to the case of a free boundary.

Accounting for thermal nonlinearity alters not only the wave amplitude (which has a square-law dependence on absorbed energy), but also the profile form of the wave: it no longer is purely exponential. Figure 5.1 shows the profile of a wave described by solution (5.8) for different values of q. It is clear that when the thermal nonlinearity is accounted for, the duration of the optoacoustic pulse diminishes (the more strongly heated regions of the medium generate sound more efficiently), although with a short laser pulse the signal remains symmetrical.

If we account for the finite laser pulse duration (Fig. 5.2), the thermal nonlinearity will increase sound generation efficiency by the end of the pulse and will induce asymmetry of the acoustic signal.[1] For example, with a free boundary, the amplitude of the rarefactional phase exceeds the amplitude of the compressional phase.

Figure 5.3 shows the experimental amplitude relation of an OA signal excited by a CO_2 laser in water.[3] As we see, the experimental data accurately confirm theoretical calculation (5.8).

With long laser pulses (for $\alpha c_0 \tau_L \gg 1$) solution (5.6) can be simplified in the same manner as in the case $\beta = \text{const}$:

$$v_r(\tau) = \frac{\beta_0 \mathscr{E}_0}{\rho_0 c_p} \left[f(\tau) + 2q f(\tau) \int_{-\infty}^{\tau} f(\xi) d\xi \right], \tag{5.9}$$
$$v_f(\tau) = (\alpha c_0)^{-1} dv_r/d\tau.$$

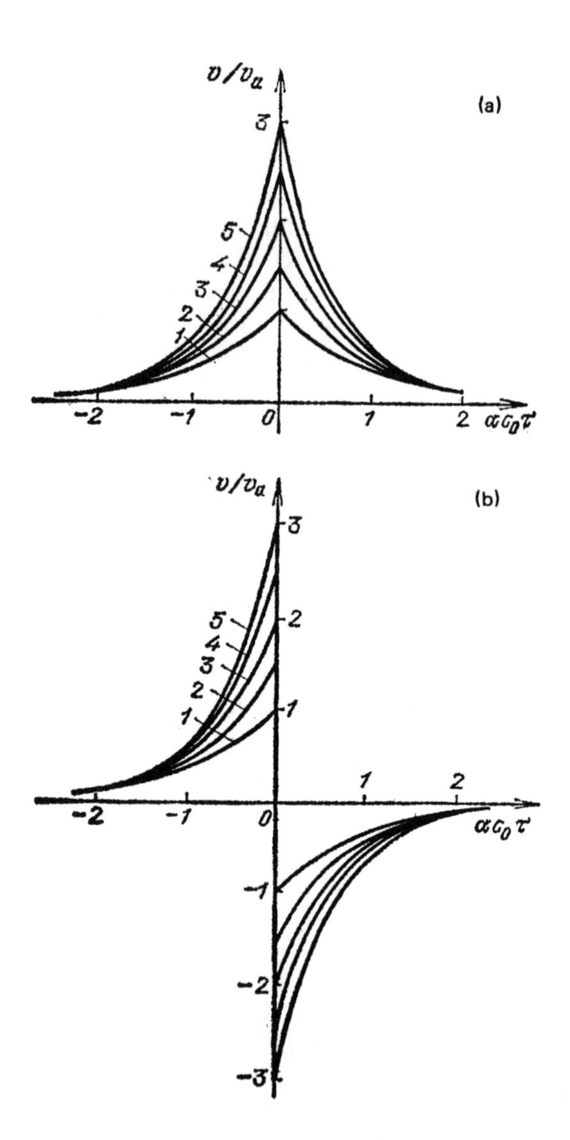

FIG. 5.1. Change in OA-signal wave form from accounting for the thermal nonlinearity for a rigid (a) and free (b) boundary. The thermal nonlinearity parameter q is equal to 0 (1), 0.5 (2), 1 (3), 1.5 (4), 2 (5).

The increase in the effect of the thermal nonlinearity at the end of the pulse with increasing heating is clearly evident from Eqs. (5.9). Solutions (5.9) explain the tripolar optoacoustic signal wave forms observed in water in Ref. 4 in the vicinity of temperature $T_0 \lesssim 4\,°\mathrm{C}$.

5.2. Influence of Nonlinear Recombination of Electron-Hole Plasma on the OA Effect in Semiconductors

The results of Chap. 4 indicate the strong dependence of the spectral transfer functions of OA conversion on the bulk recombination time τ_R of photogenerated

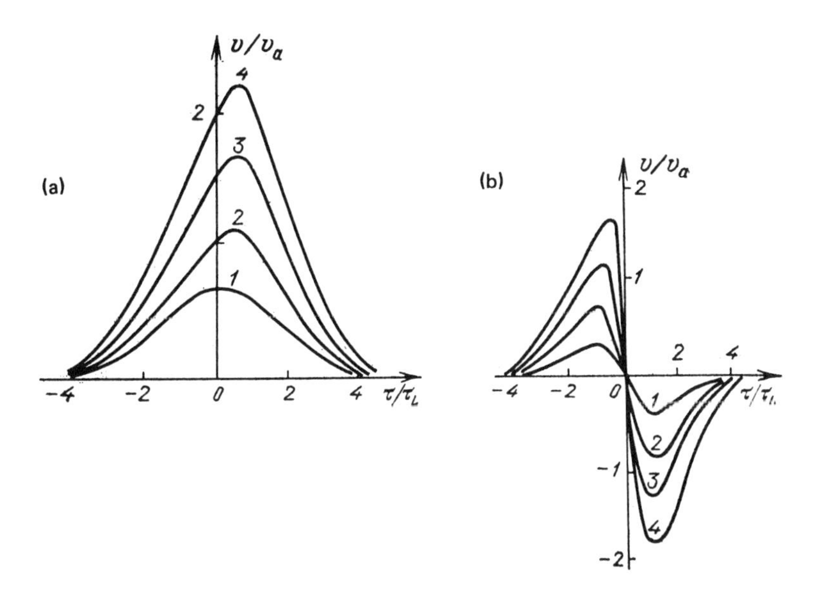

FIG. 5.2. OA-signal wave form under conditions of the thermal nonlinearity for a Gaussian laser pulse ($\alpha c_0 \tau_L = 0.6$) and rigid (a) and free (b) boundaries. Parameter q is equal to 0 (1), 0.5 (2), 1 (3), 1.5 (4).

EH pairs. In fact, the quantity $\omega\tau_R$ determines the predominance of the deformation or the thermoelastic sound generation mechanism at frequency ω. The characteristic times of the nonlinear recombination processes are essentially dependent on the concentration of the photoexcited EH plasma. In the case of nonradiative Auger recombination $\tau_R \sim n^{-2}$ (4.34) and $\tau_R \sim n^{-1}$ in the case of radiative bimolecular recombination. Hence, characteristic nonequilibrium carrier concentrations may be achieved under sufficiently intense laser irradiation such that car-

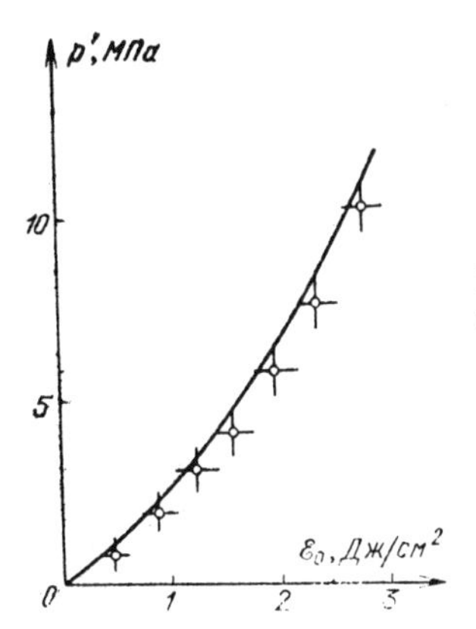

FIG. 5.3. Amplitude relation of an OA signal in water excited by a CO_2 laser. Solid curve, calculation by Eq. (5.5); circles, experiment.

rier lifetimes will be determined by nonlinear recombination. Sound excitation in this case is essentially dependent on optical radiation intensity.

Accounting for the effect of EH plasma concentration on the physical parameters of the semiconductor requires solving nonlinear inhomogeneous diffusion equations (4.2). Numerical methods[5,6] must be used for a detailed description of carrier dynamics in the general case. Hence, the development of approximate methods of qualitative analysis of the nonlinear OA effect is of timely interest as is an analysis of the simplest model cases that permit an analytical solution.

The results of linear theory (Chap. 4) can be used for a qualitative analysis of the effect on acoustic wave generation of the dependence of semiconductor parameters on the EH plasma concentration. The characteristic values dependent on optical radiation intensity I_0 an be used as the physical parameters in these solutions. For example, it is necessary to use $\tau_R=(\gamma n^2)^{-1}$ (4.34) as the nonradiative recombination time τ_R; in this relation the dependence of the characteristic carrier concentration n on I_0 is determined by Eqs. (4.34)–(4.36). Therefore by using $\tau_R=\tau_R(I_0)$ it is possible to analyze on a qualitative level the effect of an increase in optical radiation intensity on the OA effect.

We consider the effect of diminishing EH pair lifetime (increased light intensity) on the spectra of longitudinal acoustic waves excited in the $m_{D0}\ll 1$ regime (Sec. 4.1). Retaining in Eq. (4.23) only the part of the thermoelastic component that can compete with the deformation component, we obtain

$$\tilde{v}\approx\tilde{f}(\omega)(-1)\left(\frac{-i\omega}{\tau_R^{-1}-i\omega}-B\frac{\hbar\omega_L}{E_g}\right)\frac{-i\omega/\omega_a}{1+(\omega/\omega_a)^2}. \tag{5.10}$$

In the general case the spectrum of acoustic pulse (5.10) contains three characteristic frequencies: τ_L^{-1}, $\omega_a=\alpha c_L$, and $\omega_R=\tau_R^{-1}$.

In the case of slow bulk recombination $\omega_R\ll\tau_L^{-1},\omega_a$ the deformation sound generation mechanism predominates and the acoustic pulse spectrum takes the form

$$\tilde{v}\approx\tilde{f}(\omega)(-1)\frac{-i\omega/\omega_a}{1+(\omega/\omega_a)^2}. \tag{5.11}$$

If we assume $\tau_L^{-1}\gg\omega_a$ for definiteness, the profile of a pulse with spectral composition (5.11) will appear as represented by curve 1 in Fig. 4.5. The duration of the leading and trailing edges is of the order of ω_a^{-1} while the duration of the polarity crossover range is of the order of τ_L. Carrier recombination begins to have an effect on OA transfer initially at lower frequencies (for $\tau_R^{-1}\lesssim\omega_a$) and then at higher frequencies (for $\tau_R^{-1}\lesssim\tau_L^{-1}$) with increasing optical irradiation intensity.

In the case of fast bulk recombination $\omega_R\gg\tau_L^{-1},\omega_a$ the acoustic pulse spectrum takes the form

$$\tilde{v}\approx\tilde{f}(\omega)(-1)\left(-i\omega\tau_R-B\frac{\hbar\omega_L}{E_g}\right)\frac{-i\omega/\omega_a}{1+(\omega/\omega_a)^2}. \tag{5.12}$$

Here since $\tau_R^{-1}\gg\omega_a$ the signal generated by the deformation mechanism is a unipolar pulse

$$\tilde{v}_n\approx\omega_a\tau_R\tilde{f}(\omega). \tag{5.13}$$

The reduction in recombination time significantly transforms the profile of the pulse excited by the deformation mechanism. We note that the duration of the acoustic pulse drops from $\tau_a \sim \omega_a^{-1}$ to $\tau_a \sim \tau_L$ in the crossover from (5.11) to (5.13). Since the thermoelastic mechanism begins to dominate only at $\omega_R \gg \tau_L^{-1}(B\hbar\omega_L/E_g)^{-1}$, we can expect in regime $(B\hbar\omega_L/E_g)\tau_L \lesssim \tau_R \lesssim \tau_L$ that the duration of the total signal will also be determined by the laser irradiation time.

The possibility for steepening of acoustical pulses by increasing the optical radiation intensity gives semiconductor materials an advantage for use as media for generating ultrashort acoustic pulses.[7] The thermoelastic mechanism cannot be used to excite a pulse with a characteristic duration shorter than the transit time of sound through the light absorption range ($\tau_a \gtrsim \omega_a^{-1}$). This occurs when the deformation generation mechanism is used according to Eq. (5.13). This is due to the difference between the dynamics of the electron-hole and the phonon subsystems of the crystal. While the temperature diminishes solely due to heat conduction, the concentration of the photogenerated EH pair diminishes due to both pair diffusion and recombination. It is precisely acceleration of the nonlinear recombination processes by increasing the EH plasma concentration that permits a rapid (over times $\tau_R \lesssim \tau_L$) "switching off" of deformation sound sources independent of the spatial scale of the photoexcitation region. This may lead to generation of acoustic pulses whose profile mimics laser pulse envelope (5.13).[7]

A generation regime of acoustic pulses with spectral composition (5.11) was implemented experimentally in Ref. 8. This study was the first to identify the deformation mechanism of longitudinal acoustic wave excitation. However, the use of only low-intensity irradiation in this experiment $I_0 \lesssim 2$ MW/cm^2 (at $\tau_L \simeq 25$ ns) made it impossible to observe nonlinear effects. This is consistent with the estimates at the end of Sec. 4.1 according to which $\omega_R \sim 2\times 10^5\ \mathrm{s}^{-1} \ll \omega_a \sim 10^7\ \mathrm{s}^{-1} < \tau_L^{-1} \sim 5\times 10^7\ \mathrm{s}^{-1}$ are achieved in the case $\tau_L \simeq 20$ ns, $I_0 \simeq 1$ MW/cm^2. In the absence of diffusion effects, the carrier concentration for $\tau_L \lesssim \tau_R$ grows linearly with increasing intensity and with $n \propto I_0$ (4.35) while $\tau_R \propto I_0^{-2}$ (4.34). Hence, I_0 need only be increased by an order of magnitude to achieve the essentially nonlinear sound generation regime $\omega_a < \omega_R < \tau_L^{-1}$.

The nonlinear OA effect has been observed experimentally in Si in Ref. 9 ($\tau_L \simeq 20$ ns). Its primary results are reflected in Figs. 5.4, and 5.5. The symmetrical wave form of the pulse profile was conserved, while the wave amplitude grew linearly with increasing radiation intensity through a fluence of 60 mJ/cm^2. In this case the pulse begins with a rarefactional phase (Fig. 5.4), which indicates the manifestation of the deformation sound generation mechanism[8] ($d > 0$ in Si).

The rate of growth of the rarefactional wave amplitude slows upon further increases in intensity, and for $\mathscr{E}_0 \approx Q_0 \approx 250$ mJ/cm^2 its amplitude goes to saturation (Fig. 5.5). The acoustic pulse profile becomes manifestly asymmetrical for $\mathscr{E}_0 \gtrsim Q_0$ [Fig. 5.4(a)]. It is also determined that unlike the rarefactional pulse, whose duration remains nearly constant, the duration of the compressional pulse diminishes with increasing optical irradiation intensity.

The primary physical cause of the nonlinear transformation of the acoustic pulse profile observed in experiment[9] is a change of spectral transfer functions of the semiconductor (5.10) with diminishing EH pair recombination time. It is useful to identify the role of nonlinear recombination in a field description (in a time do-

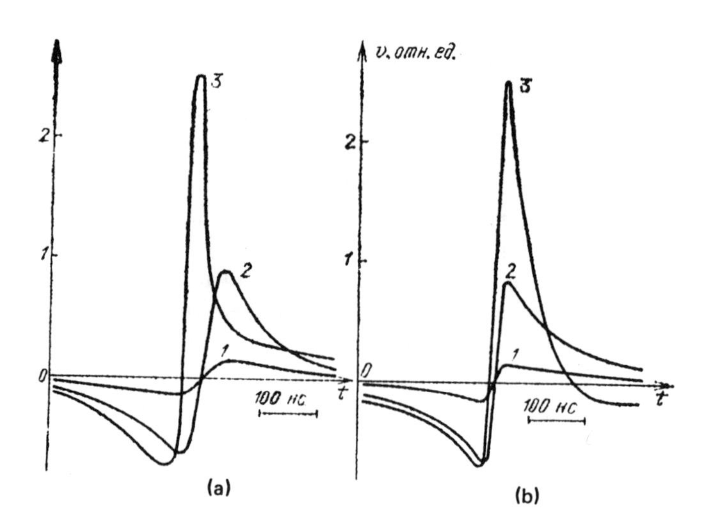

FIG. 5.4. Vibrational velocity profiles of an acoustic wave: (a) experiment [$\mathscr{E}_0 = I_0\tau_L = 0.014$ (1), 0.14 (2), 0.53 J/cm^2 (3)], (b) theory [$R \ll 1$ (1), $R = 0.1$ (2), 1 (3)].

main) as well. In the case of a Gaussian laser pulse envelope $f(t) = \exp[-(2t/\tau_L)^2]$ it is possible to obtain the following analytic description of acoustic signal (5.10) profile $\hbar\omega_L \approx E_g$):[10]

$$v_n = \frac{\sqrt{\pi}}{4} A e^{A^2/4}\left\{ -\frac{1}{1+R} e^{A\tau}\left[1 - \operatorname{erf}\left(2\tau + \frac{A}{4}\right)\right] + \frac{1}{1-R} e^{-A\tau}\left[1 + \operatorname{erf}\left(2\tau - \frac{A}{4}\right)\right]\right\} - \frac{\sqrt{\pi}}{4} A \frac{2R^2}{1-R^2} e^{-(\tau_L/4\tau_R)^2} \times e^{-\tau/\tau_R}\left[1 + \operatorname{erf}\left(2\tau - \frac{\tau_L}{4\tau_R}\right)\right], \tag{5.14}$$

$$v_T = \frac{\sqrt{\pi}}{4} B A e^{A^2/4}\left\{ e^{A\tau}\left[1 - \operatorname{erf}\left(2\tau + \frac{A}{4}\right)\right] - e^{-A\tau}\left[1 + \operatorname{erf}\left(2\tau - \frac{A}{4}\right)\right]\right\}.$$

Here the previous dimensionless parameters $R \equiv \omega_R/\omega_a = (\alpha c_L \tau_R)^{-1}$ (Sec. 4.1) and $A \equiv \alpha c_L \tau_L$ are used; the traveling wave time is normalized to the optical pulse duration ($\tau = \tau/\tau_L$). Representing the signal as (5.14) and additionally assuming that τ_R diminishes with increasing I_0 provides a qualitative description of experimental observations.[9]

1. A wave of symmetrical profile is excited at low optical irradiation intensities ($R \ll 1$) according to Eq. (5.14) [Fig. 5.4(b), profile 1]

$$v \approx v_n \approx -(\sqrt{\pi}/4) A \exp(A^2/4)\{\exp(A\tau)[1 - \operatorname{erf}(2\tau + A/4)] - \exp(-A\tau)[1 + \operatorname{erf}(2\tau - A/4)]\}. \tag{5.15}$$

The dependence on recombination time in Eq. (5.15) vanishes, and the wave amplitude is proportional to intensity (Fig. 5.5). The parameter $A \approx 0.2$ and light

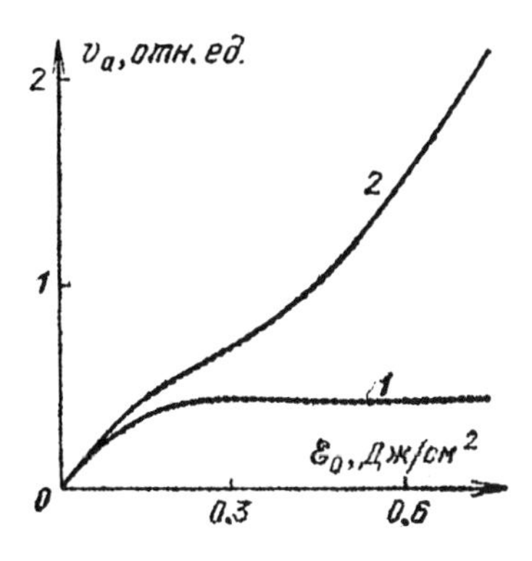

FIG. 5.5. Amplitude of rarefactional pulse (1) and compressional pulse (2) plotted as a function of incident radiation fluence.

absorptivity $\alpha \approx 10$ cm^{-1} can be estimated from the exponential edges of the pulse by comparing Eq. (5.15) to experiment [Fig. 5.4(a)]. The latter is consistent with results of optical measurements[6] (Table 4.1). Consequently, a linear generation regime is achieved for $\tau_R^{-1} \ll \omega_a \sim 10^7$ s^{-1}.

2. The thermoelastic contribution to the total signal can be neglected with good accuracy through $\tau_R \approx \tau_L (R \lesssim A^{-1} \approx 5)$. The wave profile becomes asymmetrical [Fig. 5.4(b)], and the zero point of the profile shifts towards $\tau < 0$ with diminishing τ_R, which was also observed experimentally [Fig. 5.4(a)]. The compressional pulse duration diminishes.

3. For $\tau_R \ll \tau_L (R \gg 5)$ the leading edge ($\tau \lesssim -1/2$) of the rarefactional pulse is described by the relation

$$v \approx (\sqrt{\pi}/4) A (B - R^{-1}) \exp(A\tau)[1 - \text{erf}(2\tau)], \qquad (5.16)$$

which permits the following treatment of the amplitude saturation effect of the rarefactional pulse. For $\tau_R \lesssim \tau_L$: $n \propto I_0^{1/3}$, $R^{-1} \propto \tau_R \propto I_0^{-2/3}$ according to Eqs. (4.34)–(4.36). Hence, consistent with Eq. (5.16) the contribution of the deformation mechanism to the acoustic wave continues to grow sublinearly ($v_n \propto I_0^{1/3}$ in dimensional units). Apparently, within a certain intensity range, this growth process can be entirely compensated by an increase in the thermoelastic component. In any case, theory predicts saturation at $\tau_R \sim \tau_L$. Assuming $\tau_R \sim \tau_L$, $\mathscr{E}_0 \sim Q_0$ in Eqs. (4.34)–(4.36) it is possible to obtain a rough estimate of the Auger-recombination constant in Si: $\gamma \approx 5 \times 10^{-31}$ cm^6 s^{-1}, which is in agreement with estimates obtained by other methods[6] (Table 4.1).

4. Identifying the predominant recombination mechanism as an Auger mechanism in turn makes it possible to estimate $\tau_R \gtrsim 0.5\tau_L \sim 10$ ns; $R \lesssim 10$ under experimental conditions in Ref. 9. This permits neglecting the contribution of thermoelastic generation to the compressional pulse. Specifically, the decay ($\tau > 1/2$) of the compressional pulse for $5 \leqslant R \leqslant 10$ is described by the relation

$$v \approx v_n \approx (\sqrt{\pi}/4) A \exp[(\tau_L/4\tau_R)^2 - \tau/\tau_R][1 + \text{erf}(2\tau - \tau_L/4\tau_R)].$$

It is evident that the characteristic duration of the decay of the compressional pulse is equal to the carrier recombination time and hence diminishes with increasing

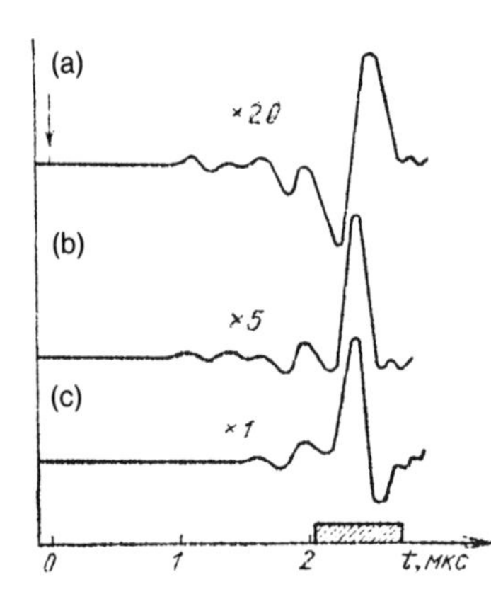

FIG. 5.6. SAW signal recorded by a piezoelectric detector. Optical radiation energy density $\mathscr{E}_0$ is equal to 0.1 (a), 0.3 (b), 0.5 J/cm^2 (c).

irradiation intensity. This explains the excitation of compressional pulses of duration $\tau_a \sim \tau_L$ in Ref. 9 for $\mathscr{E}_0 \gtrsim 2Q_0$ [Fig. 5.4(a)], since $\tau_R \lesssim \tau_L$ at these radiation intensities.

This analysis neglected a number of factors that must be taken into account in a more exact quantitative processing of the experimental results. Specifically, additional light absorption by the photoexcited free carriers was neglected:

$$\alpha_{\text{FCA}} = \sigma n, \tag{5.17}$$

α_{FCA} is the free carrier absorptivity, σ is the absorption cross section. The presence of nonlinear absorption causes only an $\alpha/(\alpha + \alpha_{\text{FCA}})$ fraction of the radiation to be expended in photogeneration of EH pairs, while the remaining fraction of radiation goes directly to heating the lattice. The spatial distribution of optical radiation changes simultaneously: $I \sim \exp\{-\alpha z - \int_0^z \alpha_{\text{FCA}}(z')dz'\}$. (The designation α is conserved here for interband light absorptivity.)

For example, $\sigma \approx 5 \times 10^{-18}$ cm^2 (Ref. 6) in silicon at room temperature for radiation at $\lambda_1 \approx 1.06$ μm. Light absorption by free carriers is comparable to interband absorption when the photoexcited EH plasma reaches a concentration $n \approx 2 \times 10^{18}$ cm^{-3}. When laser pulses with $\tau_L \approx 20$ ns are used, such carrier concentrations are achieved at the end of optical irradiation at intensities as low as $I_0 \gtrsim 3$ MW/cm^2 according to Eqs. (4.34)–(4.36). Therefore it is necessary to account for free carrier absorption as early as fluences of $\mathscr{E}_0 \gtrsim 0.1$ J/cm^2 in Si.

We carry out a qualitative analysis of the simultaneous manifestation of nonlinear recombination and nonlinear absorption (5.17) in discussing results from an experiment on laser generation of SAW's.[11] This study identified a nonlinear transformation of the profile of a Rayleigh wave excited by $\lambda_1 \approx 1.06$ μm ($\tau_L \approx 20$ ns) laser radiation in Si with increasing light intensity. Unlike experiment[12] (Sec. 4.3) the optical fluence was raised significantly (to $\mathscr{E}_0 \lesssim 0.5$ J/cm^2), in spite of a certain degree of laser beam defocusing. The width of the photoexcitation strip at half amplitude of the intensity distribution was $a \approx 1.44$ mm at $1/e$ level – $a \approx 1.7$ mm.

Figure 5.6 shows profiles of the vibrational velocity of SAW pulses recorded at different optical radiation intensities. The arrow indicates the origin of the oscillograph sweep corresponding to the instant of optical irradiation. The region corresponding to SAW detection is shaded on the time scale. Control experiments make it possible to identify the precursor on the leading edge of the SAW pulse with a bulk transverse acoustic pulse.

The negative polarity phases of the Rayleigh wave in Fig. 5.6 correspond to positive values of the vibrational velocity component normal to the surface ($z=0$), i.e., the profile in Fig. 5.6(a) corresponds to the formation of indentations on the Si surface from laser irradiation. This is consistent with experimental results[12] (Sec. 4.3) with dominance of the concentration-deformation SAW generation mechanism at low optical irradiation intensities. However, a further increase in the fluence in fact serves to invert the SAW profile [Figs. 5.6(b) and 5.6(c)].

In order to explain the experimental observations on a qualitative level we employ the results of linear theory, modifying this theory to account for supplementary light absorption by free carriers (5.17). In describing the dynamics of EH plasma and the temperature field it is possible to use Eqs. (4.2) in which

$$g(z)\equiv -\alpha \exp\left\{-\alpha z-\int_0^z \alpha_{\rm FCA}(z')dz'\right\}$$

and to carry out a substitution in the heating equation:

$$\frac{\hbar\omega_L-E_g}{\hbar\omega_L}\rightarrow\frac{\hbar\omega_\tau-E_g}{\hbar\omega_L}+\frac{\alpha_{\rm FCA}}{\alpha}.$$

Then, assuming τ_R and $\alpha_{\rm FCA}$ are constant and modifying Eqs. (4.62) and (4.65), we obtain the following representation for the spectra of the deformation and thermoelastic SAW components:

$$\begin{aligned}\tilde{w}_n&\approx\tilde{f}(\omega)\tilde{H}\left(\frac{\omega}{c_R}\right)\frac{\alpha}{\alpha+\alpha_{\rm FCA}}\frac{-i\omega}{c_R}\frac{-i\omega}{\omega_R-i\omega}\frac{\omega_a}{\omega_a+|\omega|},\\ \tilde{w}_T&\approx B\tilde{w}_n+\tilde{f}(\omega)\tilde{H}\left(\frac{\omega}{c_R}\right)(-B)\frac{\alpha}{\alpha+\alpha_{\rm FCA}}\frac{-i\omega}{c_R}\left(\frac{\hbar\omega_L}{E_g}+\frac{\alpha_{\rm FCA}}{\alpha}\right)\frac{\omega_a}{\omega_a+|\omega|}.\end{aligned}\tag{5.18}$$

Here $\omega_a\equiv(\alpha+\alpha_{\rm FCA})c_R^*\geqslant 5\times10^6\ {\rm s}^{-1}$. At the same time $\tilde{H}$ (ω/c_R) effectively establishes an upper limit on the spectrum of excited acoustic waves at $\omega\lesssim c_R/a\sim 3\times10^6{\rm s}^{-1}$. In this frequency range the spectrum of the laser pulse envelope is nearly constant: $\tilde{f}$ $(\omega)\approx\tau_L$, while modulation of the spectrum resulting from the distribution of radiation throughout the bulk is insignificant: $0.6\lesssim\omega_a(\omega_a+|\omega|)^{-1}\lesssim 1$. Hence the fundamental physical mechanisms can be identified by further simplifying Eq. (5.18):

$$\begin{aligned}\tilde{w}_n&\approx\frac{\alpha}{\alpha+\alpha_{\rm FCA}}\frac{\tau_L(-i\omega)}{c_R}\frac{-i\omega}{\omega_R-i\omega}\tilde{H}\left(\frac{\omega}{c_R}\right),\\ \tilde{w}_T&\approx B\tilde{w}_n+(-B)\frac{\alpha}{\alpha+\alpha_{\rm FCA}}\frac{\tau_L(-i\omega)}{c_R}\left(\frac{\hbar\omega_L}{E_g}+\frac{\alpha_{\rm FCA}}{\alpha}\right)\tilde{H}\left(\frac{\omega}{c_R}\right).\end{aligned}\tag{5.19}$$

A comparison of Eq. (5.19) to Eq. (4.62) shows that light absorption by free carriers will increase the thermoelastic component of SAW's relative to the deformation component. Specifically, the efficiencies of the thermoelastic and deformation mechanisms of Rayleigh wave excitation become comparable at higher frequencies $[\omega\sim B(\hbar\omega_L/E_g+\alpha_{\rm FCA}/\alpha)\omega_R]$, than in the case $\alpha_{\rm FCA}=0$. Using Eq.

(5.19) in conjunction with Eqs. (4.34)–(4.36) and (5.17) it is possible to provide the following description of the nonlinear SAW generation regime in silicon.

1. At low radiation fluences $\mathscr{E}_0 < 0.2$ J/cm^2 the recombination frequency ω_R is significantly below the characteristic frequency ω of the excited SAW's ($\omega \sim c_R/a$). In other words, the concentrations of photoexcited carriers are insignificant, τ_R exceeds the duration τ_a of the Rayleigh wave $\tau_a \sim a/c_R$, and the deformation generation mechanism is predominant. In this case a SAW will be efficiently generated only during photogeneration of the EH pairs. Rayleigh wave profile (5.19) is the first derivative of the light intensity distribution along the x axis [Fig. 5.6(a)]:

$$\widetilde{w} \approx \widetilde{w}_n \approx \frac{\alpha}{\alpha + \alpha_{\text{FCA}}} \left(\frac{\tau_L}{\tau_R} \right) (-i\omega) \widetilde{H} \left(\frac{\omega}{c_R} \right).$$

The thermal mechanism makes no significant contribution to total signal $\widetilde{w}$.

2. In the range 0.2 J/cm^2 $\mathscr{E}_0 < 0.4$ J/cm^2 the Rayleigh wave profile assumes a tripolar wave form [Fig. 5.6(b)], since n increases with increasing I_0, τ_R diminishes and the excitation of SAW from EH pair recombination becomes increasingly efficient. In fact for $\omega_R \gtrsim \omega \sim \tau_a^{-1}$ the profile of the SAW deformation component is the second derivative of the transverse light intensity distribution:

$$\widetilde{w}_n \approx \frac{\alpha}{\alpha + \alpha_{\text{FCA}}} \frac{\tau_L}{c_R \omega_R} (-i\omega)^2 \widetilde{H} \left(\frac{\omega}{c_R} \right).$$

The profile of the thermoelastic component is the first derivative:

$$\widetilde{w}_T \approx -B \frac{\tau_L}{c_R} (-i\omega) \widetilde{H} \left(\frac{\omega}{c_R} \right).$$

Here and henceforth $\hbar\omega_L/E_g \approx 1$ is accounted for under the conditions of experiment.[11]

3. For $\mathscr{E}_0 > 0.4$ J/cm^2 the SAW pulse again becomes a bipolar pulse [Fig. 5.6(c)]; the polarity is inverted compared to the case of low irradiation intensities [Fig. 5.6(a)]. This is due to the increasing role of the thermoelastic mechanism with growth of I_0, n, α_{FCA}, and ω_R. SAW pulses of the type shown in Fig. 5.6(c) will be excited not only from the dominance of thermoelastic generation but also by competition between the two mechanisms analyzed above.

The latter conclusion is based on an analytic description of the SAW profiles of spectral composition (5.19) in the case $H(x/a) = \exp[-(2x/a)^2]$:

$$w_n \approx \frac{\sqrt{\pi}}{4} \frac{\alpha}{\alpha + \alpha_{\text{FCA}}} \frac{\tau_L}{\tau_R} \frac{1}{R} \exp[(R/4)^2] \frac{\partial^2}{\partial \tau^2} \left[\exp(-R\tau) \operatorname{erfc} \left(\frac{R}{4} - 2\tau \right) \right], \tag{5.20}$$

$$w_T \approx B w_n - B \frac{\tau_L}{\tau_R} \frac{1}{R} \frac{\partial}{\partial \tau} \exp[-(2\tau)^2].$$

Here we introduce the dimensionless parameter $R = a/c_R\tau_R$ equal to the ratio of the surface wave transit time through the absorption region and the recombination time, analogous to the case of bulk waves. The time is normalized to the characteristic SAW pulse duration: $\tau \equiv \tau c_R/a$.

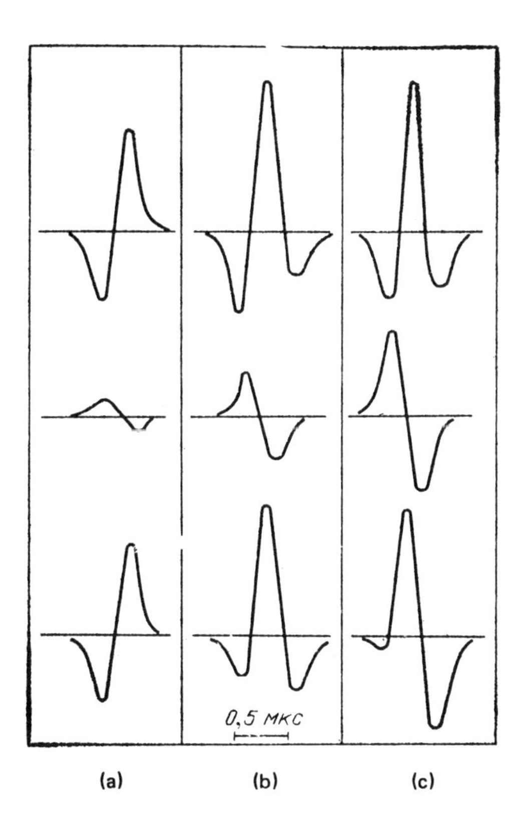

FIG. 5.7. Calculated SAW pulse profiles for different optical energy densities $\mathscr{E}_0$, free-carrier absorptivities α_{FCA} and recombination times τ_R: (a) $\mathscr{E}_0 = 0.1$ J/cm^2, $\alpha_{FCA} = 9$ cm^{-1}, $\tau_R = 770$ ns; (b) 0.3 J/cm^2, 22 cm^{-1}, 140 ns; (c) 0.5 J/cm^2, 30 cm^{-1}, 70 ns.

Field description (5.20) in cases 1 and 2 makes it possible to arrive at the same conclusions as spectral description (5.19). Specifically, $w_n \sim \partial[\exp(-(2\tau)^2)]/\partial\tau$ according to Eq. (5.20) for $R \ll 1$ and $w_n \sim \partial^2[\exp(-(2\tau)^2)]/\partial\tau^2$ for $R \gg 1$. This also permits a convenient representation of SAW profile generation in case 3 of competition between the generation mechanisms.

Equations (4.34) and (5.17) for $\langle n \rangle \sim n/2$ were used as the characteristic values τ_R and α_{FCA} in Eq. (5.20), while n was estimated by means of Eqs. (4.34)–(4.36).[11] The first upper horizontal level of Fig. 5.7 represents the evolution of w_n, while the second shows the evolution of w_T and the third shows the resulting SAW profile $w = w_n + w_T$.

It is clearly evident that, indeed, in case 3 the generation of a bipolar signal is not related to dominance of thermoelastic generation. In fact, acceleration of energy thermalization processes will cause the thermoelastic contribution to compensate the first phase of the deformation component [Fig. 5.7(c)].

An analysis of even the particular cases, which permit derivation of analytical results, represent an area of indisputable interest in analyzing nonlinear sound generation regimes.

An analytic description of the profiles of longitudinal acoustic pulses that precisely accounts for the nonlinearities $\tau_R = (\gamma n^2)^{-1}$ (4.34) and $\alpha_{FCA} = \alpha n$ (5.17) is possible when light absorption by free carriers has the predominant effect on spatial localization of the photoexcitation region:[13]

$$\alpha_{FCA} \gg \alpha. \tag{5.21}$$

We assume plane longitudinal acoustic wave regime $m_{D0} \ll 1$ is realized (Sec. 4.1). (Of course, the parameter m_{D0} is overdetermined under conditions (5.21) $m_{D0} = \omega_{D0}/\omega_a$, $\omega_a = \alpha_{FCA} c_L$.) Then, in describing fields $n(t,z)$ and $T(t,z)$ it is possible to neglect carrier diffusion and phonon heat conduction:

$$\begin{aligned} \frac{\partial n}{\partial t} &= \frac{\alpha I_0}{\hbar\omega_L} \exp\left[-\sigma \int_0^z n(t,z')dz'\right] f(t) - \frac{n}{\tau_R(n)}, \\ \frac{\partial T}{\partial t} &= \sigma n \frac{I_0}{\rho_0 c_p} \exp\left[-\sigma \int_0^z n(t,z')dz'\right] f(t) + \frac{E_g}{\rho_0 c_p} \frac{n}{\tau_R(n)}. \end{aligned} \tag{5.22}$$

Neglecting lattice heating in Eq. (5.22) from instantaneous relaxation of photogenerated electrons and holes to the bottom of the energy bands is possible when $\alpha_{FCA}/\alpha \gg (\hbar\omega_L - E_g)/E_g$. The latter is in fact due to inequality (5.21) since satisfaction of (5.21) is possible only when the optical radiation energy is insignificantly higher than the band gap.

We consider that during light irradiation, carrier recombination can be neglected:

$$\tau_R(n) \gtrsim \tau_L. \tag{5.23}$$

Then, by modeling the laser pulse envelope by the rectangle $f(t) = \theta(t) - \theta(t - \tau_L)$, it is possible to separate the description of n and T (5.22) into two stages.

Carrier photogeneration and lattice heating by free-carrier absorption of light occur over the time interval $0 \leqslant t \leqslant \tau_L$:

$$\begin{aligned} \frac{\partial n}{\partial t} &\approx \frac{\alpha I_0}{\hbar\omega_L} \exp\left[-\sigma \int_0^z n(t,z')dz'\right], \\ \frac{\partial T}{\partial t} &\approx \sigma n \frac{I_0}{\rho_0 c_p} \exp\left[-\sigma \int_0^z n(t,z')dz'\right] = \frac{\sigma\hbar\omega_L}{\alpha\rho_0 c_p} n \frac{\partial n}{\partial t}. \end{aligned} \tag{5.24}$$

Carrier recombination and recombination heating follow laser pulse irradiation $(t \gtrsim \tau_L)$:

$$\begin{aligned} \frac{\partial n}{\partial t} &= -\gamma n^3, \\ \frac{\partial T}{\partial t} &= \left(\frac{E_g}{\rho_0 c_p}\right)\gamma n^3 = -\left(\frac{E_g}{\rho_0 c_p}\right)\frac{\partial n}{\partial t}. \end{aligned} \tag{5.25}$$

A simplified system of equations easily permits estimation of maximum carrier concentrations $n_L \equiv \alpha I_0 \tau_L (\hbar\omega_L)^{-1}$ [compare to (4.35)] and reformulation of the applicability conditions of model (5.21), (5.23) as constraints on the optical radiation intensity.[13]

It is convenient for the analysis below to go over to dimensionless variables and functions:

$$t = t/\tau_L, \ z = z/c_L\tau_L, \ n = n/n_0,$$
$$v = v\rho_0 c_L/dn_0, \ n_0 = (\sigma c_L \tau_L)^{-1}. \tag{5.26}$$

Physically, n_0 is the carrier concentration where the optical radiation absorption length $\alpha_{\mathrm{FCA}}^{-1}$ is equal to the distance traveled by the acoustic wave during optical irradiation. The space-time evolution of the nonequilibrium carrier field using the designations of (5.26) and the acoustic wave sources $G(t,z)$ in wave equation (2.29) are described as follows:

$$\frac{\partial n}{\partial t} = N \exp\left[-\int_0^z n(t,z')dz'\right], \quad G = -n + \frac{Cn^2}{2}, \ 0 \leqslant t \leqslant 1,$$
$$\frac{\partial n}{\partial t} = -\Gamma n^3, \ G = -(1+B)n, \quad t \geqslant 1, \tag{5.27}$$

where we introduce the dimensionless parameters

$$N = n_L/n_0, \ \Gamma = \tau_L/\tau_R(n_0) = \tau_L/[\tau_R(n_L)N^2],$$
$$B = (K\beta E_g/d\rho_0 c_p), \ C = (\hbar\omega_L/AE_g)B, \ A = \alpha c_L \tau_L.$$

We note that the highest sound excitation efficiency from photogeneration of EH pairs is realized within the framework of linear theory (Sec. 4.1) for $A \approx 1$. We also note that the range of the parameter Γ has an upper limit due to inequality (5.23): $\Gamma \leqslant N^{-2}$.

The first term in Eqs. (5.27) for sound sources $G(t,z)$ determines the deformation contribution v_n while the second term determines the thermoelastic contribution v_T to the acoustic wave: $v = v_n + v_T$. However, from the methodologic viewpoint, it is more interesting to consider separately the profiles of acoustic waves generated by the deformation and thermoelastic mechanisms during optical irradiation, i.e., for $0 \leqslant t \leqslant 1$, $V_n \equiv v_n$, $V_T \equiv v_T/C$, as well as the pulse profile generated during EH pair recombination for $t \geqslant 1$ and by lattice heating by the energy contributed during the recombination process $V_R \equiv (v_n + v_T)/(1+B)$.

In our designations the exact solution of problem (5.27) takes the form

$$n = \begin{cases} 2Nt/(2+Ntz), & 0 \leqslant t \leqslant 1, \\ 2N/[(2+Nz)^2 + 8N^2\Gamma(t-1)]^{1/2}, & t \gtrsim 1. \end{cases}$$

This solution makes it possible to obtain an analytic description of the acoustic field by means of Eq. (2.36) outside the generation region ($\tau = t - z$ is the comoving coordinate):

$$V_n(\tau \leqslant 0) = \frac{2}{\mathscr{T}}\left\{\frac{2-\tau}{2+N(1-\tau)} - \frac{|\tau|}{2}\right.$$

$$
+\frac{2}{N}\left[\begin{array}{ll} \dfrac{1}{\sqrt{\mathcal{T}}}\ln\left|\dfrac{(2-\tau-\sqrt{\mathcal{T}})(|\tau|+\sqrt{\mathcal{T}})}{(2-\tau+\sqrt{\mathcal{T}})(|\tau|-\sqrt{\mathcal{T}})}\right| & \mathcal{T}>0, \\ \dfrac{2}{\sqrt{-\mathcal{T}}}\left(\arctan\dfrac{\sqrt{-\mathcal{T}}}{|\tau|}-\arctan\dfrac{\sqrt{-\mathcal{T}}}{2-\tau}\right) & \mathcal{T}<0, \end{array}\right]\Bigg\}
$$

$$
V_n(0\leqslant\tau\leqslant 1)=V_n(\tau\leqslant 0)+\frac{2}{\mathcal{L}}\left[\tau-\frac{4}{N\sqrt{\mathcal{L}}}\ln\left|\frac{\tau-\sqrt{\mathcal{L}}}{\tau+\sqrt{\mathcal{L}}}\right|\right],
$$

$$
V_n(\tau\geqslant 1)=\frac{2}{\mathcal{L}}\left[\frac{\tau}{2}-\frac{\tau-2}{2+N(\tau-1)}-\frac{2}{N\sqrt{\mathcal{L}}}\times\ln\left|\frac{(\tau-\sqrt{\mathcal{L}})(\tau-2+\sqrt{\mathcal{L}})}{(\tau+\sqrt{\mathcal{L}}(\tau-2-\sqrt{\mathcal{L}})}\right|\right];
$$

$$
V_T(\tau\leqslant 0)=\frac{1}{\mathcal{T}}\left\{\frac{2(4-N\tau)}{[2+N(1-\tau)]^2}-2+3\tau V_n(\tau\leqslant 0)\right\},
$$

$$
V_T(0\leqslant\tau\leqslant 1)=V_T(\tau\leqslant 0)+\frac{N\tau^2}{2\mathcal{T}}-\frac{1}{\mathcal{L}}\left\{\frac{N\tau^2}{2}+3\tau[V_n(0\leqslant\tau\leqslant 1)-V_n(\tau\leqslant 0)]\right\}, \tag{5.28}
$$

$$
V_T(\tau\geqslant 1)=\frac{1}{\mathcal{L}}\left\{2-\frac{2(4+N\tau)}{[2-N(1-\tau)]^2}-3\tau V_n(\tau\geqslant 1)\right\};
$$

$$
V_R(\tau\leqslant 1)=\frac{2\Gamma N^2}{[2+N(2\Gamma+1-\tau)][2+N(1-\tau)]},
$$

$$
V_R(\tau\geqslant 1)=\left[\frac{1+2\Gamma N}{[1+2\Gamma N^2(\tau-1)]^{1/2}}-1\right]\frac{N}{2[2+N(2\Gamma+1-\tau)]}
$$

$$
+\left[\frac{1-2\Gamma N}{[1+2\Gamma N^2(\tau-1)]^{1/2}}+\frac{4\Gamma N}{2+N(\tau-1)}-1\right]\frac{N}{2[2-N(2\Gamma+1-\tau)]}.
$$

Here $\mathcal{T}=\tau^2-8N^{-1}$, $\mathcal{L}=\tau^2+8N^{-1}$.

Figure 5.8 shows the profiles of acoustic waves generated from the increase in nonequilibrium carrier concentration (i.e., for $0\leqslant t\leqslant 1$). As the optical energy absorbed by the semiconductor (the parameter N) increases the acoustic pulses undergo steepening, due to the diminishing of penetration depth of laser light into the crystal as light absorption by free carriers (whose concentration increases) rises.

The diminishing amplitude of the pulses shown in Fig. 5.8 corresponds to an actual rise of the acoustic signal with increasing parameter N, although this process obeys a sublinear law. Therefore in this regime the efficiency of sound generation by the deformation mechanism diminishes with increasing optical radiation intensity. We obtain for the characteristic vibrational velocity in this component of the acoustic wave, using Eq. (5.28), $V_n(\tau=0,N\ll 1)\approx -N/2$; $V_n(\tau=0,N\gg 10)\approx -\pi/4(N/2)^{1/2}$; $V_n(\tau=1,N\ll 1)\approx N/2$, $V_n(\tau=1,N\gg 10)\approx 2$. Therefore the amplitude of a pulse of positive polarity is saturated as $N\to\infty$.

Figure 5.8(b) shows profiles of acoustic pulses excited by thermal expansion of a crystal heated by energy absorbed by free carriers (i.e., for $0\leqslant t\leqslant 1$). A

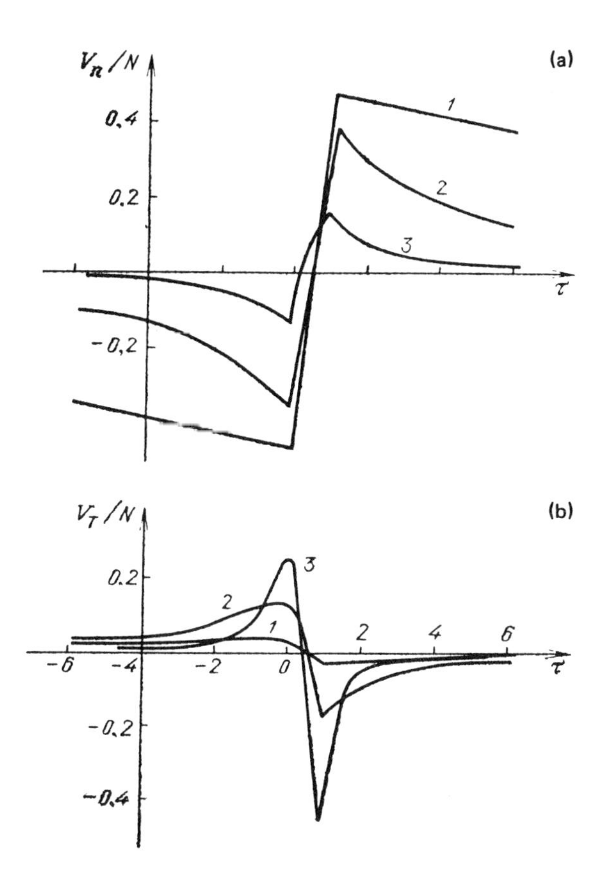

FIG. 5.8. Profiles of acoustic pulses excited by the deformation (a) and thermoelastic (b) mechanisms during optical pulse irradiation. The parameter N is equal to 0.1 (1), 1 (2), 10 (3); $\Gamma = 1$.

comparison to Fig. 5.8(a) shows that the duration of pulses $V_T(\tau)$ excited by the thermoelastic mechanism diminishes with increasing N more rapidly than the duration of pulses $V_n(\tau)$. This is due to the fact that the rising light pulse energy is accompanied not only by a diminishing size of the heating zone but also an increase in the fraction of absorbed optical energy expended directly in heating the crystal lattice. Consequently, the characteristic vibrational velocities in wave $V_T(\tau)$ initially grow superlinearly $[V_T(\tau = 0, N \ll 1) \approx (N/2)^2;\ V_T(\tau = 1, N \ll 1) \approx -(N/2)^2]$, and then at very high values of N the thermoelastic generation efficiency halts its growth $[V_T(\tau = 0, N \gg 10) \approx N/4;\ V_T(\tau = 1, N \gg 10) \approx -N/2]$.

Figure 5.9 shows the profiles of acoustic pulses excited after the termination of light irradiation, i.e., for $t \geqslant 1$). The duration of the positive swing of the pulse remains unchanged at a fixed optical intensity ($N = \text{const}$) with increasing parameter Γ, although the crossover to a pulse of negative polarity [Fig. 5.9(a)] diminishes. This is directly related to a reduction of the carrier recombination time. With a fixed parameter Γ, both the duration of the crossover region and the overall duration of the pulse diminish with increasing N [Fig. 5.9(b)]. This occurs since the

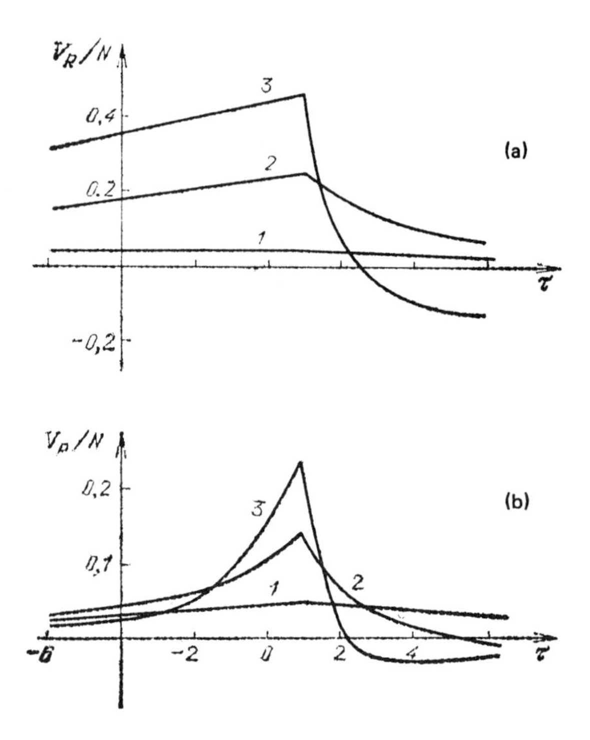

FIG. 5.9. Profiles of acoustic pulses excited by EH plasma recombination after termination of light irradiation: (a) $N = 0.1$; $\Gamma = 1$ (1); 10 (2), 100 (3); (b) $\Gamma = 1$; $N = 0.1$ (1), 0.4 (2), 1 (3).

rise of carrier concentration not only increases the carrier recombination rate but also reduces the depth of the photoexcitation region.

Figure 5.10 shows sample results from the calculation of the profile of a longi-

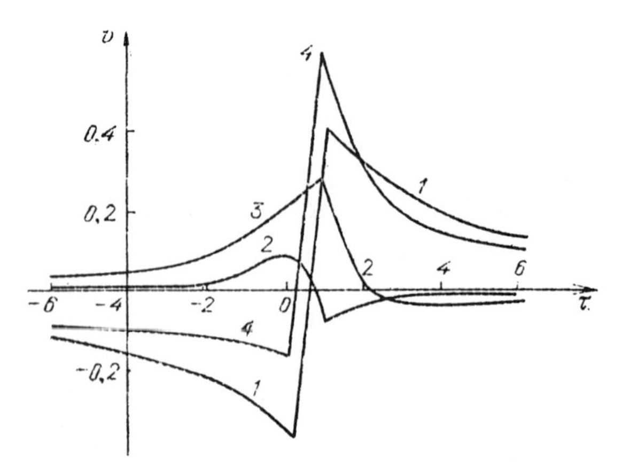

FIG. 5.10. Different acoustic signal components [1—V_n, 2—CV_T, 3—$(1+B)V_R$] and the resulting vibrational velocity profile [4—$v = V_p + CV_T + (1+B)V_R$].

tudinal acoustic wave excited in silicon by radiation $\lambda_1 \approx 1.06$ μm, $\tau_L \approx 20$ ns, $\mathscr{E}_0 \approx 0.25$ J/cm^2, which corresponds to actual experimental conditions.[9] According to our estimates, the corresponding calculated parameters are $B \approx 0.1$, $C \approx 0.5$, $N \approx 1$, $\Gamma \approx 1$. Analysis of Fig. 5.10 reveals that in this case, even when light absorption by free carriers is taken into account, the thermoelastic contribution to the acoustic wave is much less than the deformation mechanism, while the asymmetry of the resulting signal profile is caused by the wave excited by carrier recombination. Accounting for the recombination contribution to the acoustic wave makes it possible to describe the significant elevation of the compressional pulse amplitude over the rarefactional pulse amplitude, the preferential steepening of the compressional pulse, as well as the shift of the zero point of the profile towards $\tau < 0$. All these effects were observed experimentally in Ref. 9.

In conclusion, we note that these theoretical models make it possible, by comparison to results from an optoacoustic experiment, to estimate such semiconductor parameters as γ, σ, d, etc.

5.3. Sound Excitation in the Process of Fast Nonlinear Diffusion of Electron-Hole Plasma

Sound generation by fast expansion of a photoexcited EH plasma was analyzed in a linear approximation in Sec. 4.2. It was noted that the dependence of the diffusion coefficient on the photogenerated carrier concentration n may be significant in analyzing the motion of a degenerate EH plasma: $D = D(n)$. This nonlinearity can be accounted for analytically[14] in the case where carrier diffusion is limited by bulk recombination over the actual frequency range

$$\omega_S \ll \omega \ll \omega_R, \omega_n, \omega_a. \tag{5.29}$$

Inequality (5.29) differs from (4.42) only in the altered relation between ω and ω_R. This is unavoidable at sufficiently high optical irradiation intensities, since even an increase in the diffusion coefficient with increasing n cannot, according to Eqs. (4.34)–(4.36) limit growth of the plasma concentration. Nonlinear recombination processes $\tau_R = \tau_R(n)$ hence begin to dominate and ω_R jumps sharply.

Before analyzing the nonlinear problem we will determine the results of linear theory in regime (5.29). We recall that when the possibility of transonic diffusion is accounted for, the acoustic pulse spectra differ from Eq. (4.9) solely in the substitution of the multiplier in front of the square brackets in $\widetilde{v}_n$: $(\omega_R - i\omega)^{-1} \to [\omega^2/\omega_{D0} + (\omega_R - i\omega)]^{-1}$.

In spectral range (5.29) the transfer function of the deformation mechanism of sound generation takes the form

$$K_n(\omega) = (-1)\frac{-i\omega}{\omega_R}\frac{-i\omega(\omega_{D0}\omega_R)^{1/2}}{\omega^2 + \omega_{D0}\omega_R}. \tag{5.30}$$

The characteristic frequency $(\omega_{D0}\omega_R)^{1/2}$ may therefore appear in the acoustic signal spectrum. This requires that the frequency fall within the range defined by inequalities (5.29). The right-hand side of Eq. (5.29) holds if

$$\omega_{D0} \ll \omega_R \ll m_{D0}^2 \omega_{D0},\ m_{D0}^4 \omega_{D0}.$$

Consequently, the regime $m_{D0} \gg 1$ will be realized, which in practice requires strong (surface) light absorption. We note that when $\omega_{D0} \ll \omega_R$ holds, frequency ω_D (4.6) at which the wave vectors of the acoustic and concentration waves are comparable, adopts a value $(\omega_{D0}\omega_R)^{1/2}$. This reveals the physical meaning of the characteristic frequency in Eq. (5.30).

Unlike regime (4.42), in the case of fast bulk recombination in regime (5.29) it is necessary to also account for thermoelastic sound excitation. For purposes of simplicity we assume slow phonon heat conduction (substantially subsonic), and a thermally thin light absorption region:

$$\omega \ll \omega_\chi, \omega_T. \tag{5.31}$$

Then the vibrational velocity spectra of the acoustic wave can be represented as

$$\tilde{v}_T(\omega) \approx B \frac{-i\omega(\omega_{D0}\omega_R)^{1/2}}{\omega^2 + \omega_{D0}\omega_R} \tilde{f}(\omega), \quad \tilde{v}_n = -\frac{1}{B}\frac{-i\omega}{\omega_R}\tilde{v}_T. \tag{5.32}$$

The frequency $(\omega_{D0}\omega_R)^{1/2}$ falls within spectral range (5.31), if $m_{D0}m_\chi \gg 1$ and $(\omega_{D0}\omega_R)^{1/2} \ll \omega_\chi$. The latter inequality holds in the case of fast diffusion.

According to Eq. (5.32), the thermoelastic sound generation mechanism is dominant in the frequency range $\omega \ll B\omega_R$. A comparison of Eq. (5.32) to Eq. (5.12) shows that generation in test regime (5.29), (5.31) occurs in the same manner as in the case of a thermally and diffusively thick absorption region ($m_{D0} \ll 1$). The difference is that the characteristic dimensions of the heating zone are not determined by the depth of penetration of light α^{-1} (characteristic frequency $\omega_a = \alpha c_L$), but rather by the diffusion length $l_D = (D\tau_R)^{1/2}$ [characteristic frequency $c_L/l_D = (\omega_{D0}\omega_R)^{1/2}$]. Physically, the frequency $(\omega_{D0}\omega_R)^{1/2}$ is the inverse of the sound travel time through the EH plasma localization region.

The profiles of acoustic pulses with a spectral composition similar to Eq. (5.32) have already been analyzed in Secs. 2.2 and 5.2. It is important for our purposes that in the case $\tau_L^{-1} \gg (\omega_{D0}\omega_R)^{1/2}$ the profile of the thermoelastic component of the signal will acquire characteristic bipolar form (2.28) where the durations of the leading and trailing edges will be determined by the diffusion length $l_D/c_L \sim (\omega_{D0}\omega_R)^{-1/2}$. Therefore, the acoustic wave profile bears information on the spatial distribution of EH plasma.

An analysis carried out within the framework of linear model (4.2) provides the physical meaning for the mathematical techniques utilized in analyzing nonlinear diffusion.[14] Under conditions where the carrier recombination time τ_R is significantly less than the laser irradiation duration τ_L, the EH pair generation equation

$$\frac{\partial n}{\partial t} = -\frac{\partial}{\partial z}(nv_d) - \frac{n}{\tau_R(n)} - \frac{I_0}{\hbar\omega_L}\frac{\partial}{\partial z}\exp(-\alpha z)f(t)$$

adopts a quasistationary form

$$\frac{\partial}{\partial z}(nv_d) + \frac{n}{\tau_R(n)} = -\frac{I_0}{\hbar\omega_L}\frac{\partial}{\partial z}\exp(-\alpha z)f(t), \quad nv_d|_{z=0} = 0. \tag{5.33}$$

Here v_d is the hydrodynamic velocity of the EH pairs [which is not substantially different from the drift velocity in the assumptions traditionally made in deriving Eq. (5.33)].[15]

The equation of motion of the EH pairs

$$Mn\left(\frac{\partial v_d}{\partial t}+v_d\frac{\partial v_d}{\partial z}+\frac{v_d}{\tau_d}\right)=-\frac{\partial p}{\partial z} \tag{5.34}$$

includes the viscous term v_d/τ_d, which phenomenologically accounts for carrier scattering processes responsible for loss of the total plasma momentum.[15] The characteristic damping time of hydrodynamic motion is of the order of the minimum carrier scattering time for the case where carriers are scattered by defects, ionized impurities, and phonons.

We assume that in the low-temperature case examined here, scattering by ionized inclusions predominates. The fact that τ_d is independent of temperature $(\tau_d \sim n)$ is of fundamental importance.[16] This makes it possible to uncouple the equation of motion of the nonequilibrium carriers and the equation of phonon heat conduction. In Eq. (5.34) M is the mass of the EH pair, p is the internal pressure of EH plasma $[p=n^2(\partial F/\partial n)_T$, where F is the free energy of the EH pair]. EH plasma expansion at supersonic velocities requires a plasma concentration far exceeding the concentration of EH liquid.[17] Hence the kinetic part, $F\approx an^{2/3}$, where a is a constant, predominates in the EH pair free energy.[18] Then, assuming that the internal pressure forces are in equilibrium with the viscosity, i.e.,

$$\left|\frac{1}{vp}\frac{\partial v_d}{\partial t}\right|,\quad \left|\frac{\partial v_d}{\partial z}\right| \ll \tau_d^{-1},$$

the quasistationary relation of v_d to the plasma concentration gradient is determined from Eq. (5.34),

$$v_d \approx -\frac{\tau_d}{M}\frac{1}{n}\frac{\partial p}{\partial z}\approx -\frac{10}{9}\frac{a\tau_d}{M}n^{-1/3}\frac{\partial n}{\partial z}.$$

The last relation in conjunction with Eq. (4.41) provides a representation for the diffusion coefficient of the degenerate EH plasma:[19] $D(n)\equiv(10/9)(a\tau_d/M)n^{2/3}$. Using this designation, Eq. (5.33) in the case of surface light absorption is conveniently recast as

$$\begin{gathered}\frac{\partial}{\partial z}\left[D(n)\frac{\partial n}{\partial z}\right]-\frac{n}{\tau_R(n)}=0,\\ D(n)\frac{\partial n}{\partial z}\bigg|_{z=0}=-\frac{I_0}{\hbar\omega_L}f(t)\equiv -J_0 f(t).\end{gathered} \tag{5.35}$$

Therefore the carrier diffusion equation in approximation (5.29) is a homogeneous equation, although it has changed boundary conditions. According to the estimates derived above, in describing the temperature field under conditions (5.31) $(\omega \sim \tau_L^{-1})$ it is possible to neglect phonon heat conduction and only account for recombination heating of the crystal:

$$\frac{\partial T}{\partial t}=\left(\frac{E_g}{\rho_0 c_p}\right)\frac{n}{\tau_R(n)}. \tag{5.36}$$

Equation (5.36) describes the component of the temperature field making the primary contribution to thermoelastic sound generation.

The thermoelastic acoustic wave excitation mechanism is dominant in regime $\tau_R \ll B\tau_L$. The time dependence of carrier concentration for a rectangular laser pulse $[f(t)=\Theta(t) - \Theta(t-\tau_L)]$ will be a replica of the pulse envelope. In this model case, general description (2.36) of the profile of an acoustic pulse excited near the free surface is transformed by means of Eqs. (5.35) and (5.36) to

$$v=\frac{K\beta E_g}{2\rho_0^2 c_L^2 c_p} D(n)\frac{\partial n}{\partial z}\Bigg|_{c_L|\tau|}^{c_L|\tau-\tau_L|}. \tag{5.37}$$

Here $n=n(z)$ is the solution of problem (5.35) in the stationary case $[f(t) \equiv 1]$, while the vertical line denotes the difference of the substitutions of the upper and lower argument into the functions. Solution (5.37) determines the dependence of the acoustic wave profile on the spatial distribution of the EH pairs and the laser irradiation time. Equation (5.37) describes the decay of n with distance from boundary $z=0$ (specifically this is a consequence of ignoring surface recombination). Hence, in the case of interest to us $\partial D/\partial n>0$ relation (5.37) defines a symmetrical bipolar pulse with fixed peak position: $(v(\tau) = -v(\tau_L-\tau)$, $v_a=v(\tau=0) = -v(\tau=\tau_L)$ $(\beta>0$ for definiteness).

As an example we find the spatial distribution of a degenerate EH plasma scattered by ionized impurities $[D(n) = \delta n^{5/3}$, δ is a constant] in the case of plasma Auger recombination (4.34). With these dependences of the diffusion coefficient and lifetime of the EH plasma on pair concentration, the exact solution of (5.35) takes the form

$$n=(1+z/10)^{-6}, \tag{5.38}$$

where n is normalized to its surface value

$$n=n/n_0,\ n_0=n(z=0) = (17J_0^2/6\gamma\delta)^{3/17}{\sim}I_0^{6/17},$$

while the spatial coordinate is normalized to the diffusion length:

$$z=z/z_0,\ z_0=\sqrt{1.02 l_D(n_0)}\approx[D(n_0)\tau_R(n_0)]^{1/2}{\sim}I_0^{-1/17}.$$

Solution (5.38) describes a twofold reduction in plasma concentration at the diffusion length $z\approx 1$ from the surface. The increase in carrier concentration n_0 near the surface is predominantly limited by Auger recombination of carriers. This is evident from the fact that the relation $n_0{\sim}I_0^{6/17}$ is very close to $I_0^{1/3}$ which derives from the solution of Eq. (5.33) neglecting carrier diffusion.

The depth of penetration of the plasma is weakly dependent on the laser radiation intensity $(l_D{\sim}I_0^{-1/17})$, since the rise in the diffusion coefficient is compensated by the diminishing lifetime of the EH pairs. This characteristic anomaly of the examined plasma motion regime can be utilized for its identification purposes. We note that problem (5.37) in the case $f \equiv 1$ permits an analytic solution with any

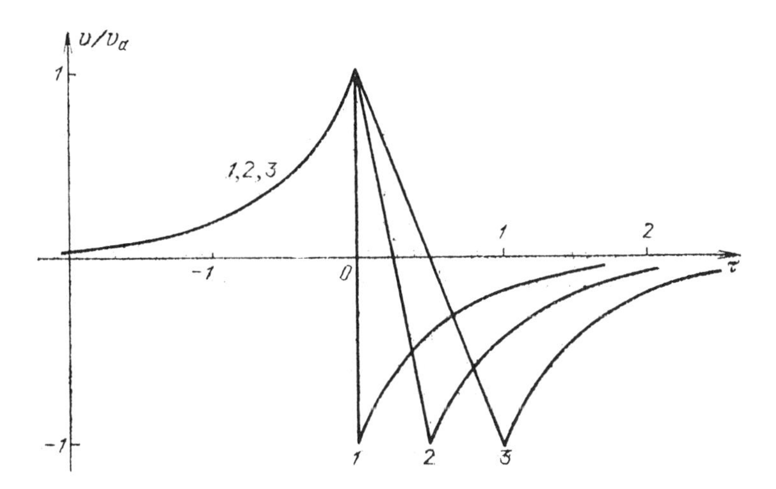

FIG. 5.11. Variation of acoustic pulse profiles with increasing optical irradiation duration for different values of the parameter A: 0 (1), 0.5 (2), 1 (3).

approximations of $D(n)$, $\tau(n)$ by power functions, even when nonlinear surface recombination $[s=s(n)]$ is additionally accounted for.[20]

Utilizing Eq. (5.38), we transform the acoustic pulse profile to

$$v=(1+z/10)^{-17}\,\Big|{}^{|\tau|}_{|\tau-A|}, \tag{5.39}$$

where

$$\tau=c_L\tau/z_0,\;\; v=v/v_0,$$

$$v_0\equiv(3K\beta E_g\delta n_0^{8/3}/10\rho_0^2c_L^2c_pz_0)\sim I_0, \tag{5.40}$$

$$A\equiv c_L\tau_L/z_0\sim I_0^{1/17}.$$

According to Eq. (5.40) deviations from a linear transformation of the acoustic wave profile with increasing photoexcitation intensity are related solely to the variation of the parameter A which determines the efficiency of acoustic pulse generation. However, this parameter changes little due to the nearly constant geometry of the heating zone. Hence, the dependence of the amplitude of the acoustic pulses on laser intensity can only be used to identify this plasma expansion regime. The acoustic signal profile contains information on the spatial distribution of EH plasma. Specifically, for $A\lesssim 1$ nonequilibrium carrier diffusion has an effect on the duration of the acoustic pulses.

Figure 5.11 provides profiles of acoustic pulses for different ratios of the laser pulse irradiation duration τ_L and the transit time of sound across the diffusion length $l_D/c_L(A\approx c_L\tau_L/l_D)$. The acoustic wave profile is most easily related to the spatial carrier distribution in the limit $A\ll 1$:

$$v/v_a=-(1+|\tau|/10)^{-18}\,\mathrm{sgn}\,\tau.$$

The regime $\tau_R\ll\tau_L$ occurs in many low-temperature experiments employing nanosecond optical radiation.[21–23] The results of experiments on CdTe (Ref. 23) indicate the possibility for thermoelastic generation of acoustic waves where the duration of the acoustic pulse is determined by the transit time of sound over a distance of the order of the EH pair diffusion length. According to the author's estimates, the parameter values were as follows: $\tau_L \approx 10$ ns, $\tau_R \approx 0.3$ ns, $v_d \approx 1.2 \times 10^7$ cm/s. In this case $D \sim v_d^2 \tau_R \sim 4 \times 10^4$ cm^2/s, $l_D \sim v_d \tau_R \approx 35$ μm, $c_L \approx 4 \times 10^5$ cm/s, $c_L \tau_L \approx 40$ μm. Therefore, $A \sim 1$ and can be further reduced by reducing the optical radiation time. We emphasize that $A \lesssim 1$ at appreciably supersonic EH plasma diffusion velocities: $V_D \gtrsim c_L(\tau_L/\tau_R)^{1/2} \gg c_L (M_D \gg 1)$.

In conclusion, we point out that in describing thermoelastic sound generation by Eqs. (2.29) and (5.36) we have neglected the nonlinearity associated with the temperature dependence of the coefficient of volume expansion and specific heat. At the same time, the relations $\beta = \beta(T)$ and $c_p = c_p(T)$ are stronger at low temperatures. However, in the Debye solid-state model β and c_p vary identically with temperature: $\beta \propto c_p \propto T^3$, which results in mutual compensation of nonlinear effects in phonon heat conduction equation (5.36) and wave equation (2.29).

In actual crystals such a situation will hardly occur uniformly. For example, we can assume $\beta \propto T^3$ in Si only at $T \lesssim 14$ K,[24] while Si compresses upon heating in the range $18 \lesssim T \lesssim 120$ K.[25] Certain semiconductors do not even have regions $\beta > 0$ at low temperatures.[26] Clearly, when analyzing sound generation under such conditions it is necessary to simultaneously account for concentration nonlinearities, as done in the present section as well as thermal nonlinearities (Sec. 5.1).

5.4. The Change in OA-Generation Efficiency in Nonlinear Light Absorption Regimes

The method of recording pulsed acoustic signals is widely used to analyze nonlinear light absorption in matter. For example, the results of Ref. 27 indicate a relation between radiation intensity nonlinear distortions to the acoustic pulse profile and the change in light reflectivity in metal upon heating [$R = R(T)$]. The deviation from linear behavior of the dependence of the acoustic signal amplitude on light intensity has already been traditionally employed to determine the multiphoton light absorptivities [$\alpha = \alpha(I)$] in semiconductors and dielectrics.[28–30]

In the majority of cases (including those in Refs. 29 and 30) the cross section of multiphoton processes and light absorption by photogenerated carriers are determined under conditions of rather weak total absorption. This permits effective analysis in a "transverse" scheme, where the acoustic wave detector is placed on the lateral face of the sample parallel to light propagation.[30] In this situation a change in optical absorption will only alter the intensity of the thermoelastic sound sources and will have no effect in practice on their spatial distribution. However, there have been experiments[31] in which a deviation from cylindrical geometry was observed with increasing light absorption.

A "longitudinal" experimental scheme is most appropriate for analyzing nonlinear absorption of optical radiation in nontransparent (strongly absorbing) samples. In this configuration the acoustic pulse detector is placed on the face of the sample opposite the irradiated surface.[27,32] In this case, nonlinear distortions of the acoustic signals may be related not only to the nonlinear variation of the energy contribution near the surface, but also to the change in sound generation efficiency resulting from transformation of the geometry of the heating zone.

In this connection we recall that the light absorptivity has a direct effect on the multiplier $\omega\omega_a/(\omega^2+\omega_a^2)$ in the spectral transfer function ($\omega_a=\alpha c_L$) according to linear theory (Secs. 2.2 and 4.1) in the absence of carrier diffusion or heat conduction. In the same manner, a change in light absorptivity will cause both a frequency shift of the region of most efficient sound generation ($\omega\sim\omega_a$), and will also alter the width $\Delta\omega$ of the spectral band of efficient generation: $\Delta\omega\sim\omega_a$ (Sec. 2.2).

The role of light absorption by photogenerated free carriers under essentially nonstationary conditions $\tau_R(n)\gtrsim\tau_L$ (5.23) was identified in Sec. 5.2. Nonlinear light absorption in this case will affect both the concentration-deformation and the thermoelastic sound generation mechanisms. Here we first analyze the regime where thermoelastic excitation of acoustic waves ($\tau_R\ll B\tau_L$) dominates (Sec. 4.1). In this case a quasistationary distribution of photoexcited EH plasma is set up under laser irradiation since $\tau_R\ll\tau_L$. As we have seen (Sec. 5.3) the latter fact substantially simplifies the description of nonlinear sound generation regimes.

We limit the analysis to the nonlinearity mechanisms that are not related to changes in temperature of the material caused by optical irradiation. The possibility for neglecting the temperature dependences $\alpha=\alpha(T)$ must, of course, be confirmed by numerical estimates in each specific case.[13] For definiteness we analyze sound generation from two-photon interband light absorption in a semiconductor ($\hbar\omega_L<E_g<2\hbar\omega_L$). We also neglect radiative recombination channels of the nonequilibrium EH plasma compared to Ref. 33, and we use to describe the fields $n(t,z)$, $T(t,z)$, and $I(t,z)$ the system of equations

$$\frac{\partial n}{\partial t}=k_2\frac{I^2}{2\hbar\omega_L}-\frac{n}{\tau_R(n)}, \tag{5.41}$$

$$\rho_0c_p\frac{\partial T}{\partial t}=\left(\alpha_0+\sigma n+\frac{2\hbar\omega_L-E_g}{2\hbar\omega_L}k_2I\right)I+E_g\frac{n}{\tau_R(n)},$$
$$\frac{\partial I}{\partial z}=-(\alpha_0+\sigma n+k_2I)I,\ I|_{z=0}=(1-R)I_Sf(t)\equiv I_0f(t), \tag{5.42}$$

where α_0 is the residual linear absorptivity (which, specifically, may be due to the presence of free carriers in the absence of laser irradiation), and k_2 is the two-photon light absorptivity. It is assumed that the light absorption region is diffusively and thermally thick (regime $m_{D0}\ll1$ is realized) (Sec. 4.1).

When $\tau_R(n)\ll\tau_L$ holds, the solution of Eq. (5.41) describes the quasistationary distribution of photogenerated carriers $n=n(I)$:

$$n=\tau_R(n)k_2I^2/2\hbar\omega_L. \tag{5.43}$$

This makes it possible to reduce (5.42) to the form

$$\rho_0c_p\frac{\partial T}{\partial t}=\alpha(I)I=-\frac{\partial I}{\partial z}, \tag{5.44}$$

where

$$\alpha(I) = \alpha_0[1 + \sigma n(I)/\alpha_0 + k_2 I/\alpha_0]. \tag{5.45}$$

Let us use a simple model for the laser pulse envelope $f(t) = \theta(t) - \theta(t-\tau_L)$. In this case $I(t,z) \equiv I(z)f(t)$ and, using Eqs. (2.36) and (5.44), it is possible to describe analytically the profile of the vibrational velocity pulse excited by the thermoelastic effect for an arbitrary distribution $I(z)$:

$$v = \frac{K\beta}{2\rho_0^2 c_L^2 c_p} I(z) \Bigg|_{c_L|\tau - \tau_L|}^{c_L|\tau|}. \tag{5.46}$$

Acoustic signal (5.46) is a symmetrical, bipolar signal $[v(\tau = \tau_L/2) = 0, v(\tau) = -v(\tau_L - \tau)]$. If the material expands upon heating ($\beta > 0$), the sound pulse begins with a compressional phase. If the energy delivery diminishes with distance from the surface $z = 0$ ($\partial^2 T/\partial t \partial z < 0$), for which

$$\frac{\partial[\alpha(I)I]}{\partial I} > 0 \tag{5.47}$$

must hold, the acoustic signal grows monotonically for $\tau < 0$ and $\tau > \tau_L$. The sound signal will always decay ($\partial v/\partial \tau < 0$) for $0 < \tau < \tau_L$. The following representation will therefore be valid for the acoustic pulse amplitude under condition (5.47):

$$v_a = |v(0)| = |v(\tau_L)| = K\beta(2\rho_0^2 c_L^2 c_p)^{-1}[I(0) - I(c_L \tau_L)]. \tag{5.48}$$

Analysis of relation (5.48) reveals that the simplest relation between the amplitude of an excited acoustic wave and the light absorptivity characteristic of the transverse experimental scheme is realized in the longitudinal scheme only in the limiting case $\alpha^{-1} \gg c_L \tau_L$. Here the light absorption length far exceeds the distance traveled by the sound during optical irradiation. Utilizing the expansion $I(c_L \tau_L) \approx I(0) + c_L \tau_L \, \partial I/\partial z(0)$, Eqs. (5.46) and (5.48) can be recast as

$$v_a \approx K\beta\tau_L(2\rho_0^2 c_L c_p)^{-1}\alpha(I_0)I_0,$$

$$v \approx -K\beta\tau_L(2\rho_0^2 c_L^2 c_p)^{-1}\alpha[I(c_L|\tau|)]I(c_L|\tau|)\,\mathrm{sgn}\,\tau.$$

In the opposite limiting case $\alpha^{-1} \ll c_L \tau_L$ the second term in Eq. (5.48) can be ignored:

$$v_a \approx K\beta(2\rho_0^2 c_L^2 c_p)^{-1} I_0.$$

Consequently, in the case of strong absorption the wave amplitude grows linearly with increasing radiation intensity in spite of the change in absorption. This is due to the diminishing sound generation efficiency which occurs as the spectral range of efficient acoustic signal transmission ($\omega \sim \omega_a \sim \alpha c_L$, $\Delta\omega \sim \omega_a$) shifts increasingly away from the characteristic frequencies of the excited sound ($\omega \lesssim \tau_L^{-1}$).

It is only necessary to know the spatial distribution of light in the sample to describe the acoustic pulse by means of (5.46) and (5.48). We calculate $I(z)$ for certain specific cases of nonlinear light absorption. If linear carrier recombination $[\tau_R(n) = \mathrm{const}]$ predominates, then according to Eq. (5.43) $n \approx k_2 \tau_R (2\hbar\omega_L)^{-1} I^2$ and

$$\alpha(I) \approx \alpha_0 + k_2 I + \sigma k_2 \tau_R (2\hbar\omega_L)^{-1} I^2. \tag{5.49}$$

In the case of nonradiative Auger recombination (4.34) $n \approx (k_2/2\hbar\omega_L\gamma)^{1/3} I^{2/3}$ and

$$\alpha(I) \approx \alpha_0 + \sigma(k_2/2\hbar\omega_L\gamma)^{1/3} I^{2/3} + k_2 I. \tag{5.50}$$

The following representation of the dependence of light absorptivity on intensity may be convenient for modeling light absorption:

$$\alpha(I) = \alpha_0(1 + I/I_{cr})^{\delta}, \tag{5.51}$$

where I_{cr} is the critical intensity for manifestation of nonlinear effects; δ is a constant. For $\delta > 0$ Eq. (5.51) describes the rise in absorption, while for $\delta < 0$ it describes the induced transparency effect. Specifically, a value $\delta = -\frac{1}{2}$ corresponds to light-induced transparency of the inhomogeneously broadened transition; $\delta = -1$ corresponds to saturation of the homogeneously broadened absorption line.[34] We note that Eq. (5.51) satisfies Eq. (5.47) for all I if $\delta \geqslant -1$. From the known relation $\alpha = \alpha(I)$ the spectral distribution of radiation is given, consistent with Eq. (5.44) by the integral

$$z = \int_I^{I_0} [\alpha(I')I']^{-1}\, dI'. \tag{5.52}$$

All cases (5.49)–(5.51) listed here permit analytic integration (5.52). It is true that in most cases the relation $I(z)$ is expressed in the implicit form $z = z(I)$, which makes it easier to use the graphic plots of $I(z)$ to generate the profiles of acoustic pulses (5.46) and determine their amplitudes (5.48).

As an example that permits explicit description of the light intensity distribution, we consider model (5.51) for $\delta = 1$, which also corresponds to neglecting light absorption by free carriers in Eqs. (5.49) and (5.50). In dimensionless variables $z = \alpha_0 z$, $I = I/I_{cr}$, $I_0 = I_0/I_{cr}$ distribution (5.52) can be represented as

$$I = I_0 \exp(-z)(1 + I_0[1 - \exp(-z)])^{-1}.$$

We obtain for the dimensionless wave amplitude $v_a = v_a/v_0$

$$v_a = I_0 - I(A) = I_0(1 + I_0)/(I_0 + [1 - \exp(-A)]^{-1}). \tag{5.53}$$

Here $v_0 = K\beta(2\rho_0^2 c_L^2 c_p)^{-1} I_{cr}$, $A \equiv \alpha_0 c_L \tau_L$ is the parameter characterizing OA-generation efficiency at low laser intensities ($\alpha \approx \alpha_0$). According to Eq. (5.53) $v_a \approx [1 - \exp(-A)] I_0$ for $I_0 \ll 1$ and $v_a \approx I_0$ for $I_0 \gg [1 - \exp(-A)]^{-1}$.

Deviations from a linear relation are observed within a limited intensity range of $1 \leqslant I_0 \leqslant [1 - \exp(-A)]^{-1}$. Hence in order to identify such variations in experiment it is necessary to reduce parameter A. This can be achieved by reducing the laser pulse duration τ_L. The results obtained here, however, are valid only when $\tau_L \gg \tau_R/B$. In the case of nonlinear recombination such an inequality imposes an additional lower limit on the range of usable laser intensities.[33]

Figure 5.12 shows relation (5.53) (v_a plotted as a function of I_0 for different values of the parameter A), which confirms our qualitative arguments. The profiles of OA pulses normalized to pulse amplitudes are shown in Fig. 5.13; these reveal steepening of the acoustic signals with increasing I_0. Such steepening may be significant only for $A \ll 1$.

Superlinear growth of acoustic pulse amplitude and steepening of pulse leading edges were observed experimentally in Ref. 32 in GaSb irradiated by a yttrium-

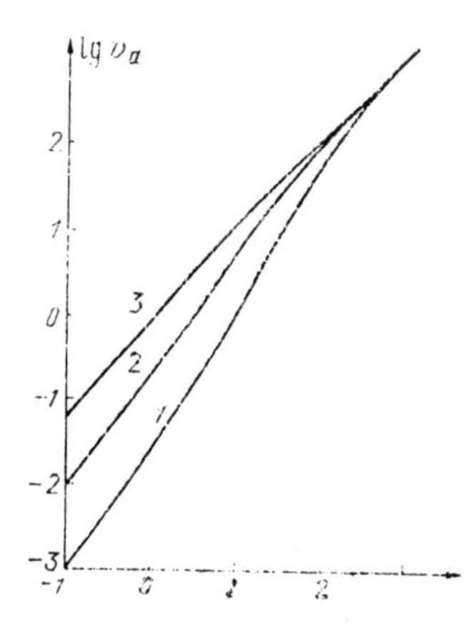

FIG. 5.12. OA-signal amplitude plotted as a function of light intensity in the nonlinear absorption regime for different values of the parameter A: 0.01 (1), 0.1 (2), 1 (3).

erbium-aluminum garnet laser ($\lambda \approx 2.92\ \mu$m, $\tau_L \approx 65$ ns). Nonlinear light absorption appeared in this study against a background of strong linear absorption ($\alpha_0 \approx 25\ \text{cm}^{-1}$). The authors determined[32] that regime (5.50) was realized, where the superlinear rise in temperature in the surface region of the crystal was primarily caused by energy absorption by the additional free carriers created from two-photon light absorption ($2\hbar\omega_L \approx 0.85$ eV $> E_g \approx 0.72$ eV). The nonlinear growth of signal amplitude was enhanced by the increasing efficiency of sound generation with rising absorption, since $A \approx 0.7 < 1$.

We consider Eq. (5.51) for $\delta = -1$ as an example of an implicit description of intensity distribution. With use of the same dimensionless variables,

$$z = I_0 - I - \ln(I_0/I).$$

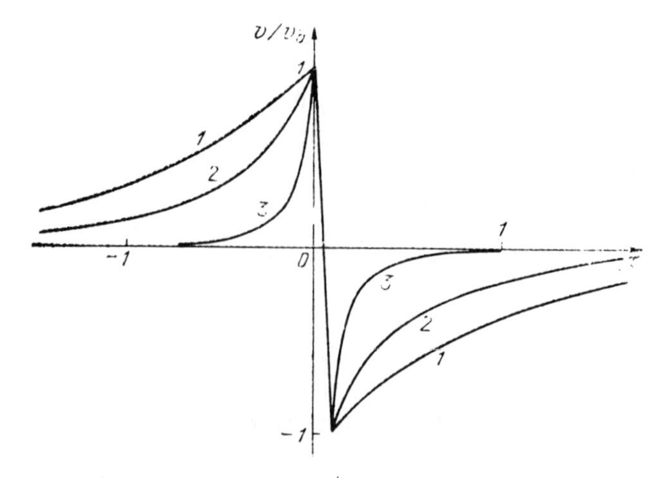

FIG. 5.13. Variation of acoustic pulse profiles with increasing laser radiation intensity I_0 equal to 0.1 (1), 1 (2), 10 (3); $A = 0.1$, $\tau = \tau\alpha c_L = A\tau/\tau_L$ — dimensionless time.

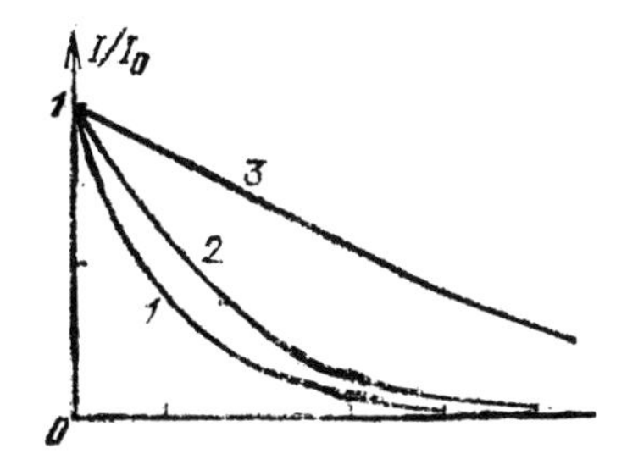

FIG. 5.14. Variation in the spatial distribution of light in the sample with increasing light intensity at the boundary $I_0 =$ 0.1 (1), 1 (2), 10 (3), $A = 1$.

This spatial distribution was plotted for a number of values of I_0 in Fig. 5.14. This figure is employed to draft the acoustic pulse profiles in accordance with Eq. (5.46) (Fig. 5.15). An implicit expression can be derived for the wave amplitude

$$I_0 = v_a[1 - \exp(v_a - A)]^{-1}. \qquad (5.54)$$

An implicit intensity dependence of the amplitude similar to Eq. (5.54) can also be obtained in the general case. According to Eqs. (5.48) and (5.52) it takes the form

$$A = \int_{I_0 - v_a}^{I_0} \alpha_0[\alpha(I)I]^{-1}\, dI.$$

Consistent with Eq. (5.54) at low intensities $I_0 \ll A[1 - \exp(-A)]^{-1}$ the OA-wave amplitude grows linearly: $v_a \approx [1 - \exp(-A)]I_0$. At high intensities $I_0 \gg A$ it goes to saturation at $v_a \approx A$. This is determined by narrowing of the spectral domain of efficient sound generation ($\Delta\omega \sim \omega_a \approx \alpha c_L$) during quasistationary light-induced transparency of the medium. Transient regimes may be observed upon increasing laser intensity only in the case if the initial intensities lie in the range $I_0 \lesssim A$.

Figure 5.16 shows the OA-signal amplitude (5.54) plotted as a function of I_0 for different values of the parameter A.

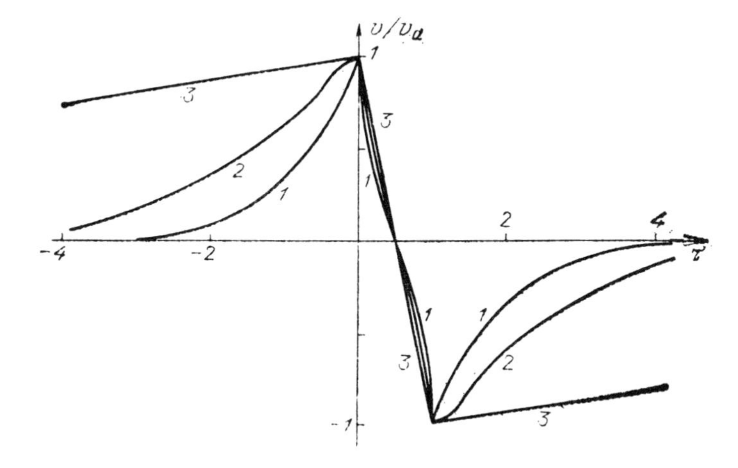

FIG. 5.15. Behavior of the acoustic pulse profiles for quasistationary light-induced transparency of a crystal at intensity I_0 of 0.1 (1), 1 (2), 10 (3); $A = 1$, $\tau = A\tau/\tau_L$.

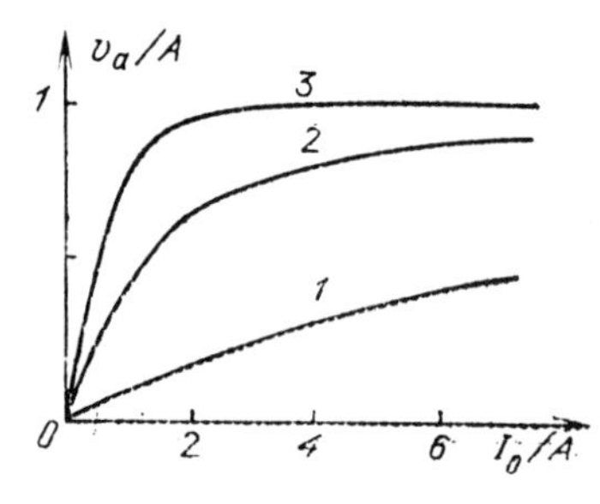

FIG. 5.16. Saturation of OA-signal amplitude from light-induced transparency of the medium $A = \alpha_0 c_L \tau_L = 0.1$ (1), 1 (2), 10 (3).

The sublinear rise in amplitude of the acoustic pulses with increasing laser intensity was observed experimentally during transparency of water by an erbium laser ($\tau_L \approx 60$–100 ps, $\alpha_0 \approx 1.3 \times 10^4$ cm^{-1}).[35] Calculations[33] confirm the conclusions of the authors of Ref. 35 that the retarded growth of v_a was caused by the diminishing sound generation efficiency with increasing size of the energy delivery region.

Thermoelastic sound generation from nonstationary induced transparency of liquids was first analyzed in Ref. 36. The elastic wave excitation process was described by a complete system of nonlinear equations of hydrodynamics, which made possible only numerical computer calculations of the acoustic wave profiles. The authors of Ref. 36 did not differentiate distortions to the acoustic pulse profile caused by absorption saturation and an acoustic nonlinearity. Reference 35 has analyzed the excitation of acoustic waves in water and glycerin under conditions where induced transparency of the surface due to breaking of the hydrogen bonds is possible. When the surface of the absorbing liquid was covered with a quartz plate, unipolar acoustic pulses were recorded in this experiment. Steepening of the leading edge at high laser intensities was attributed by the authors of Ref. 35 exclusively to the acoustic nonlinearity in the propagation of these pulses.

We use a simple model of nonstationary self-induced transparency to obtain an analytic description of the transformation of acoustic pulse profiles caused by the change in the geometry of the optical irradiation region over time.[37] When necessary the nonlinear acoustic effects can be described by the stage-by-stage approach (Chap. 2). The methods of accounting for the acoustic nonlinearity in the acoustic wave generation stage are examined in Chap. 6.

The transparency of semiconductors can be effectively induced by the dynamical Burstein-Moss effect[38] while that of dye solutions by reducing the number of absorbing bonds.[35] In both cases the light absorptivity diminishes with increasing photoexcited elements in the absorber (EH pairs in the semiconductor and broken bonds in the dye). Therefore a fundamental characteristic of the absorber is the dependence of α on the concentration n of photoexcited states: $\alpha = \alpha(n)$, $\partial\alpha/\partial n < 0$. For definiteness we shall henceforth assume that n refers to the concentration of EH plasma.

Let the absorption saturation effect be described by the following model:

$$\alpha(n) = \alpha_0[1 - \theta(n - n_{\mathrm{cr}})], \tag{5.55}$$

whereby the absorptivity drops to zero when plasma concentration reaches the critical value n_{cr}.

We utilize the following equations to describe the space-time evolution of fields $n(t,z)$ and $I(t,z)$:

$$\frac{\partial n}{\partial t}=\frac{\alpha(n)I}{\hbar\omega_L},\ n|_{t\to-\infty}=0,$$
$$\frac{\partial I}{\partial z}=-\alpha(n)I,\ I|_{z=0}=I_0 f(t). \tag{5.56}$$

One fundamental difference between model (5.56) and (5.41) is the EH pair recombination over the optical irradiation time is neglected ($\tau_R \gg \tau_L$). Dye relaxation was also treated as negligibly slow by Ref. 36. In this approximation the carrier concentration is not determined by the optical radiation intensity at the given instant and at the given point in space [compared to Eq. (5.43)] but rather is dependent on the entire history of laser irradiation. The induced transparency is essentially a nonstationary process; the spatial distribution of radiation and the EH plasma are largely transformed the interaction of light with the material.

As usual, analytic calculations are possible for the model $f(t)=\theta(t)-\theta(t-\tau_L)$. Equation system (5.56) in dimensionless variables $n=n/n_{\rm cr}$, $I=I/I_{\rm cr}$, $I_{\rm cr}=\hbar\omega_L c_L n_{\rm cr}$, $z=z/c_L\tau_L$, $t=t/\tau_L$ takes the form

$$\frac{\partial n}{\partial t}=A[1-\theta(n-1)]I,$$
$$\frac{\partial I}{\partial z}=-A[1-\theta(n-1)]I,\ I|_{z=0}=I_0[\theta(t)-\theta(t-1)], \tag{5.57}$$

where $A=\alpha_0 c_L\tau_L$ is our traditional parameter. The solution of system (5.57) demonstrates that critical carrier concentrations ($n=1$) are not achieved even on the surface of the medium at intensities $I_0<A^{-1}$, i.e., in this regime the absorber is not saturated. Light absorption and growth of plasma concentration occur throughout the sample:

$$\frac{\partial n}{\partial t}=-\frac{\partial I}{\partial z}=AI_0\exp(-Az)[\theta(t)-\theta(t-1)]. \tag{5.58}$$

If $I_0\gtrsim A^{-1}$, the surface $z=0$ will become transparent as early as $t_{\rm cr}=(\mathrm{AI}_0)^{-1}\leqslant 1$ and the absorption region begins to move into the bulk of the medium:

$$\frac{\partial n}{\partial t}=-\frac{\partial I}{\partial z}=AI_0\exp(-Az)[\theta(t)-\theta(t-t_{\rm cr})],\ t\lesssim t_{\rm cr},$$
$$\frac{\partial n}{\partial t}=-\frac{\partial I}{\partial z}=AI_0\exp\{-A[z-I_0(t-t_{\rm cr})]\}\theta[z-I_0(t-t_{\rm cr})]$$
$$\times[\theta(t-t_{\rm cr})-\theta(t-1)],\quad t\geqslant t_{\rm cr}. \tag{5.59}$$

According to solution (5.59) radiation is not absorbed in the time interval $t_{\rm cr}<t<1$ for $0<z<I_0(t-t_{\rm cr})$. The boundary of this region (the transparency edge) travels into the bulk of the medium at a constant velocity I_0 (in dimensionless variables). Since the coordinate is normalized to $c_L\tau_L$, while time is normalized

to τ_L, the velocity is normalized to the sound speed c_L. Consequently, physically I_0 is the Mach number of the transparency front (the velocity of the edge $v_e = I_0 c_L$). Thus, for $I_0 > 1$ a supersonic regime of the transparency is realized ($v_e > c_L$), while a subsonic regime ($v_e < c_L$) is realized for $I_0 < 1$.

The deformation sound generation mechanism is dominant in semiconductors in the regime $m_{D0} \ll 1$ for $\tau_L \ll \tau_R$. According to Eqs. (2.29) and (2.36), the acoustic wave source is the nonstationary field of carrier concentration $n(t,z)$ ($\partial n/\partial t \neq 0$). Consistent with the description of EH plasma dynamics (5.58), (5.59) the sound sources have a constant amplitude during laser irradiation. For $I_0 < A^{-1}$ their spatial distribution remains fixed, while for $I_0 > A^{-1}$ and $t > t_{\rm cr}$ the sources travel into the bulk of the absorbing medium.

It is assumed in analyzing thermoelastic sound generation during light-induced transparency[35,36] that the laser energy expended in bond breaking thermalizes instantaneously, although the bonds themselves are not restored. Equation system (5.56) is supplemented by a description of liquid heating:

$$\rho_0 c_p \frac{\partial T}{\partial t} = \alpha(n) I = \hbar\omega_L \frac{\partial n}{\partial t}.$$

This relation demonstrates that thermoelastic sound sources in the dye are linearly related to concentration-deformation sources in a semiconductor [see Eqs. (5.58) and (5.59)]. Hence the results obtained below for a semiconductor are also valid for dyes accurate to the redesignations.

We use Eq. (2.36) to calculate the profiles of acoustic waves excited at a rigid boundary. We introduce the dimensionless sound velocity of an acoustic wave $v = v/v_0$ where $v_0 = -dn_{\rm cr}/2\rho_0 c_L$ in a semiconductor and $v_0 = K\beta\hbar\omega_L n_{\rm cr}/2\rho_0^2 c_L c_p$ in a liquid.

It is advisable to calculate separately the profiles of waves excited by fixed and moving sources. A calculation of the signal v_s from fixed sources yields the following result:

$$v_s = I_0(1 - \exp(-At_s))\exp(A\tau), \quad \tau \leqslant 0,$$

$$v_s = I_0(2 - \exp[A(\tau - t_s)] - \exp(-A\tau)), \quad 0 \leqslant \tau \leqslant t_s, \tag{5.60}$$

$$v_s = I_0(1 - \exp(1 - At_s)\exp[-A(\tau - t_s)], \quad \tau > t_s.$$

The characteristic time t_s in relations (5.60) represents the operating time of the fixed sources, i.e., it is identical to the laser pulse duration $t_s = 1$, if transparency is absent ($I_0 \leqslant A^{-1}$) and is equal to the onset time of transparency $t_s = t_{\rm cr}$ if transparency has occurred ($I_0 \geqslant A^{-1}$). Analysis of Eq. (5.60) reveals that an acoustic pulse excited by fixed sources has a characteristic bell shape and is symmetrical relative to time $\tau = t_s/2$. Typical profiles of pulses excited by fixed sources for the case $A = 1$ and different intensity levels are shown in Fig. 5.17(a).

The analytic description of profiles of acoustic pulses excited by moving sources is differentiated by subsonic ($I_0 < 1$) and supersonic ($I_0 > 1$) velocities. In the case $A^{-1} < 1$ the onset of transparency is accompanied by subsonic velocities of the transparency front. In this case ($A^{-1} < I_0 < 1$) profiles v_M of acoustic waves excited by a moving front over characteristic time intervals

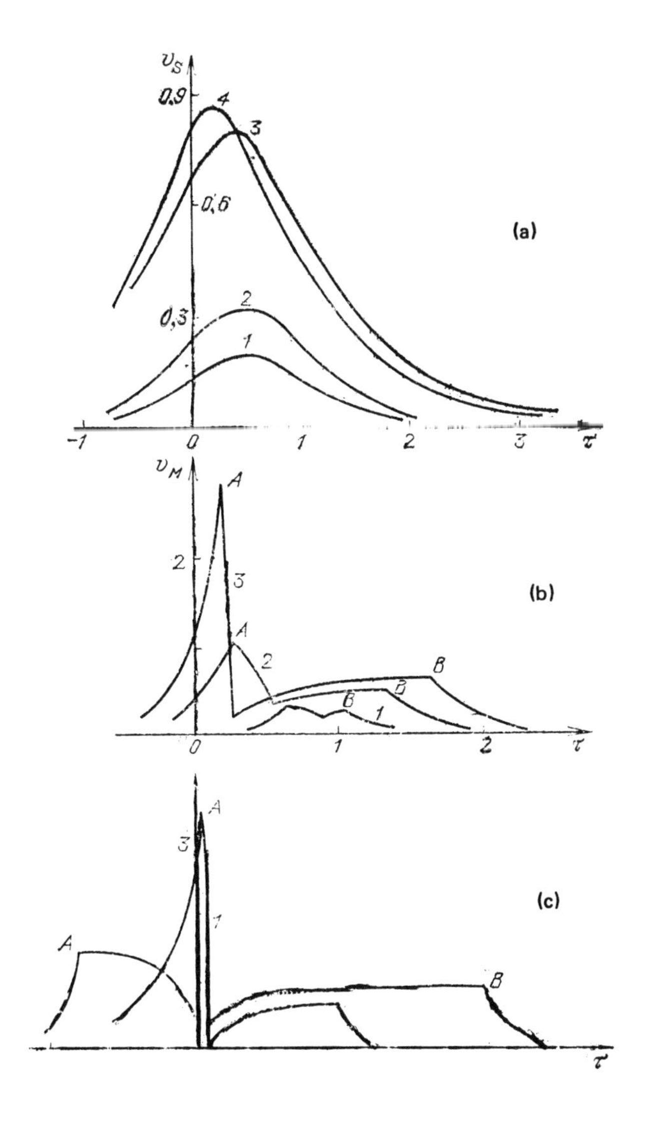

FIG. 5.17. Profiles of acoustic pulses excited in different regimes of source motion. (a) Fixed sources: $A = 1, I_0 = 0.2$ (1), 0.4 (2), 1.2 (3), 2 (4); (b) subsonic sources: $A = 5, I_0 = 0.3$ (1), 0.6 (2), 0.9 (3); (c) supersonic sources: $A = 5, I_0 = 1.1$ (1), 3 (2).

1) $\tau \leqslant t_{cr}$,
2) $t_{cr} < \tau \leqslant 1 - I_0(1 - t_{cr})$,
3) $1 - I_0(1 - t_{cr}) < \tau \leqslant 1 + I_0(1 - t_{cr})$,
4) $\tau > 1 + I_0(1 - t_{cr})$

are described by the relations

$$
\begin{aligned}
&1)\ v_M = (I_0/(1-I_0))\{1-\exp[-A(1-I_0)(1-t_{\mathrm{cr}})]\}\exp[A(\tau-t_{\mathrm{cr}})],\\
&2)\ v_M = I_0\{(1-I_0)^{-1}[1-\exp][A(\tau-t_{\mathrm{cr}})-A(1-I_0)(1-t_{\mathrm{cr}})]]\\
&\qquad +(1+I_0)^{-1}[1-\exp[-A(\tau-t_{\mathrm{cr}})]]\},\\
&3)\ v_M = (I_0/(1+I_0))\{1-\exp[-A(\tau-t_{\mathrm{cr}})]\},\\
&4)\ v_M = (I_0/(1+I_0))\{1-\exp[-A(1+I_0)(1-t_{\mathrm{cr}})]\}\\
&\qquad \times\exp[-A(\tau-1-I_0(1-t_{\mathrm{cr}}))].
\end{aligned}
\tag{5.61}
$$

A physical analysis of the generation process reveals that a contribution to the acoustic signal proportional to $(1-I_0)^{-1}$ make the acoustic waves propagating from the generation site in the same direction as the traveling transparency edge (i.e., away from the boundary of the medium) (Sec. 1.3). Waves propagating against the direction of the sources and towards the detector after reflection off the boundary of the medium make a contribution to the acoustic signal proportional to $(1+I_0)^{-1}$. According to Eq. (5.61) only waves that have not been reflected by the boundary will be recorded for $\tau<t_{\mathrm{cr}}$, while for $\tau>1-I_0(1-t_{\mathrm{cr}})$ only reflected waves will be recorded. One feature of generation at subsonic velocities of the transparency front is that both "direct" and reflected waves reach the detector at specific times $[t_{\mathrm{cr}}<\tau<1-I_0(1-t_{\mathrm{cr}})]$.

Typical profiles calculated by Eq. (5.61) are shown in Fig. 5.17(b). The evident asymmetry of these profiles can easily be attributed to the fact that the acoustic pulses propagating in the same direction as the sources are more efficiently excited than waves propagating opposite this direction (Sec. 1.3). Equations (5.61) can be used to obtain a rough estimate of the amplitude ratio of the absolute and local maxima denoted by A and B in Fig. 5.17 (b):

$$
\frac{v_A}{v_B}=\frac{1+I_0}{1-I_0}\,\frac{1-\exp[-(1-I_0)(A-I_0^{-1})]}{1-\exp[-(1+I_0)(A-I_0^{-1})]}\,.
$$

As demonstrated by an analysis of solutions (5.61), the amplitude of a wave generated by moving sources is limited even when the sources travel at sound speed (i.e., in the synchronous sound generation regime):

$$
v_A(I_0=1)=I_0A(1-t_{\mathrm{cr}}).
$$

This limit on a synchronously excited wave is attributable to the finite time during which the sources are active.[39]

A supersonic transparency edge regime is realized for $I_0>1$. In the case $A>1$ a crossover from subsonic to supersonic velocities is realized with increasing intensity. In the case $A<1$ the sources begin to travel at supersonic velocities immediately after the initiation of transparency (since $I_0>1$ follows automatically from $I_0>A^{-1}$ for $A<1$). In both cases the analytic representation for acoustic pulses over characteristic time intervals

$$
\begin{aligned}
&1)\ \tau\leqslant 1-I_0(1-t_{\mathrm{cr}}),\\
&2)\ 1-I_0(1-t_{\mathrm{cr}})<\tau\leqslant t_{\mathrm{cr}},\\
&3)\ t_{\mathrm{cr}}<\tau\leqslant 1+I_0(1-t_{\mathrm{cr}}),\\
&4)\ \tau>1+I_0(1-t_{\mathrm{cr}})
\end{aligned}
$$

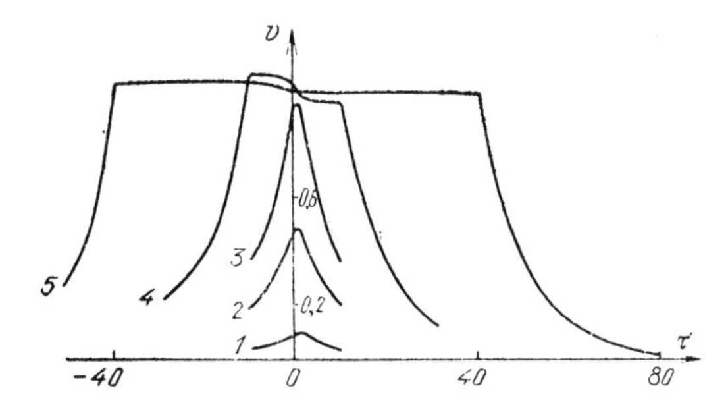

FIG. 5.18. Transformation of an OA-pulse profile with increasing laser radiation intensity I_0 equal to 1 (1), 5 (2), 10 (3), 20 (4), 50 (5). $A = 0.1$.

takes the form

$$\begin{aligned}
&1)\ v_M = (I_0/(I_0 - 1))\{1 - \exp[-A(I_0 - 1)(1 - t_{cr})]\} \\
&\qquad \times \exp[A(\tau - 1 + I_0(1 - t_{cr}))], \\
&2)\ v_M = (I_0/(I_0 - 1))\{1 - \exp[A(\tau - t_{cr})]\}, \\
&3)\ v_M = (I_0/(I_0 + 1))\{1 - \exp[-A(\tau - t_{cr})]\}, \\
&4)\ v_M = (I_0/(I_0 + 1))\{1 - \exp[-A(I_0 + 1)(1 - t_{cr})]\} \\
&\qquad \times \exp[-A(\tau - 1 - I_0(1 - t_{cr}))].
\end{aligned} \tag{5.62}$$

Unlike the subsonic regime, at supersonic velocities of these sources, "direct" and reflected waves will always arrive at the detector at different points in time according to solution (5.62): the direct waves arrive for $\tau < t_{cr}$; boundary-reflected waves arrive for $\tau > t_{cr}$ while no signal is present for $\tau = t_{cr}$. Figure 5.17(c) provides profiles of acoustic waves excited by sources traveling at supersonic velocities.

Analysis of Eq. (5.62) demonstrates that the arrival time of the peak of a pulse excited by moving sources for $I_0 > 1$ diminishes linearly with increasing intensity:

$$\tau_A = 1 - I_0(1 - t_{cr}) = 1 + A^{-1} - I_0. \tag{5.63}$$

The following representation is valid for the pulse amplitude:

$$v_A = (I_0/I_0 - 1))\{1 - \exp[-A(I_0 - 1)(1 - t_{cr})]\},$$

whereby the amplitude of the excited pulses goes to saturation with increasing intensity: $v_A \to 1$ for $I_0 \gg \max\{1,A\}$. The saturation is due to the fact that the increasing intensity will reduce sound generation efficiency [in proportion to $(I_0 - 1)^{-1}$] due to the linearly increasing shift away from phase matching.

Figure 5.18 shows the resulting acoustic pulse profiles ($v = v_S + v_M$) calculated by means of Eqs. (5.60)–(5.62). The analytic results obtained here on the whole confirm the conclusions of Ref. 36 that absorption saturation may broaden the excited acoustic pulses and reduce the efficiency of sound generation. At the same time, at transonic edge velocities, there may be a significant increase in the acoustic wave excitation efficiency. It should be noted that according to the results obtained here (and as clearly indicated from the profiles) induced transparency of the me-

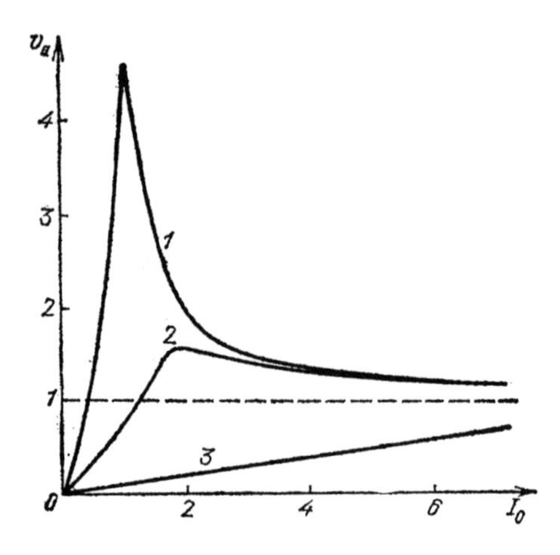

FIG. 5.19. Acoustic pulse amplitude plotted as a function of laser radiation intensity for different values of the parameter A: 5 (1), 1 (2), 0.1 (3).

dium will always generate asymmetry of the excited wave profiles. The shift of the pulse peak arising from induced transparency will cause steepening of the leading edge of the pulse and will broaden the trailing edge. Consequently, correct accounting of wave distortions resulting from absorption saturation may be quite important in a proper estimation of nonlinear acoustic effects that may also result in a similar type of transformation of pulse profiles.[35]

The analytic representations of the acoustic pulse profile obtained here permit calculation of the pulse peak position and pulse amplitude as a function of the radiation intensity I_0 and parameter A.[37] These calculations have demonstrated that the pulse peak is delayed to some degree compared to the case of fixed sources immediately following the onset of induced transparency. However, the pulse peak begins to appear earlier than the peak from fixed sources upon further increase in intensity. In the range $\tau < 0$ the peak of the resulting signal is always determined by the arrival time of pulse peak (5.63) excited by moving sources.

Figure 5.19 shows the acoustic pulse amplitude plotted as a function of laser radiation intensity for different values of the parameter A. This relation confirms the increase in wave amplitude near the synchronous generation regime noted above ($I_0 \approx 1$) as well as a constant amplitude value at high intensities. We note that the intensity shift that generates the highest amplitude pulse relative to $I_0 = 1$ (Fig. 5.19, curve 2) is related to the fact that the change in I_0 not only alters the velocity of the sources but also alters their time of motion.

These calculations have demonstrated that the appearance of moving acoustic wave sources from induced transparency of a medium will lead to asymmetric transformation of the excited acoustic pulses. We estimate at what intensities absorption saturation in water may lead to a profile transformation similar to the version attributed to a nonlinear acoustic effect by the authors of Ref. 35. The asymmetry of the pulse reported by the authors of Ref. 35 in glycerin can qualitatively be characterized by the fact that the leading edge of the pulse is approximately one-half the duration of the trailing edge. For water experiments at

$\tau_L \approx 80$ ps the estimate yields $A = 0.156$ (i.e., the acoustic pulse duration τ_a is determined by the transit time of sound across the absorption region $\tau_a \approx A^{-1}$). With this value of the parameter A, the acoustic pulse is transformed in the manner shown in Fig. 5.18 for $A = 0.1$ with increasing intensity.

In order for the leading edge to be one-half the size of the trailing edge, the pulse peak need only be shifted by a time of the order of the pulse duration, i.e., $\tau_{\text{peak}}(I_0) - 1/2 \approx -A^{-1}$, where $\frac{1}{2}$ is the peak position of a pulse produced by fixed sources. Hence

$$\tau_{\text{peak}}(I_0) \approx 1/2 - A^{-1}. \tag{5.64}$$

In the case $A \ll 1$ the peak will appear in the range $\tau < 0$ according to Eq. (5.64) where Eq. (5.63) will always hold for the position of the peak. A joint solution of Eqs. (5.64) and (5.63) yields $I_0 \approx 2A^{-1}$. However, $I_0 = A^{-1}$ is the critical intensity for the onset of induced transparency.

Consequently, such a significant profile transformation is possible already for intensities only twice higher than the transparency threshold ($I_{\text{cr}} \approx 3$ GW/cm^2 in H_2O). This is due to the fact that the sources travel at velocities far exceeding the sound speed during induced transparency in the case $A \ll 1$ (indeed, in the case $A = 0.156$ the Mach number of the transparency front I_0 will always exceed $A^{-1} = 6.4$). It may be important to account for the predicted distortions of the acoustic pulse profile in order to properly assess nonlinear effects.

The nonlinearities of the optoacoustic effect manifested in the light absorption region were examined in this chapter. An acoustic nonlinearity also appears on the OA-signal propagation path. These processes will be analyzed in Chap. 6.

REFERENCES

1. L. V. Burmistrova, A. A. Karabutov *et al.*, Akust. Zh. **25**, 616 (1979) [Sov. Phys. Acoust. **25**, 348 (1979)].
2. T. A. Dunina, S. V. Egerev *et al.*, Akust. Zh. **25**, 622 (1979) [Sov. Phys. Acoust. **25**, 353 (1979)].
3. V. M. Gordienko, A. A. Karabutov, and S. B. Nemirovskiy, *Tez. dokl. V Vsesoyuzn. sobeshch. po nerezonansnomu vzaimodeystviyu opticheskogo izluchniya s veshchestvom* [*Conference Proceedings of the Fifth All-Union Conference on Nonresonant Interaction of Optical Radiation with Matter*] (Nauka, Leningrad, 1981), p. 367.
4. M. W. Sigrist, J. Appl. Phys. **60**, R83 (1986).
5. M. I. Gallant and H. M. Driel, Phys. Rev. B **26**, 2133 (1982).
6. M. G. Grimaldi, P. Baeri and E. Rimini, Appl. Phys. A, **107** (1984).
7. S. M. Avanesyan and V. E. Gusev, Kvantovaya Electron. **13**, 1241 (1986) [Sov. J. Quantum Elektron. **16**, 812 (1986)].
8. W. B. Gauster and D. H. Habing, Phys. Rev. Lett. **18**, 1058 (1967).
9. S. M. Avanesyan, V. E. Gusev, and N. I. Zheludev, Appl. Phys. A **40**, 163 (1986).
10. S. M. Avanesyan and V. E. Gusev, Izv. Akad. Nauk. SSSR Ser. Fiz. **51**, 248 (1987) [Bulletin of the Academy of Sciences of the USSR. Physical Series, **51**, 37 (1987)].
11. S. M. Avanesyan, V. E. Gusev *et al.*, Pis'ma Zh. Tekh. Fiz. **12**, 1067 (1986) [Sov. Tech. Phys. Lett. **12**, 442 (1986)].
12. S. M. Avanesyan, V. E. Gusev *et al.*, Akust Zh. **32**, 562 (1986) [Sov. Phys. Acoust. **32**, 356 (1986)].
13. V. E. Gusev, Akust. Zh. **32**, 778 (1986) [Sov. Phys. Acoust. **32**, 486 (1986)].
14. V. E. Gusev, Vestn. Moscow State Univ. Phys. Astron. Ser. **30**, 38 (1989) [Moscow University Physics Bulletin (1989)].

15. G. Mahler and A. Fourikis, J. Lumin. **30**, 18 (1985).
16. V. F. Gantmakher and I. B. Levinson, *Rasseyanie nositely toka v metallakh i poluprovodnikakh* [*Charge Carrier Scattering in Metals and Semiconductors*] (Nauka, Moscow, 1984).
17. V. E. Gusev, Fiz. Tverd. Tela **29**, 2316 (1987) [Sov. Phys. Solid State **29**, 1335 (1987)].
18. T. Rice, G. Hensel, T. Phillips and G. Thomas, *Elektronno-dyrochnaya zhidkost' v poluprovodnikakh* [*Electron-Hole Liquid in Semiconductors, edited by T. I. Galkina and B. G. Zhurkin* (Mir, Moscow, 1980).
19. M. Combescot, J. Bok, and J. Lumin. **30**, 1 (1985).
20. L. A. Almazov, V. K. Malyutenko, and L. L. Fedorenko, Uk. Fiz. Zh. **26**, 734 (1981).
21. S. Modesti, A. Frova *et al.*, J. Lumin. **31-32**, 503 (1984).
22. K. M. Romanek, H. Nanther *et al.*, J. Lumin. **24-25**, 585 (1981).
23. H. Schweizer, E. Zielinski *et al.*, J. Lumin. **31-32**, 503 (1984).
24. K. G. Lyon, G. L. Salinger *et al.* J. Appl. Phys. **48**, 865 (1977).
25. C. A. Swenson, J. Phys. Chem. Ref. Data **12**, 179 (1983).
26. T. F. Smith and G. K. White, J. Phys. C **8**, 2031 (1975).
27. A. A. Bondarenko, A. K. Vologdin, and A. I. Kondrat'ev, Akust. Zh. **26**, 828 (1980) [Sov. Phys. Acoust. **26**, 467 (1980)].
28. Y. Bae, J. J. Song, and Y. B. Kin, J. Appl. Phys. **53**, 615 (1982).
29. P. Horn, P. Braunlich, and A. Schmid, J. Opt. Soc. Am. **2**, 1095 (1985).
30. B. G. Gorshkov, L. M. Dorozhkin *et al.*, Zh. Eksp. Teor. Fiz. **88**, 21 (1985) [Sov. Phys. JETP **61**, 12 (1985)].
31. A. A. Betin, O. V. Mitropol'skiy, *et al.*, Kvantovaya Elektron. **12**, 1856 (1985) [Sov. J. Quantum Electron. **15**, 1227 (1985)].
32. V. E. Gusev and B. V. Zhdanov, Izv. Akad. Nauk. SSSR Ser. Fis. **53**, 1157 (1989) [Bulletin of the Academy of Sciences of the USSR. Physical Series, **53**, 128 (1989)].
33. V. E. Gusev, B. V. Zhdanov, and B. G. Shakirov, Akust. zh. **34**, 463 (1988) [Sov. Phys. Acoust. **34**, 269 (1988)].
34. F. Keilmann, IEEE J. Quantum Electron. **12**, 592 (1976).
35. K. L. Vodop'yanov, L. A. Kulevskiy *et al.*, Zh. Eksp. Teor. Fiz. **91**, 114 (1986) [Sov. Phys. JETP **64**, 67 (1986)].
36. G. S. Bushanam and F. S. Barnes, J. Appl. Phys. **46**, 2074 (1975).
37. V. E. Gusev and N. M. Tyutin, Akust. Zh. **34**, 1034 (1988) [Sov. Phys. Acoust. **34**, 593 (1988)].
38. J. Shah, R. F. Leheny, and C. Lin, Solid State Commun., **18**, 1035 (1976).
39. F. V. Bunkin, A. I. Malyarovskiy *et al.*, Kvantovaya Elektron. **5**, 457 (1978) [Sov. J. Quantum Electron. **8**, 270 (1978)].

Chapter 6

Quasisynchronous Sound Excitation Under Acoustic Nonlinearity Conditions

Laser excitation of sound may be initiated by sources moving at transonic velocities. Quasisynchronous excitation of acoustic waves can be initiated by, for example, laser beam scanning (Secs. 1.3 and 3.3), mixing of counterpropagating light waves (Sec. 3.4), ballistic phonon heat conduction (Sec. 3.6), fast expansion of a photo-generated electron-hole plasma (Sec. 4.2), and transonic propagation of the induced-transparency edge of the medium (Sec. 5.4). Long-term travel of the sources at velocity $V(t)$ near the sound speed c_L will lead to efficient amplification of the comoving acoustic wave whose amplitude can grow without limit over time in a linear approximation in the case of exact phase matching: $V(t) = c_L$. The effect of an acoustic nonlinearity on quasisynchronous sound generation is investigated in this chapter.

6.1. Nonlinear Limiting of the Efficiency of a Moving Optoacoustic Antenna

Interest in research on acoustic wave emission by sources traveling at transonic velocities has derived from the capability to employ lasers to generate sound sources traveling at an arbitrary velocity along a predetermined trajectory.[1] Distributed thermal sources traveling at transonic velocities can, specifically, be implemented from displacement of light foci[2] and by absorption saturation of dye molecules (Sec. 5.4). However, the development of controlled broadband sources employing light-beam scanning over the surface of an absorbing medium at a velocity near that of the eigenwaves (by a transonic material flow around the heating zone) has attracted the most interest.[3–8] A substantial number of theoretical[6–13] and experimental[14–17] studies have been devoted to analyzing the properties of such an optoacoustic radiator for the case where the light energy converted into sound is quite small in magnitude and linear waves are excited. We note that such research is of interest for analyzing the thermal self-action of light pulses in gas flows.[4,18,19]

Generation of acoustic waves of finite amplitude by distributed moving thermal sources has been an area of special interest.[20–27] This is primarily due to the fact that the quasisynchronous sound generation regimes in which an acoustic nonlinearity is possible are the most promising for generating perturbations of the highest amplitude and shortest duration[14,15] and for action of the heating sources on high-speed bodies.[28]

The theory has beyond doubt been successful in describing the nonlinear limit on the acoustic wave amplitude from source motion at the sound speed.[20,21] The shift of the maximum of the resonant curve towards the supersonic range[20,21,24] (Fig. 1.8) has been qualitatively confirmed by numerical calculations for strong shock waves.[23,25] The development of the theory of generation of acoustic waves of finite amplitude by distributed sources has directly employed asymptotic and qualitative methods of analyzing quasilinear partial differential equations.[28–32]

We carry out a detailed analysis of the effect of an acoustic nonlinearity on quasisynchronous thermoelastic sound generation by a moving light beam. In order to describe the plane longitudinal nonlinear acoustic waves in liquids or gases in the absence of viscosity and heat conduction we employ continuity equation (2.1) and equation of motion (2.2) in Euler form:

$$\frac{\partial \rho}{\partial t} + \frac{\partial(\rho v)}{\partial x} = 0,$$

$$\rho\left(\frac{\partial v}{\partial t} + v\frac{\partial v}{\partial x}\right) = -\frac{\partial p'}{\partial x}. \tag{6.1}$$

In analyzing weakly nonlinear acoustic disturbances it is necessary, in addition to thermoelastic stresses (I.5), to account for the quadratic-nonlinear term in the state equation[33,21]

$$p' = c_0^2\rho' + (c_0^2/\rho_0)(\varepsilon - 1)\rho'^2 + c_0^2\rho_0\beta T'. \tag{6.2}$$

The rise in temperature T' of the medium is caused by its heating by a thermal source (scanning laser beam) traveling at velocity V_0 (Sec. 1.3):

$$\frac{\partial T'}{\partial t} = \frac{\alpha I_0}{\rho_0 c_p} H\left(\frac{x - V_0 t}{a}\right). \tag{6.3}$$

We recall that the function $H(x/a)$ describes the transverse distribution of light intensity within the optical radiation layer (Fig. 1.5).

According to linear theory (Sec. 1.3), when the sources travel at a transonic velocity ($|V_0 - c_0|/c_0 \ll 1$) the acoustic wave comoving the energy delivery region is amplified more effectively. By modifying the slowly varying profile method (the Khokhlov method[29]), we assume that the profile of this wave is slowly distorted due to source action and the nonlinear acoustic effects in the coordinate system traveling at velocity V_0.[21,28] In fact, transformation of the excited acoustic pulse profile is assumed to be insignificant over a period of the order of its duration. Assuming $\rho' = \mu\rho'(\mu t, \xi = x - V_0 t)$, $v = \mu v(\mu t, \xi = x - V_0 t)$, $T' = \mu^2 T(\mu t, \xi = x - V_0 t)$, in Eqs. (6.1) and (6.3), we only conserve terms up through second order in the small parameter $\mu \ll 1$ ($\mu \sim |\rho'|/\rho_0 \sim |v|/c_0$). Using equation of state (6.2) and collecting terms of first order of smallness ($\sim\mu$) on the left-hand sides of these equations and those of second order of smallness ($\sim\mu^2$) on the right-hand sides, we find

$$\rho_0 \frac{\partial v}{\partial \xi} - V_0 \frac{\partial \rho'}{\partial \xi} = -\frac{\partial(\rho' v)}{\partial \xi} - \frac{\partial \rho'}{\partial t}, \tag{6.4}$$

$$-\rho_0 V_0 \frac{\partial v}{\partial \xi} + V_0^2 \frac{\partial \rho'}{\partial \xi} = V_0 \rho' \frac{\partial v}{\partial \xi} - \rho_0 v \frac{\partial v}{\partial \xi} - 2c_0^2(\varepsilon - 1)\rho_0^{-1}\rho' \frac{\partial \rho'}{\partial \xi} + (V_0^2 - c_0^2)\partial\rho'/\partial\xi + (c_0^2 \beta \alpha I_0 / V_0 c_p) H(\xi). \tag{6.5}$$

The term $(V_0^2 - c_0^2)\partial\rho'/\partial\xi$ on the right-hand side of Eq. (6.5) has an order of smallness μ^2 at transonic velocities of the laser beam ($|V_0 - c_0|/c_0 \sim \mu$).

When Eq. (6.4) is multiplied by V_0 and summed with Eq. (6.5), terms of first order of smallness cancel. The relation of the vibrational velocity to the density increments in a linear propagating acoustic wave $v/c_0 = \rho'/\rho_0$ can be used with sufficient accuracy in the remaining terms of the equation. Explicitly accounting for the fact that the generation regime is only slightly asynchronous (we set $V_0/c_0 = 1 + \mu\Delta_0$, $\Delta_0 = (V_0 - c_0)/c_0$ is the relative velocity mismatch), we finally obtain

$$\frac{\partial v}{\partial t} - c_0 \Delta_0 \frac{\partial v}{\partial \xi} + \varepsilon v \frac{\partial v}{\partial \xi} = \left(\frac{c_0 \beta \alpha I_0}{2\rho_0 c_p}\right) H(\xi). \tag{6.6}$$

Equation (6.6) is called an inhomogeneous equation of simple waves with velocity mismatch.[21,24] It is the simplest equation that provides an adequate description of quasisynchronous generation of weakly nonlinear waves in dispersionless media.[34,35] It is clearly evident that the solution of linearized equation (6.6) in the case of precise velocity matching ($\Delta_0 = 0$) describes an unlimited growth of the acoustic wave proportional to time. Consequently, it is fundamentally necessary to account for the nonlinearity with small mismatches, while the quasisynchronous sound excitation regime is essentially nonlinear.

It is convenient for the subsequent analysis to reduce Eq. (6.6) to dimensionless form. After setting

$$\xi = \xi/a, \ v = v/v_0, \ t = t/t_0, \ t_0 = a/\varepsilon v_0,$$
$$v_0 = c_0(\alpha a N/2\varepsilon)^{1/2}, \ N = \beta I_0/\rho_0 c_0 c_p, \tag{6.7}$$

we obtain

$$\frac{\partial v}{\partial t} - \Delta \frac{\partial v}{\partial \xi} + v \frac{\partial v}{\partial \xi} = H(\xi). \tag{6.8}$$

The characteristic scales classified in accordance with (6.7) have a clear physical meaning: t_0 is the time in which a discontinuity is formed in a freely propagating pulse of amplitude v_0 and length a; the scale v_0 for the vibrational velocity defines the maximum possible variation of v over time t_0 in the linear problem. Scale v_0 is related to two dimensionless combinations: the quantity αa characterizing light absorption over a length of the order of the beamwidth, and the number N, dependent on the external source intensity and defining the acoustic Mach number ($M_a \equiv |v|/c_0$) of the excited wave. The characteristic wave amplitude has a nonlinear dependence on the laser radiation intensity: $v_0 \sim I_0^{1/2}$.

As follows from Eq. (6.8), the process is characterized by the single parameter Δ equal to the ratio of the velocity mismatch, detuning of the source and sound $V_0 - c_0$ to the nonlinear variation in the sound speed εv_0:

$$\Delta = (V_0 - c_0)t_0/a = (V_0 - c_0)/\varepsilon v_0.$$

The parameter Δ represents the relative effect of the asynchronicity of the sources and the nonlinearity of the acoustic wave. If we have a small dimensionless mismatch Δ ($|\Delta| \lesssim 1$) the influence of nonlinear acoustic effects will be significant. The values $\Delta < 0$ correspond to subsonic source velocities, while $\Delta > 0$ correspond to supersonic velocities.

Equation (6.8) is a homogeneous quasilinear first-order partial differential equation. We utilize the simple initial condition $v(t=0, \xi) = 0$ to investigate the problem of sound generation. It is particularly convenient to represent Eq. (6.8) as

$$\frac{\partial w}{\partial t} + w\frac{\partial w}{\partial \xi} = H(\xi), \quad w = v - \Delta. \tag{6.9}$$

Then the parameter Δ will only enter into the initial conditions:

$$w(t=0,\xi) = -\Delta. \tag{6.10}$$

Problems (6.9) and (6.10) can be solved by quadratures for any form of the relation $H(\xi)$ describing the spatial distribution of the thermal sources. Indeed, we write for this equation the characteristic relations[36]

$$dt = \frac{d\xi}{w} = \frac{dw}{H(\xi)}. \tag{6.11}$$

The last of these equalities defines the integral

$$w^2/2 + \mathscr{H}(\xi) = C_1, \tag{6.12}$$

where $-\mathscr{H}(\xi)$ is the antiderivative of the function $H(\xi)$($\partial\mathscr{H}(\xi)/\partial\xi = -H(\xi)$). Using Eq. (6.12) and the first relation from (6.11), we find one additional integral of the system:

$$t = \pm \int^{\xi} \{2[C_1 - \mathscr{H}(\xi')]\}^{-1/2}\, d\xi' + C_2. \tag{6.13}$$

Initial condition (6.10) defines the relation between the constants C_1 and C_2:

$$0 = \pm \int^{\mathscr{H}^{-1}(C_1 - \Delta^2/2)} \{2[C_1 - \mathscr{H}(\xi')]\}^{-1/2}\, d\xi' + C_2,$$

which upon substitution of Eq. (6.12), (6.13) makes it possible to represent the solution of problem (6.9), (6.10) in implicit form:

$$t = \pm \int_{\mathscr{H}^{-1}[\mathscr{H}(\xi) + w^2/2 - \Delta^2/2]}^{\xi} \left\{2\left[\mathscr{H}(\xi) - \mathscr{H}(\xi') + \frac{w^2}{2}\right]\right\}^{-1/2} d\xi'. \tag{6.14}$$

The possibility for obtaining an analytic solution is determined by whether or not it is possible to express quadrature (6.14) by known functions, i.e., the form of the distribution $\mathscr{H}(\xi)$. However, even if the process can be described analytically,[20,21,37–39] there are still some difficulties in analyzing the dependence of the solution on the parameters and identifying physically different excitation regimes.

The method of analyzing Eq. (6.9) by means of characteristic diagrams[40] utilizes relation (6.13) to describe the motion of the points of the initial disturbance on the (ξ, t) plane. Using a notation of the initial condition $w(t=0) = -\Delta$, $\xi(t=0) = \xi_0$ we obtain from integrals (6.12), (6.13) the expression

$$t = \pm \int_{\xi_0}^{\xi} \left\{ 2\left[\mathscr{H}(\xi_0) - \mathscr{H}(\xi') + \frac{\Delta^2}{2} \right] \right\}^{-1/2} d\xi', \tag{6.15}$$

describing the characteristic passing at initial time $(t=0)$ through $\xi = \xi_0$.

Using the language of characteristics it is convenient to write the formation condition of weak shock waves (discontinuities in the profile of acoustic disturbances). As in the case of homogeneous simple waves, the intersection condition of the characteristics takes the form $\partial\xi/\partial\xi_0 = 0$.[41] However, this relation can only be used effectively in the case of an analytic description of the characteristics, i.e., when there exist also a description of the solution of the overall problem using tabular functions [compare quadratures (6.14) and (6.15)]. The primary drawback of this method lies in the fact that the values of the disturbances are not conserved on the characteristics for a nonhomogeneous equation of simple waves[41] and hence the (ξ,t) plane does not contain all information required for describing the sound excitation process.

References 24, 34, and 42 propose analyzing the excitation of weakly nonlinear sound by distributed moving sources on the (ξ,w) plane (coordinate, velocity), i.e., on the phase plane.[43,44] In order to identify the physical meaning of this approach we write system of characteristic equations (6.11) as

$$\frac{dw}{dt} = -\frac{d\mathscr{H}}{d\xi}, \tag{6.16}$$

$$\frac{d\xi}{dt} = w. \tag{6.17}$$

Equation system (6.16), (6.17) is autonomous, which makes possible its effective analysis on the phase plane (ξ,w).[43,44] The relation describing phase trajectories is obtained by deleting time from the problem (6.16), (6.17) and is identical to the second equation in (6.11). Therefore, the phase portrait is described by integral (6.12).

Physically, $\mathscr{H}(\xi)$ is a potential function[43] proportional to the potential energy of the liquid element at the given point in space; the function $w^2/2$ is related to kinetic energy; the parameter C_1 characterizes total energy. The representative points travel along equienergy lines on the phase plane defined by law of conservation (6.12) at a velocity whose projection onto the ξ axis is equal to w according to Eq. (6.17). Here, the phase plane (ξ,w) contains all information required to describe the displacement of the representative point (the liquid element). Phase portrait (6.12) can be plotted for any form of the function $\mathscr{H}(\xi)$ represented analytically or graphically.[38]

It is possible to describe the motion of a distributed physical system (wave profile evolution) on the phase plane as follows. An energy distribution

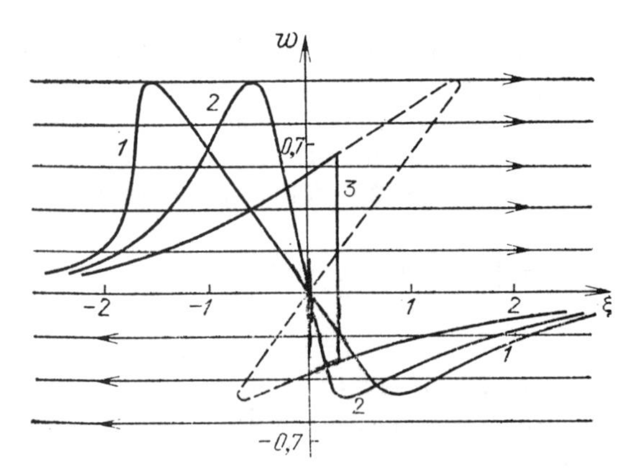

FIG. 6.1. Analysis of the evolution of a freely propagating acoustic pulse on a phase plane.

$C_1(\xi) = w_0^2 \ (\xi)/2 + \mathscr{H}(\xi)$ is assigned to the initial disturbance profile $w(t = 0,\xi) \equiv w_0(\xi)$ i.e., to each point on the initial profile corresponds a parameter $C_1(\xi)$ of equienergy line (6.12) passing through it. Hence, each of the liquid elements (each representative point) travels along its own equienergy line. The instantaneous state of the distributed physical system under test is determined by the entire collection of representative points (the representative profile).

When ambiguity arises in the profile of the wave, the leading front follows the "equivalent dissected area rule"[33,41] which accounts for the law of conservation of momentum from energy absorption by the discontinuity. This method of formulating possible wave forms of a disturbance is characteristic of hyperbolic waves.[41] At the same time there are many physical problems that also permit a phase plane analysis where no such constraint on the nature of collective motion of the elements within the distributed system exists.[45–47]

Before proceeding directly to analyzing quasisynchronous sound excitation, we consider on a phase plane the evolution of the initial acoustic disturbance in the absence of external sources $H(\xi) = 0$, $\Delta = 0$, $w = v$. Then according to Eq. (6.12) the horizontal lines $w = \pm C_1$ represent the phase trajectories. The phase portrait of the system is shown in Fig. 6.1. The arrows indicate the direction of the representative points motion along the trajectories. According to Eq. (6.17) the regions $w = v > 0$ of the acoustic wave travel at a higher velocity than the sound speed, while the regions $w = v < 0$ travel more slowly than acoustic waves of infinitely small amplitude.[33]

Figure 6.1 also shows profiles 2 and 3 of an acoustic pulse propagating in the positive direction of the z axis plotted at successive times $t = 1$, 3 by displacement of the points of initial profile 1 $[w(\xi,t = 0) = w_0(\xi)]$ along the phase trajectories as $\xi = \xi_0 + w_0(\xi_0)t$, $w = w_0(\xi_0)$. In the absence of external sources, the velocity of representative points is independent of time.

The appearance of external sources at transonic velocities will cause the horizontal phase trajectories to bend. In the case of a bell distribution of light intensity in the laser beam cross section $H(\xi)$ the potential function $\mathscr{H}(\xi) = \int_\xi^\infty H(\xi')d\xi'$ takes the form of a smoothed step from which liquid would flow in the direction of

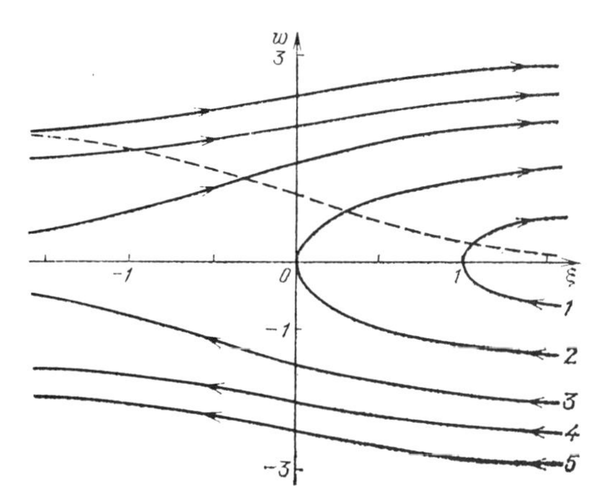

FIG. 6.2. Phase portrait of quasisynchronous excitation of acoustic pulses for a step-type potential function (dashed curve).

increasing ξ (Fig. 6.2, dashed curve). We note that the potential function in Fig. 6.2 is similar to the distribution of the temperature increment T' in the accompanying coordinate system. The spatial inhomogeneity of the potential energy of the liquid elements from activation of the external sources drives the elements in the direction of diminishing potential. Their vibrational velocity is altered in accordance with the law of conservation of energy (6.12) and the acoustic wave is thus excited and amplified.

The phase portrait of the system for this potential is shown in Fig. 6.2. The arrows indicate direction of point motion along the phase trajectories. Trajectories corresponding to parameter values $0 < C_1 < \mathscr{H}(-\infty)$ have a turning point $\xi = \mathscr{H}^{-1}(C_1)$ (for example, curves 1 and 2). The representative points in this region are "reflected" by the heating layer; their "kinetic energy" $w^2/2$ is less than the potential barrier. The representative points in the range $C_1 \geqslant \mathscr{H}(-\infty)$ do not change their direction over time (curves 3–5). It is evident that the initial profile (6.10), which is a horizontal line on the plane (ξ, w) is unstable, since it is intersected by phase trajectories. The acoustic wave is excited in the system.

Since equienergy lines (6.12) correspond to stationary solutions of Eq. (6.9) (for $\partial w/\partial t = 0$) the wave profile that is established at $t \to \infty$ $[w(t = \infty, \xi) \equiv w_\infty]$ will either match one of the phase trajectories or will consist of parts of different phase trajectories and contain a discontinuity. Hence, the phase portrait of the system makes possible a rather easy determination of the wave form of the stationary waves for different values of the parameter Δ. We represent the initial profiles $w = -\Delta$ on the phase portrait for different dimensionless mismatches (Fig. 6.3).

In a subsonic source regime ($\Delta < 0$, line 1 in Fig. 6.3), the stationary vibrational velocity distribution established as $t \to \infty$ coincides with the trajectory of point $A(-\infty, -\Delta)$:

$$w_\infty = \{\Delta^2 + 2[\mathscr{H}(-\infty) - \mathscr{H}(\xi)]\}^{1/2}.$$

The acoustic wave amplitude $v_a = v_{\max} - v_{\min} = w_{\max} - w_{\min}$ in this case

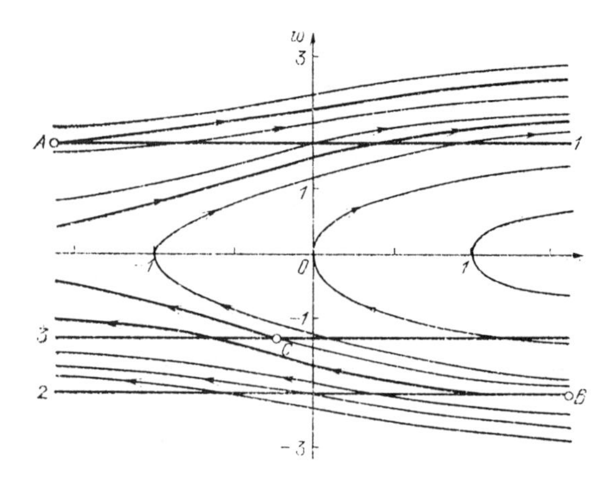

FIG. 6.3. Formation of the stationary profile of an acoustic wave in subsonic (1), supersonic (2), and mixed (3) regimes.

$$v_a = \Delta + (\Delta^2 + \Delta_{cr}^2)^{1/2}, \tag{6.18}$$

where we use the designation $\Delta_{cr} = [2\mathscr{H}(-\infty)]^{1/2}$.

The regime where sources travel at velocity $\Delta > \Delta_{cr}$ will be called the totally supersonic regime. In this case the "kinetic energy" of the elements of the initial profile $w_0^2/2 = \Delta^2/2$ exceeds the potential barrier $\mathscr{H}(-\infty)$. The initial profile $w_0 = -\Delta$ lies in the region of phase trajectories with no turning points (curve 2 in Fig. 6.3). The trajectory of the point $B(+\infty, -\Delta)$ is a stationary profile:

$$w_\infty = -[\Delta^2 - 2\mathscr{H}(\xi)]^{1/2}, \; v_a = \Delta - (\Delta^2 - \Delta_{cr}^2)^{1/2}. \tag{6.19}$$

In the intermediate case $0 < \Delta < \Delta_{cr}$ (line 3 in Fig. 6.3) some of the initial profile points are reflected by the heating layer. The steady-state velocity distribution is identical to the trajectory of point C after its turn at an infinite point $(-\infty, 0)$, i.e., to one of the separatrixes that bisects the phase plane regions where the trajectories have and do not have turning points:

$$w_\infty = \{2[\mathscr{H}(-\infty) - \mathscr{H}(\xi)]\}^{1/2}, \; v_a = \Delta + \Delta_{cr}. \tag{6.20}$$

The dependence of acoustic wave amplitude on velocity of the quasisynchronous sources as determined by (6.18)–(6.20) is plotted in Fig. 1.8. Accounting for the acoustic nonlinearity led to a shift of the maximum of the resonant curve towards the supersonic velocity range $(\Delta > 0)$. This is due to the fact that the compressional waves $(\rho'/\rho_0 = v/c_0 > 0)$ of finite amplitude propagate at a velocity exceeding the linear sound speed. Hence, when the source velocity only slightly exceeds the linear sound speed $(0 < \Delta \leqslant \Delta_{cr})$ their synchronous gain conditions are enhanced.

In accordance with this discussion, the development of an acoustic disturbance with time can be roughly described as follows on a phase plane: we represent the small interval Δt; using the wave form $w(t,\xi)$ known at time t as well as equality (6.17), we determine the displacement $d\xi$ of a sufficient number of profile points; implementing this displacement along trajectory (6.12), we plot the profile of the disturbance $w(t+dt,\xi)$, etc. In this case the representative points travel discretely along the exact point trajectories. Characteristic results of such a plot for a phase portrait similar to that shown in Fig. 6.1 are reported in Ref. 35. We also note that in the particular case $H(\xi)=1/ch^2\xi$ implicit solution (6.14) of problem (6.9), (6.10) can be expressed by elementary functions.[20,21]

However, another approach that utilizes a continuous displacement of the profile points along approximate phase trajectories is also possible for describing generation on a phase plane. In order to use this approach it is necessary to appropriately approximate the distribution function of the sources $H(\xi)$ (for example, a piecewise-linear approximation) so that Eq. (6.17) permits exact integration along the trajectory. The analytic results obtained from this approach essentially serve as estimates for an exact solution.

We consider a model of a square rectangular intensity distribution in the laser-beam cross section $H(\xi)=\theta(\xi+1)-\theta(\xi-1)$. This model corresponds to a potential function of the type

$$\mathscr{H}(\xi)=\begin{cases}2, & \xi<-1,\\ 1-\xi, & |\xi|<1,\\ 0, & \xi>1.\end{cases} \tag{6.21}$$

This simple approximation of the light beam shape form makes it possible to identify all primary characteristics of the process of acoustic wave excitation by broadband sources.

The phase portrait of Eq. (6.9) in the neighbourhood of the traveling elevated temperature front (6.21) is represented in Fig. 6.4. In the heating zone ($|\xi|<1$) the phase trajectories take the form of a parabola; the representative points travel along the ξ axis under constant acceleration and travel at a fixed, constant velocity along the w axis. Outside the heating zone the representative points travel rectilinearly and the disturbance profile is distorted in the same manner as in a simple wave (Fig. 6.1).

The following characteristic stages of the generation process can be identified. The disturbance begins to grow at the activation point of the sources ($t=0$), and at time $t=t_2$ the amplitude of the disturbance reaches its maximum value. The nonlinearity simultaneously magnifies the curvature of the profile, and a discontinuity forms at $t=t_1$. Subsequently, the discontinuity grows and reaches maximum wave amplitude at time $t=t_3$. The discontinuity accelerates over the interval $t_1<t<t_3$, and travels at a constant speed for $t>t_3$. The dynamics of this process are essentially dependent on the source motion regime (the parameter Δ).

The development of a disturbance in the subsonic scanning regime $\Delta<0$ is shown in Fig. 6.4(a). In this case all points of the initial profile overtake the light beam with time. The maximum wave amplitude is reached when the point $A(-1,\ -\Delta)$ lying along the left boundary of the heating zone exits this region. The time of point travel along this trajectory can be found by Eq. (6.15); the time to achieve maximum amplitude in the subsonic case is $t_2=\Delta+(\Delta^2+4)^{1/2}(\Delta\leqslant 0)$.

For $t<t_2$ the amplitude of the disturbance grows in proportion to time: $v_a=t$. The pulse amplitude then remains constant (for $t>t_2$). The maximum amplitude is

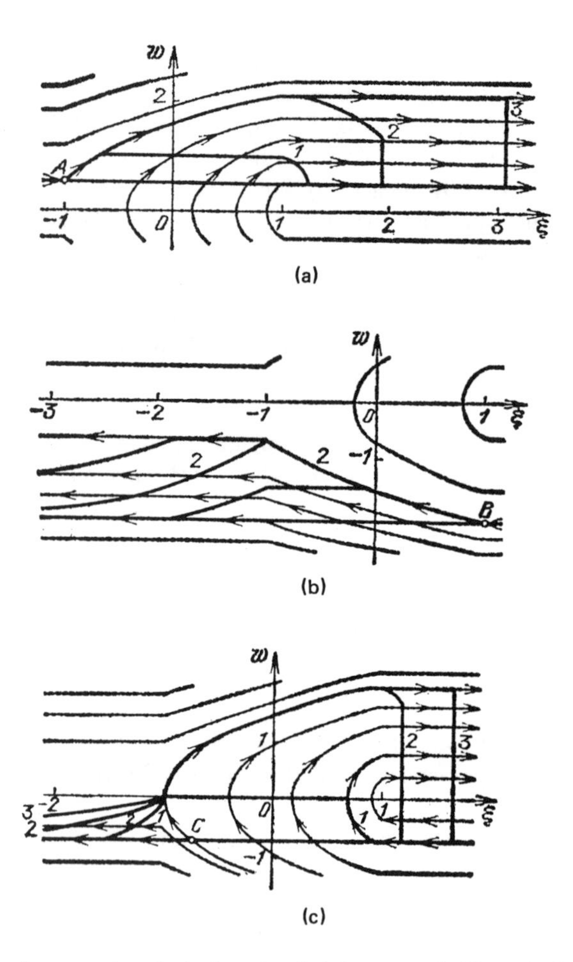

FIG. 6.4. Excitation of an acoustic pulse in the subsonic (a), supersonic (b), and mixed (c) regimes of the heating layer motion. The wave profiles are plotted at characteristic times t_1 (1), t_2 (2), and t_3 (3).

achieved increasingly rapidly (t_2 diminishes) with detuning from linear phase matching ($\Delta = 0$), while the amplitude itself diminishes and the discontinuity formation time grows.

The flow profile is parabolic outside the heating zone: $\xi = 1 + w(t - w)$ $(-\Delta \leqslant w \leqslant w_{\max})$. Therefore the discontinuity is initially formed on the line $w = -\Delta$ at time $t_1 = -2\Delta$. The stabilization time of the discontinuity is found by means of the integral relation

$$\int_{-\infty}^{\infty} v\, d\xi = \int_{-\infty}^{\infty} (w + \Delta) d\xi = \mathscr{H}(-\infty)t, \tag{6.22}$$

which follows from differential equation (6.8) in the case of sound generation $[v(t=0) = 0]$ assuming that the disturbance reaches infinitely distanced points in space in infinite time ($v(\xi = \pm\infty) = 0$).

According to Eq. (6.22) the momentum of an acoustic wave grows in proportion to the source action time and irradiation intensity regardless of their velocity regime or the acoustic nonlinearity (compare to the results of Sec. 1.3). The velocity of the heating zone has an effect on the acoustic disturbance wave form (its amplitude and duration), although not on its area on the phase plane (ξ,w). Using Eq. (6.22) for $\Delta \leqslant 0$, we find $t_3 = [4(\Delta^2+4)^{1/2}+\Delta]/3$. Prior to this instant the discontinuity was accelerated and after this instant it travels at a constant velocity:

$$\frac{d\xi_e}{dt} = \begin{cases} (5\Delta - 3t)/8, & t_1 < t < t_3; \\ [\Delta - (\Delta^2+4)^{1/2}]/2, & t > t_3. \end{cases}$$

The latter formula makes it possible to find the spatial length of the disturbance as a function of time (for $t < t_1$ it is equal to $l_a = 2 - \Delta t$), although the corresponding expressions are not given due to their cumbersome size. In a subsonic flow regime, the spatial length of the generated acoustic wave can be calculated from linear theory only when the action time of the thermal sources τ_L [the light pulse duration normalized in accordance with (6.7)] does not exceed the discontinuity formation time t_1: $\tau_L \leqslant -2\Delta$. Calculation reveals that, for example, the excited pulse duration for $\Delta = 0$ calculated accounting for the acoustic nonlinearity is twice the result of linear theory as early as $\tau_L = \frac{10}{3}$.

All points of the profile lag the optical beam in the totally supersonic regime $(\Delta \geqslant \Delta_{cr} = 2)$ [Fig. 6.4(b)]. The motion of point $B(1, -\Delta)$ on the right boundary of the heating zone defines the pulse amplitude. The amplitude saturates at the moment of time $t_2 = \Delta - (\Delta^2-4)^{1/2}$; for $t < t_2$ it grows in proportion to time. In this source motion regime no discontinuity is formed and pulse length grows linearly with time: $l_a = 2 + \Delta t$.

Sound excitation has a number of anomalies in the mixed motion regime $(0 < \Delta < 2)$. The points of the initial profile in the supersonic trajectory range [the left of point C in Fig. 6.4(c)] lag the heating layer and form a pulse "tail" whose length grows in proportion to the detuning Δ and time t. The amplitude of the tail stabilizes on the level of Δ at time $t = \Delta$ when point C passes through boundary $\xi = -1$ of the heating zone. The points of the initial profile in the turning trajectories region (to the right of point C) form the pulse "head."

Since the points travel at a constant velocity along the w axis (in the heating zone), the initial profile section in this zone travels parallel to itself. The discontinuity forms at time $t_1 = \Delta$ (at the same time as point C arrives at line $w = 0$) at coordinate $\xi_e = 1 - \Delta^2/2$, i.e., within the heating zone. The smaller the value of Δ, the closer to the right boundary of the heating zone the discontinuity is formed. For $0 < \Delta < 1$ the leading edge exits the heating zone before the maximum pulse amplitude is achieved at time $t_2 = 2 + \Delta$ when point C crosses boundary $\xi = 1$. The amplitude of the discontinuity will then grow through time $t_3 = \Delta + 8/3$ when it becomes comparable to the wave amplitude; the front then moves at a constant velocity.

Point C overtakes the shock front even within the segment $[-1,1]$ for $1 < \Delta < 2$. Hence the wave amplitude stabilizes simultaneously with the exit of the shock front from the heating zone: $t_2 = [8(2-\Delta)^{-1} + \Delta]/3$ [this time can easily be determined by integral relation (6.22)]. When $\Delta \to 2$ (corresponding to excitation of a wave of maximum amplitude $v_a = 4$) the rise time of amplitude t_2 approaches ∞, since the front is decelerated and stops as it approaches the heating zone boundary.

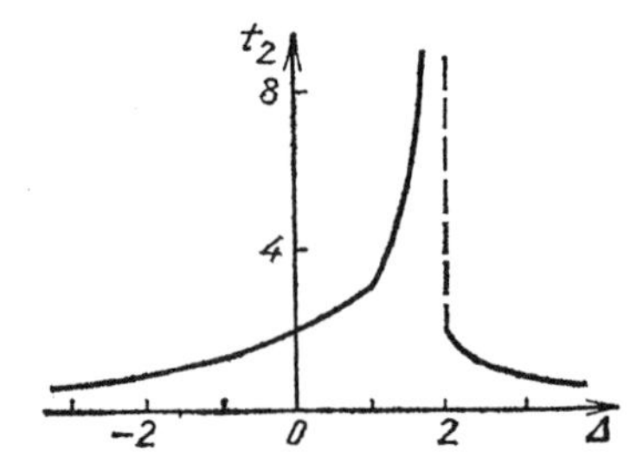

FIG. 6.5. The rise time of the acoustic pulse plotted as a function of source velocity.

In the mixed regime $0 < \Delta < 2$ the amplitude of the pulse "head" may far exceed the amplitude of the "tail." In other words, the pulse steepens and its spatial length at half amplitude may be less than the size of the heating zone. Qualitatively, this follows from relation (6.22): the pulse duration is shorter the higher the pulse amplitude for the fixed period of source action.

Figure 6.5 shows the relation $t_2(\Delta)$: the rise time to maximum pulse amplitude plotted as a function of source velocity (detuning Δ). This relation can be used to easily determine the detuning band where maximum acoustic disturbance amplitude can be achieved for a given laser pulse duration. The range of allowed mismatches is limited by the intersection points of line $t_2 = \tau_L$ with curve $t_2(\Delta)$.

Figure 6.6 shows the dependence of the time required for excitation of a wave of fixed amplitude on the velocity of the light beam (the heating layer). The interval of detunings Δ where this level can be achieved narrows with increasing wave amplitude, while the time required rises simultaneously. Curves are given for the following amplitude values: $v_a = 1$, 1.5, 2, 2.5, 3, 3.3, and 3.5. We note that the phase plane method used here permits a simple description of the interaction of waves excited by distributed sources with free acoustic disturbances propagating in the same direction as the transonic sources. Only the initial conditions change compared to the problem examined above, while the phase portrait of the system remains the same.

The velocity of the heating layer was assumed to be constant in the analysis above. At the same time, constant velocity of the light beam can be achieved in the experiments (for example, Refs. 14 and 15) only within a certain degree of accu-

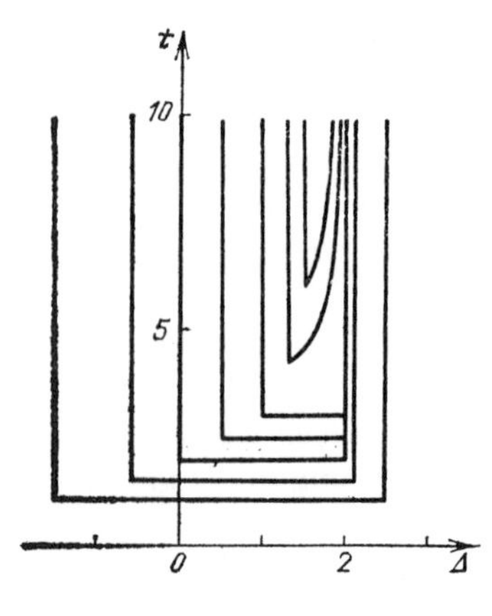

FIG. 6.6. The time required for exciting a wave of fixed amplitude plotted as a function of light-beam velocity.

racy. It is therefore interesting to investigate the effect of a change in the velocity of quasisynchronous sources on the nonlinear sound excitation process.[26]

In the noninertial coordinate system related to a light beam at a varying velocity, the transformation of the vibrational velocity profile of the acoustic disturbance is, as before, described by evolution equation (6.6). However, the relative detuning Δ_0 in this equation is dependent on time: $\Delta_0 = [V(t) - c_0]/c_0$. We note that in spite of this fact integral relation (6.22) remains valid.

We explicitly isolate the initial component in the time dependence of OA antenna velocity

$$V(t) = V_0 + \int_0^t g(t')dt',$$

where the function $g(t)$ describes the time dependence of the light-beam acceleration. Then, in dimensionless variables (6.7), Eq. (6.6) can be transformed to

$$\frac{\partial w}{\partial t} + w\frac{\partial w}{\partial \xi} = H(\xi) - g(t), \quad w = v - \Delta. \tag{6.23}$$

Here $g(t) = ag/(\varepsilon v_0)^2$ is the dimensionless acceleration of the heating region, while the dimensionless mismatch $\Delta = [V(t) - c_0]/\varepsilon v_0$ is dependent on time. The following condition is imposed on the solution of Eq. (6.23) for an undisturbed initial state of the medium $v(t=0,\xi) = 0$:

$$w(t=0,\xi) = -\Delta(t=0) = -(V_0 - c_0)/\varepsilon v_0.$$

The solution of $w(t,\xi)$ found here is used to determine the acoustic pulse profile as follows:

$$v = w + \Delta(t) = w + \Delta(t=0) + \int_0^t g(t')dt'. \tag{6.24}$$

Equation (6.23) is a quasilinear equation. Its characteristic equations take the form

$$\frac{dw}{dt} = H(\xi) - g(t), \quad \frac{d\xi}{dt} = w. \tag{6.25}$$

Equation system (6.25) is an autonomous system for the case of constant acceleration of the light beam $g(t) = \text{const} \equiv g$. In this situation Eq. (6.23) may be effectively analyzed on the phase plane (ξ,w).

The singularities of the phase plane are determined from the relations

$$H(\xi_i) = g, \quad w_i = 0.$$

Consequently, they appear only for a bounded range of values of constant source acceleration: $0 \leqslant g \leqslant 1$ [since $0 \leqslant H(\xi) \leqslant 1$ from the normalization]. If $\partial H(\xi_i)/\partial\xi < 0$, then a center singularity is realized; if $\partial H(\xi_i)/\partial \xi > 0$, a saddle-point singularity occurs. The physical meaning of the equilibrium points can be determined by solving the nonstationary problem.

The phase portrait of Eq. (6.25) in the case $g(t) = \text{const}$ is described by the relation

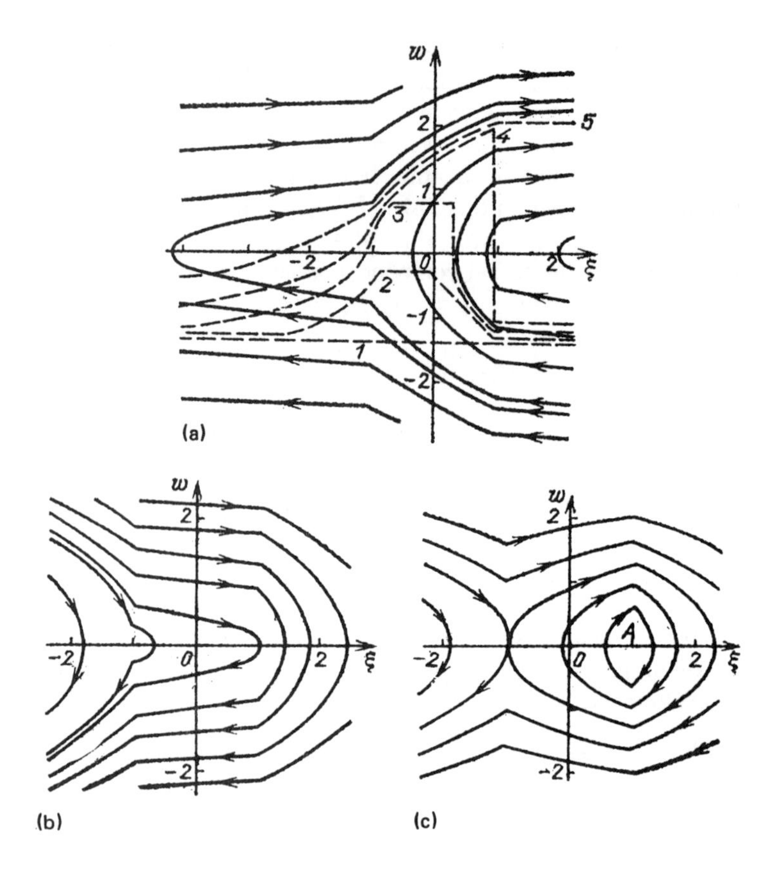

FIG. 6.7. Phase plane for the case of constant deceleration (a) and constant acceleration of (b), (c) OA-antenna motion. The wave profiles are plotted at successive times t of 0 (1), 1 (2), 2 (3), 3 (4), and 6 (5).

$$w^2/2 + \mathscr{H}(\xi) + g\xi = C_1.$$

The phase portrait is shown in Fig. 6.7 for a rectangular light intensity distribution in the beam cross section $H(\xi) = \theta(\xi + 1) - \theta(\xi - 1)$. Outside the heating zone the representative points are accelerated along the ξ axis at an acceleration $-g$, and within this zone at an acceleration $1 - g$ [since $d^2\xi/dt^2 = H(\xi) - g(t)$, which derives from system (6.25)]. In the case of constant deceleration of the beam ($g<0$) acceleration of the representative points has the same sign inside and outside the heating zone (the inertial force and the potential step run in the same direction). Therefore, sooner or later they all begin to overcome the heating layer [Fig. 6.7(a)]. Analogously, for $g>1$ the acceleration sign is also constant (the decelerating action of the inertial force on the liquid element exceeds the accelerating action of the thermal sources). The representative points therefore sooner or later lag the accelerating beam [Fig. 6.7(b)].

Closed trajectories appear on the phase plane in the intermediate case $0<g<1$ [Fig. 6.7(c)]. The elementary disturbances periodically exit the thermal layer and

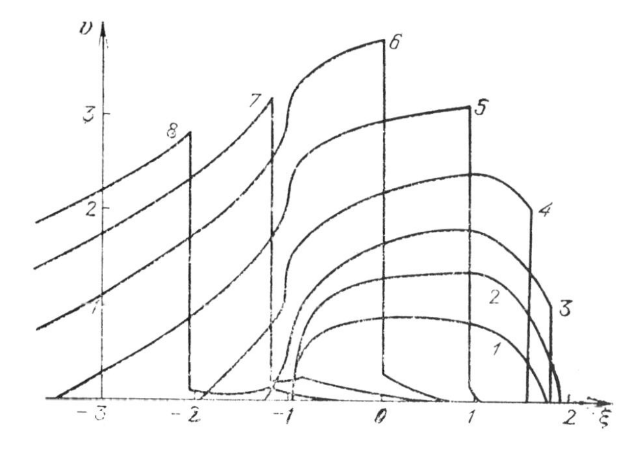

FIG. 6.8. Formation of an acoustic pulse for the case of the sound barrier crossing by the thermal layer at constant acceleration [$\Delta(t=0) = -1.4$, $g = 1.1$]. The wave profiles are plotted at successive times t of 0.7 (1), 1.2 (2), 1.8 (3), 2.5 (4), 3.4 (5), 4.2 (6), 4.8 (7), 5.2 (8).

return to the source localization region near the center-type singularity $A(1,0)$. This corresponds to the following physical process: the elementary disturbance driven by the thermal sources overcomes the beam, but it is traveling at a constant velocity outside the heating zone and the accelerating beam catches it. The sources further amplify the wave; it again exits the thermal layer, etc. The velocity of the representative point trapped by the accelerating thermal layer in the absence of collective effects grows without limit. Acceleration of a spatially distributed disturbance is limited by nonlinear attenuation: the formation of the wave front where energy dissipates.

The phase portrait provides a convenient qualitative representation of processes occurring in the system and also makes it possible to achieve an approximate graphical plot of the excited signals. In Fig. 6.7(a) the dashed lines represent acoustic wave profiles on the phase plane (ξ,w) for $g = -0.1$. Analysis reveals that the acoustic disturbance front exits the heating zone at time $t \approx 3$ and, at this instant, the signal amplitude begins to decay. Using relation (6.24), we can easily go to a representation of the profiles on the plane (ξ,v): this only requires shifting the ξ axis.

Under constant-acceleration radiator motion $(g>0)$ the excited acoustic pulse begins to lag the antenna beginning at a certain instant. Figure 6.8 shows the evolution of the acoustic wave profile for a transit crossover from subsonic scanning velocities to supersonic velocities. It is clear that the acoustic disturbance front intersects the thermal layer and exits the layer with increasing beam velocity. The maximum pulse amplitude is achieved at the point of its "separation" from the radiator, when the profile discontinuity crosses over the left boundary $(\xi = -1)$ of the heating zone. The signal formed when the thermooptical antenna crosses the sound barrier then propagates freely.[15]

We use a specific example to show how this theory can be employed to analyze the possibility of observing nonlinear effects in experiment. We consider from this viewpoint the experimental conditions of Ref. 14 which was the first to achieve sound excitation by laser-beam scanning of a water surface [$\alpha = 0.17\ \text{cm}^{-1}$, $\beta = 2.1\times10^{-4}\ \text{K}^{-1}$, $c_0 = 1.2\times10^5$ cm/s, $c_p = 4.18\text{J}/(\text{g}\times\text{K})$, $\varepsilon \approx 4$].

The space-time distribution of the laser radiation was defined by the following parameters: $\tau_L \sim 0.2$ ms, $a \approx 0.5$ cm, and a characteristic laser pulse energy $E \approx 0.2$ J. Using these data and relation (6.7) we obtain the following estimates: for the radiation intensity I_0 ($I_0 \sim E/\pi a^2\tau_L \approx 5\times10^3\ \text{W/cm}^2$), for the number N ($N \approx 1.7\times10^{-6}$) and the characteristic disturbance velocity ($v_0 \approx 20$ cm/s). Since the dimensionless light pulse duration $\tau_L = \varepsilon v_0\tau_L/a \approx 3.2\times10^{-2}$ is much less than the characteristic time to achieve maximum amplitude for the case of transonic scanning at constant velocity t_2 ($t_2 \sim 1$), the amplitude of the disturbance grows in proportion to the period of activity of the thermal sources (linear acoustic wave excitation regime): $v_a = \tau_L = 3.2\times10^{-2}$. We estimate the pressure increment of the wave at the end of the trajectory: $p'_{\text{theor}}(0) = c_0\rho_0 v = c_0\rho_0 v_0 v_a \approx 10^5\ \text{dyn/cm}^2$. The pressure was measured at a length $L = 45$ cm from the end of the light track in Ref. 14. In order to obtain a theoretical estimate we employ an approximate relation for waves from a cylindrical source: $p'_{\text{theor}}(L)L \sim p'_{\text{theor}}(0)a/2$. We finally obtain $p'_{\text{theor}}(L) \sim 500\ \text{dyn/cm}^2$, which is within an order of magnitude of the experimentally observed values of $p'_{\text{expt}}(L) \sim 300\ \text{dyn/cm}^2$.

The estimates provided above demonstrate that a linear acoustic wave excitation regime was realized in Ref. 14. The question arises as to whether or not, by increasing the laser pulse duration, it is possible to achieve nonlinear operating regimes of the thermooptical antenna. More specifically, to what degree of accuracy is it necessary to maintain constant velocity of scanning to achieve $v_a \approx 1$?

Studies of an OA antenna traveling at a nonconstant velocity have demonstrated that nonlinear waves can be excited with the following limits on the dimensionless acceleration: $|g| \lesssim 1$. Therefore, in order to achieve $v_a \sim 1$ it is not sufficient to increase the optical irradiation time τ_L by a factor of 300. The deviation of $V(t)$ from sound speed across the entire track must also satisfy the relation

$$(V(t) - c_0)/300\tau_L \sim g \lesssim (\varepsilon v_0)^2/a.$$

It turns out that constant velocity must be maintained accurate to better than 0.5%.

Estimates reveal that essentially accelerated light-beam scanning ($|g| \sim 10^3$) was realized in experiment.[14] In this case the round-trip time of the representative point through the heating zone (the maximum effective amplification time of the acoustic pulse) was of the order of $4|g|^{-1/2} \ll 1$. Over such times the wave amplitude grows linearly with time and consequently $v_a \lesssim 4|g|^{-1/2}$. Such a dependence of the excited wave amplitude on beam acceleration is related to the diminishing operating time of the antenna near velocity matching. The latter estimate demonstrates that $v_a \lesssim 10^{-1}$ must hold under the experimental conditions of Ref. 14 and, consequently, the realized values $v_a \approx 3\times10^{-2}$ are near the maximum possible values. Increasing the amplitude of the excited pulse requires simultaneously increasing the duration of the light pulse of fixed intensity and improving scanning stability.

We now investigate the effect of relaxation processes in the medium on nonlinear acoustic wave excitation by distributed synchronous sources.[27] We know that the propagation of sound waves disrupts thermodynamic equilibrium in the medium.[33,48] The system does not achieve a new equilibrium state (with parameter values altered by the wave) instantaneously, but rather after a characteristic time τ_r (the relaxation time). Hence if the acoustic pulse duration τ_a does not far exceed the relaxation time ($\tau_a \lesssim \tau_r$), it will propagate in a nonequilibrium medium.[33,49] It is fundamentally necessary to take account of the processes responsible for the equilibration delay.

Applying the phenomenological approach proposed by Mandelshtam and Leontovich (see, for example, Refs. 48 and 33) it is possible to avoid the specific nature of the relaxation processes. The relaxation mechanisms in the medium will be accounted for by introducing an "internal coordinate" r whose "motion" to the equilibrium value r_0 is described by the equation

$$\frac{dr}{dt} = -\frac{(r-r_0)}{\tau_r}.$$

In this model, the equation of state of the gas (accounting for thermoelastic stresses, quadratic nonlinear effects, and relaxation processes) can be reduced to

$$\left(\frac{d}{dt} + \tau_r^{-1}\right)[p' - c_0^2\rho' - (\varepsilon - 1)c_0^2\rho_0^{-1}\rho'^2 - c_0^2\rho_0\beta T'] = \frac{mc_0^2\, d\rho'}{dt}. \tag{6.26}$$

The parameter $m \equiv (c_\infty^2 - c_0^2)/c_0^2$ characterizes the difference in the equilibrium c_0 and "frozen" c_∞ sound speeds. The propagation velocity of acoustic waves of infinitely small amplitude is equal to c_∞ if the characteristic time of the relaxation process τ_r far exceeds the acoustic pulse duration.[33] Equation (6.26) is valid for minor deviations from equilibrium ($|r/r_0 - 1| \sim \mu$). The other equations (6.1), (6.3) of the thermohydrodynamic system describing the test phenomenon remain unchanged.

Utilizing the modification of the slowly varying profile method and further assuming that the parameter m is small ($m \sim \mu$) for these media, we can obtain the following equation for the velocity disturbance v:[27]

$$\frac{\partial v}{\partial t} + (\varepsilon v - c_0\Delta_0)\frac{\partial v}{\partial \xi} + \frac{mc_0}{2}\frac{\partial}{\partial \xi}\int_{+\infty}^{\xi} \frac{\partial v}{\partial \xi'} \exp\left(\frac{\xi - \xi'}{c_0\tau_r}\right) d\xi' = \frac{\alpha c_0^2}{2}\frac{\beta I_0}{\rho_0 c_0 c_p} H(\xi). \tag{6.27}$$

We note that the considered regime ($|\Delta_0| \sim \mu$, $m \sim \mu$) in transonic gas flow theory is called a totally transonic regime.[50]

We reduce Eq. (6.27) to dimensionless form, using the designations of Eq. (6.7):

$$\frac{\partial v}{\partial t} + (v - \Delta)\frac{\partial v}{\partial \xi} + D_S\frac{\partial}{\partial \xi}\int_{+\infty}^{\xi} \frac{\partial v}{\partial \xi'} \exp[\mathscr{R}(\xi - \xi')]d\xi' = H(\xi). \tag{6.28}$$

The dimensionless parameter $D_S = mc_0/2\varepsilon v_0 \approx (c_\infty - c_0)/\varepsilon v_0$ characterizes the relative influence of dispersion and nonlinear effects on the disturbance evolution process. For $D_S \gg 1$ a nonlinear change in sound speed is much less than the difference in the propagation velocities of the high-frequency and low-frequency disturbances. In the opposite case $D_S \ll 1$ the effect of a nonlinearity is substantially stronger than the effect of dispersion. The parameter $\mathscr{R} \equiv a/c_0\tau_r$ is determined by the ratio of the sound transit time through the heating zone $a/c_0 \sim \tau_a$ to the characteristic relaxation time τ_r. The third parameter on which the solution depends is the dimensionless detuning Δ which indicates the flow regime (scanning regime).

We consider certain limiting cases. If $\mathscr{R} \ll 1$, the particles of the medium over the relaxation time will travel a distance $c_0\tau_r$ much less than the spatial scale of the disturbance a; a flow is quasiequilibrium. Equation (6.28) in this case takes the following asymptotic form:

$$\frac{\partial v}{\partial t} + (v - \Delta)\frac{\partial v}{\partial \xi} - \left(\frac{D_S}{\mathscr{R}}\right)\frac{\partial^2 v}{\partial \xi^2} - \left(\frac{D_S}{\mathscr{R}^2}\right)\frac{\partial^3 v}{\partial \xi^3} = H(\xi). \tag{6.29}$$

A detailed analysis of inhomogeneous Korteweg-de Vries–Burgers' equation (6.29) [as is the case for Eq. (6.28)] at present evidently requires a computer for the solution (see, for example, Refs. 51 and 52) or a stage-by-stage method (see Chap. 2). We note that the high-frequency attenuation and weak dispersion do not limit the growth in the momentum of acoustic disturbance. Relation (6.22) remains valid. Therefore, although the acoustic pulse profile is deformed, the area of the disturbance on the plane (ξ, v) remains the same as in the absence of relaxation processes.

If $\mathscr{R} \ll 1$, the medium cannot "follow" the changes caused by the wave ("quasi-freezing"). When the thermal sources are concentrated in a limited spatial domain [and, consequently, $v(t, \xi = \pm\infty) = 0$], Eq. (6.28) becomes

$$\frac{\partial v}{\partial t} + (v - \Delta_\infty)\frac{\partial v}{\partial \xi} + \gamma v = H(\xi). \tag{6.30}$$

Its solution is dependent solely on two parameters: the dimensionless detuning Δ_∞ relative to the frozen sound speed: $\Delta_\infty = \Delta - D_S = (V_0 - c_\infty)/\varepsilon v_0$ and the dimensionless decay decrement $\gamma = D_S\mathscr{R}$. Analysis of Eq. (6.30) reveals that low-frequency attenuation limits the increase in the mass of gas driven by the light beam:

$$\int_{-\infty}^{\infty} v\, d\xi = \mathscr{H}(-\infty)[1 - \exp(-\gamma t)]\gamma^{-1}. \tag{6.31}$$

The higher the dimensionless decay decrement γ the lower the momentum conveyed to the wave comoving with the laser beam.

First-order inhomogeneous quasilinear equation (6.30) can be effectively analyzed on the phase plane (ξ, v). The equations of the characteristics take the form

$$\frac{dv}{dt} = -\gamma v + H(\xi), \quad \frac{d\xi}{dt} = v - \Delta_\infty. \tag{6.32}$$

Each point of the initial profile follows the characteristic curve of stationary equation (6.30):

$$\frac{dv}{d\xi} = (H(\xi) - \gamma v)/(v - \Delta_\infty). \tag{6.33}$$

Unlike the case of an ideal gas, the phase portrait of the flow of relaxing gas (6.33) is not a universal portrait, and depends on the incoming flow velocity through the parameter Δ_∞.

We can use existing graphical techniques[43,44] to plot phase trajectory family (6.33) in the case of a complex spatial distribution of the thermal source intensity $H(\xi)$. The singularities (ξ_i, v_i) on the phase portrait are determined by

$$H(\xi_i) = \gamma\Delta_\infty, \; v_i = \Delta_\infty.$$

For the heating zone $H(\xi) \geqslant 0$ and hence the phase plane will contain no singularities in the case of a subsonic inflow ($\Delta_\infty < 0$). Singularities are also absent when $\gamma\Delta_\infty > 1$ [the source distribution function $H(\xi)$ is normalized so that $\max H(\xi) = 1$]. The singularity is a node for $-(\gamma/2)^2 \leqslant \partial H(\xi_i)/\partial\xi < 0$, a saddle point for $\partial H(\xi_i)/\partial\xi > 0$, and a focus for $\partial H(\xi_i)/\partial\xi < -(\gamma/2)^2$. For $\partial H(\xi_i)/\partial\xi = 0$ a higher-order node-saddle singularity is realized.[44]

An analytic description of the phase portrait of the system is possible for a rectangular wave intensity distribution in the beam cross section [potential function of the type (6.21)]:

$$\xi = -\gamma^{-1}(v - \Delta_\infty \ln|v|) + C_1, \qquad |\xi| > 1.$$
$$\xi = -\gamma^{-1}[v - \gamma^{-1} - (\Delta_\infty - \gamma^{-1})\ln|v - \gamma^{-1}|] + C_2, \quad |\xi| \leqslant 1.$$

Here C_1, C_2 are the trajectory parameters.

Three characteristic flow regimes can be identified: subsonic regime $\Delta_\infty \leqslant 0$, totally supersonic regime ($\Delta_\infty > \Delta_{cr} \geqslant 0$), and mixed regime ($0 < \Delta_\infty \leqslant \Delta_{cr}$) where a portion of the initial profile points overtake the heating zone, while the remaining points lag this zone. In the cases noted here the disturbance will evolve differently from a qualitative viewpoint. This classification must be differentiated from the three ranges of the parameter Δ_∞ values characteristic of a relaxing medium: ($\Delta_\infty \leqslant 0$, $0 < \Delta_\infty \leqslant \gamma^{-1}$, $\Delta_\infty > \gamma^{-1}$), in which the phase portraits of the test system are qualitatively different.

Figure 6.9 shows the phase trajectories for inflow velocities that do not exceed the high-frequency "frozen" sound speed ($\Delta_\infty \leqslant 0$). There are no singularities. The level $v = \gamma^{-1}$ corresponds to a limitation on OA-antenna efficiency under velocity matching conditions ($\Delta_\infty = 0$) neglecting the nonlinearity (due to frequency-independent sound absorption). The initial profile points $[v(t = 0, \xi) = 0]$ overtakes the light beam in regime $\Delta_\infty \leqslant 0$. All elementary disturbances decay outside the heating zone. The profile of the resulting stationary (as $t \to \infty$) pulse is identical to the trajectory of point $A(-1, 0)$ situated on the left boundary of the heating zone.

The phase portrait of the system under detunings $\Delta_\infty > \gamma^{-1}$ will also contain no singularities (Fig. 6.10). All disturbances lag the heating zone and decay. Employing the present terminology, this regime is a totally supersonic regime (relative to the greatest "frozen" sound speed), and, consequently, the detuning in the crossover regime Δ_{cr} will not exceed the level of the low-frequency limit γ^{-1}.

Figure 6.11 shows a phase plane in the regime $0 < \Delta_\infty \leqslant \gamma^{-1}$. Point $B(-1, \Delta_\infty)$ on the phase plane is a saddle point, while the point $C(1, \Delta_\infty)$ is a stable focus. Trajectories below separatrix b_1 are totally supersonic. The initial profile points traveling along these trajectories lag the heating zone. If separatrix b_1

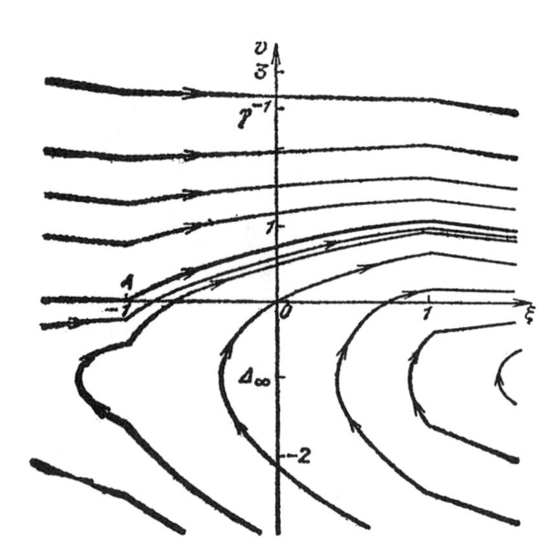

FIG. 6.9. Phase portrait of a system in the subsonic source regime ($\Delta_\infty \leqslant 0$).

does not intersect the initial profile [Fig. 6.11(a)], a totally supersonic regime ($\Delta_{cr} < \Delta_\infty < \gamma^{-1}$) is realized. We note that at flow velocities $\Delta_\infty > \Delta_{cr}$ [Figs. 6.10 and 6.11(a)] the wave form of the resulting disturbance is identical to the trajectory of the point $D(1,0)$.

In the mixed flow regime ($0 < \Delta_\infty \leqslant \Delta_{cr}$) the profile of the stationary pulse contains a discontinuity [Fig. 6.1(b)]. There will either be no flow disturbance in front of the shock edge ($\xi \geqslant \xi_e$) (for $\xi_e \geqslant 1$), or it will coincide with the trajectory

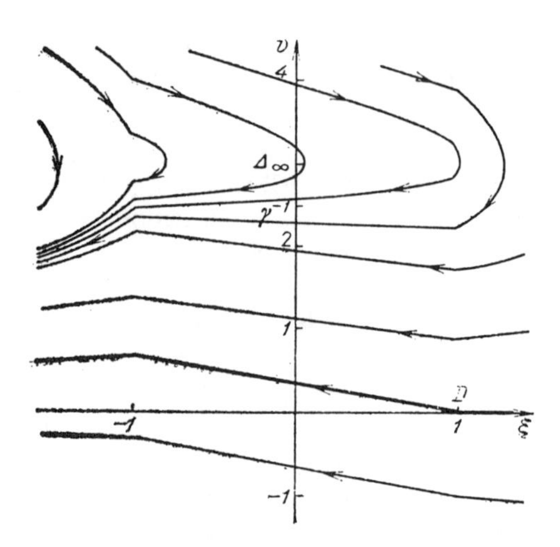

FIG. 6.10. Phase portrait of a system for supersonic velocities exceeding the low-frequency damping level ($\Delta_\infty > \gamma^{-1}$).

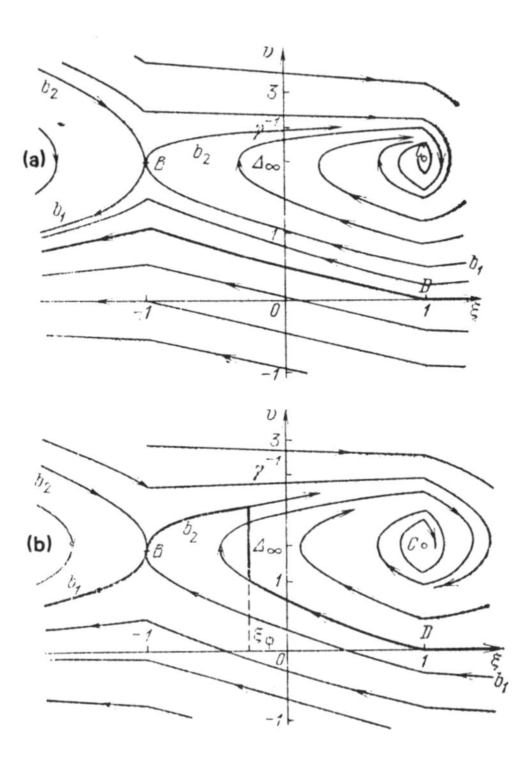

FIG. 6.11. Phase portrait of the system in the case $0<\Delta_\infty \leqslant \gamma^{-1}$; (a) supersonic regime ($\Delta_{cr}<\Delta_\infty <\gamma^{-1}$); (b) mixed regime ($0<\Delta_\infty \leqslant \Delta_{cr}$).

of point $D(1,0)$ (for $\xi_e<1$); the profile behind the shock front is described by the segments of the separatrix b_2 (for $-1 \leqslant \xi \leqslant \xi_e$) and b_1 (for $\xi \leqslant -1$). The discontinuity behaves in accordance with pulse area law (6.31). At inflow velocities $0<\Delta_\infty \leqslant \Delta_{cr}$ the elementary disturbances that correspond to the points on the initial profile to the right of separatrix b_1 are amplified to the degree that they overcome the thermal layer. However, outside the source localization region ($\xi >1$) the wave amplitude diminishes due to damping and the heating zone catches it, the wave is reamplified, etc. This process on the phase plane corresponds to a stable focus singularity.

In the crossover regime ($\Delta_\infty = \Delta_{cr}$) separatrix b_1 intersects the ξ axis (initial profile) at the point $D(1,0)$. Calculation reveals that the magnitude of the shift Δ_{cr} is related to the damping decrement γ by the relation (Fig. 6.12, curve 1)

$$2\gamma^2 = \gamma\Delta_{cr} + (1-\gamma\Delta_{cr})\ln(1-\gamma\Delta_{cr}). \tag{6.34}$$

With small decrements ($\gamma \ll 1$): $\Delta_{cr} \approx 2(1-\gamma/3)$.

We note that the stationary pulse amplitude in the crossover regime is equal to the detuning: $v_a(\Delta_{cr}) = \Delta_{cr}$.

The maximum amplitude of the disturbance is achieved at an optimum velocity $\Delta_\infty = \Delta_{max}$ ($0 \leqslant \Delta_{max} \leqslant \Delta_{cr}$), which is determined by the relation

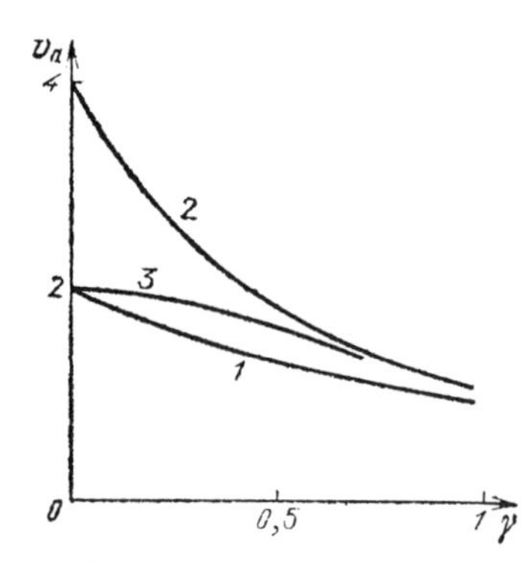

FIG. 6.12. Acoustic pulse amplitude plotted as a function of the damping decrement for different characteristic source velocities: $\Delta_\infty = \Delta_{cr}$ (1); Δ_{max} (2); 0 (3).

$$2\gamma^2 = -\gamma\Delta_{max} + (1-\gamma\Delta_{max})\ln[(1-\gamma\Delta_{max})/(1-2\gamma\Delta_{max})].$$

Specifically, $\Delta_{max} \to \Delta_{cr}$ for $\gamma \to 0$ and $\Delta_{max} \to 0$ for $\gamma \to \infty$, which corresponds to weak manifestation of nonlinear effects. The dependence of maximum acoustic pulse amplitude on the decay decrement $v_a^{max}(\gamma) = 2\Delta_{max}(\gamma)$ is shown in Fig. 6.12 (curve 2). We note that the discontinuity edge of the stationary acoustic pulse lies within the heating zone ($-1 \leqslant \xi_e \leqslant 1$) for $\Delta_{max} \leqslant \Delta_\infty \leqslant \Delta_{cr}$ and outside this zone ($\xi_e > 1$) for $0 < \Delta_\infty < \Delta_{max}$.

Using relation (6.31) it is possible to determine the dependence of the disturbance amplitude v_a on the decay decrement γ and detuning Δ_∞. We note that at flow velocities $\Delta_\infty < 0$ and $\Delta_\infty > \Delta_{cr}$ this relation is given in implicit form:

$$\Delta_\infty = \gamma^{-1} + [v_a - 2\gamma \operatorname{sgn}(\Delta_\infty)]/\ln(1-\gamma v_a).$$

The corresponding expressions for the mixed regime are not provided due to their cumbersome size. Figure 6.13 shows the dependence of wave amplitude on the velocity of wide-band sources for different damping decrements. It is clear that when the relaxation processes are accounted for, this eliminates the discontinuity in the curve that exists for an ideal medium ($\gamma = 0$). The optimum detuning Δ_{max}

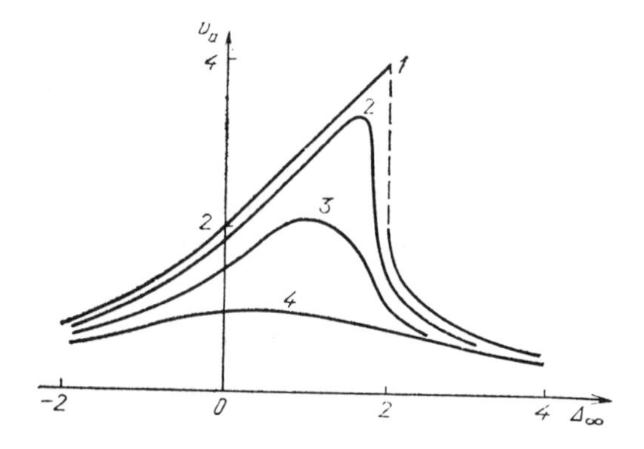

FIG. 6.13. Stationary wave amplitude plotted as a function of thermal layer velocity for different damping decrements γ of 0 (1), 0.1 (2), 0.4 (3), 1 (4).

diminishes with increasing damping and approaches $\Delta_{max} = 0$, as follows from linear theory. We note that at high decrements ($\gamma > 1$) the amplitude of the disturbance will diminish in inverse proportion to the value of γ for any source velocity. This is due to the fact that in the case of strong dispersion (and since $\mathscr{R} \ll 1$ and $\gamma > 1$, then $D_S \gg 1$) only an insignificant number of spatial harmonics of the disturbance will be resonantly excited. Here, not only the waves propagating in the direction opposite to the sources (Sec. 1.3) but also a fraction of the comoving waves will be poorly excited, since their velocity may be substantially different from that of the sources due to the strong dispersion.

With use of the characteristics of Eq. (6.32) it is possible to determine the dependence of the rise time to maximum amplitude of the disturbance t_2 in the given regime on the damping decrement. For example, for a crossover detuning $\Delta_\infty = \Delta_{cr}$:

$$t_2 = \gamma^{-1} \ln[1 - \gamma \Delta_{cr}(\gamma)],$$

while $\Delta_{cr}(\gamma)$ in turn is determined by relation (6.34). Analysis reveals that in spite of the reduction of the maximum amplitude of the disturbance, the time to achieving this amplitude grows monotonically with increasing damping decrement.

In the majority of actual physical systems it is necessary to account for the finite size of the laser beam perpendicular to the beam direction and, consequently, the influence of diffraction effects on OA-signal formation. It is most natural to first analyze the problem in a quasi-one-dimensional geometry when the transverse dimensions of the heating zone b far exceed their longitudinal dimensions a in the direction of motion: $b \gg a$ (Fig. 1.5).

We assume that all physical quantities change more slowly in the transverse direction y than along the accompanying coordinate $\xi = x - V_0 t$ [for example, the projection v_x of the vibrational velocity v onto the x axis: $v_x = \mu v(\mu t, \xi = x - V_0 t, \mu^{1/2} y]$. Then, with use of Eqs. (6.2) and (6.3), it is possible to obtain from a system of inhomogeneous equations for an ideal liquid an evolution equation for the longitudinal component of the velocity disturbance:[28]

$$\frac{\partial}{\partial \xi}\left[\frac{\partial v}{\partial t} + (\varepsilon v - c_0 \Delta_0)\frac{\partial v}{\partial \xi} - \frac{c_0 \beta \alpha I_0}{2\rho_0 c_p} H\left(\frac{\xi}{a}\right)\varphi\left(\frac{y}{b}\right)\right] = -\frac{c_0}{2}\frac{\partial^2 v}{\partial y^2}. \quad (6.35)$$

As with Eq. (6.6), this equation is valid only at transonic source velocities. The function $\varphi(y/b)$ describes the light intensity distribution perpendicular to beam motion. The transverse component of sound velocity $[v_y \equiv v_\perp = \mu^{3/2} v_\perp(\mu t, \xi = x - V_0 t, \mu^{1/2} y)]$ is given by

$$\frac{\partial v}{\partial y} = \frac{\partial v_\perp}{\partial \xi}. \quad (6.36)$$

Homogeneous equation (6.35) (i.e., in the absence of external sources) was derived in Ref. 53 and used to describe the internal and external flow around mechanical profiles (see, for example, Refs. 54 and 55). In this connection we note that nonstationary nozzle processes can be accurately analyzed by means of a type (6.6) one-dimensional equation on a phase plane.[42,56,57] From the mathematical viewpoint in the absence of sources ($I_0 = 0$) Eq. (6.35) is analogous to the

Khokhlov-Zabolotskaya equation.[58,59] Apparently Ref. 60 was the first to report analyzing solutions of the second-order inhomogeneous quasilinear equation (6.35).

Utilizing the designations of (6.7) and additionally assuming $\eta = y/b$, we reduce Eq. (6.35) to dimensionless form

$$\frac{\partial}{\partial \xi}\left[\frac{\partial w}{\partial t} + w\frac{\partial w}{\partial \xi} - H(\xi)\varphi(\eta)\right] = -D_f\frac{\partial^2 w}{\partial \eta^2}, \quad w = v - \Delta. \tag{6.37}$$

The initial conditions for Eq. (6.37) are determined by the OA-antenna motion regime: $w(t=0,\xi,\eta) = -\Delta$. The dimensionless parameter $D_f \equiv (c_0/\varepsilon v_0)\times(a/b)^2$ characterizes the relative role of diffraction and nonlinear effects. For $D_f \to 0$ ($b \to \infty$) we have a transition to a one-dimensional geometry of the problem.

It is not possible at present to derive exact analytic solutions of Eq. (6.37). Approximate methods that make it possible to establish certain fundamental mechanisms of the process will be employed below to analyze quasisynchronous excitation of acoustic waves. Primary attention will be focused on determining the dependence of the velocity of the crossover regime Δ_{cr} and the maximum wave amplitude v_a^{max} on the "diffraction/nonlinearity" parameter D_f.

We utilize the technique proposed in Ref. 61 for analyzing nonlinear paraxial sound beams to investigate the process near the plane $\eta = 0$. We generalize this method to the case of sound excitation by traveling volume sources. Following Ref. 61, we solve the problem as a series expansion in the small parameter η: the ratio of the transverse coordinate y to the characteristic width of the heating zone b:

$$w(t,\xi,\eta) = w_0(t,\xi) + (\eta^2/2!)w_2(t,\xi) + (\eta^4/4!)w_4(t,\xi) + \cdots.$$

Substituting this expansion into Eq. (6.37) and assuming $\varphi(\eta) = \exp(-\eta^2)$ for definiteness, we obtain an infinite chain of coupled equations

$$\begin{aligned}
&\frac{\partial}{\partial \xi}\left[\frac{\partial w_0}{\partial t} + w_0\frac{\partial w_0}{\partial \xi} - H(\xi)\right] = -D_f w_2,\\
&\frac{\partial}{\partial \xi}\left[\frac{\partial w_2}{\partial t} + \frac{\partial}{\partial \xi}(w_0 w_2) + H(\xi)\right] = -D_f w_4,
\end{aligned} \tag{6.38}$$

System (6.38) can be closed if we set $w_4 = w_6 = \cdots = 0$, which, of course, makes the subsequent calculations only approximate. Then, integrating once the first equation of (6.38) and carrying out double integration on the second equation with respect to the longitudinal coordinate, we obtain for the functions w_0 and

$$W_2 = \int_\infty^\xi w_2(t,\xi')d\xi'$$

a system of inhomogeneous quasilinear equations with an identical principal part[36]:

$$\frac{\partial w_0}{\partial t} + w_0\frac{\partial w_0}{\partial \xi} = H(\xi) - D_f W_2,$$

$$\frac{\partial W_2}{\partial t} + w_0 \frac{\partial W_2}{\partial \xi} = \mathscr{H}(\xi). \tag{6.39}$$

Equations (6.39) are found with the natural assumption $H(\xi \to +\infty) \to 0$. The solution of system (6.39) is dependent on the velocity of the sources (the parameter Δ) through the initial conditions

$$w_0(t=0,\xi) = -\Delta, \ W_2(t=0,\xi) = 0. \tag{6.40}$$

The system of characteristic equations corresponding to Eq. (6.39) is an autonomous system

$$\frac{dw_0}{dt} = H(\xi) - D_f W_2, \ \frac{dW_2}{dt} = \mathscr{H}(\xi), \ \frac{d\xi}{dt} = w_0, \tag{6.41}$$

and, in principle, can be analyzed in the phase space (ξ,w_0,W_2). However, it is difficult to represent a family of phase trajectories in three-dimensional space and investigate the behavior of the representative points without a computer. Hence, it is best at the outset to limit the analysis to the process in the plane of symmetry $(\eta = 0)$ whose description only requires knowing the phase trajectories on the plane $(\xi,w_0,0)$.

We utilize simple model (6.21) of the heat source distribution to make it possible to obtain certain analytic results. Then system (6.41) will describe free propagation of quasi-one-dimensional acoustic waves of finite amplitude in the range $\xi > 1$. In the range $\xi < -1$ there likewise will be no external sources, although there will be a temperature "tail" that is inhomogeneous along the transverse coordinate; this tail further distorts the wave. We note that in both regions signal propagation can be described analytically,[61] if we know its profile at the boundaries of the heating region $(\xi = \pm 1)$. In the range where the sources are active $|\xi| \leqslant 1$ the characteristic system (6.41) takes the form

$$\frac{dw_0}{dt} = 1 - D_f W_2, \ \frac{dW_2}{dt} = 1 - \xi, \ \frac{d\xi}{dt} = w_0. \tag{6.42}$$

Consistent with the classification proposed by Poincaré[62] the singularity $(\xi = 1, w_0 = 0, W_2 = D_f^{-1})$ of the phase space is a saddle focus. It is easily determined that system (6.42) has a first integral

$$(\xi - 1 + D_f^{-1/3} w_0 - D_f^{1/3} W_2 + D_f^{-2/3}) \exp(-D_f^{1/3} t) = \text{const},$$

which can be used to project the trajectories onto the plane (ξ,w_0). The following analysis differs from the preceding one solely in its cumbersome size and more difficult accessibility. It reveals the capabilities of the phase plane method for analyzing non-one-dimensional processes. However, for this specific problem it is not advisable to carry these formulations through to the end.

As noted above, the solutions of system (6.39) only approximately describe quasi-two-dimensional waves near plane $\eta = 0$. Estimates of the applicability of Eqs. (6.39) can be obtained by linearizing the system and comparing its solutions to a series expansion in η^2 of the exact solution of linearized Eq. (6.37). Such a comparison reveals that the problem contains an additional small parameter $D_f t \ll 1$. Therefore, the results of analyzing system (6.39) are valid over times $t \ll D_f^{-1}$. As noted in Ref. 61, the approximation used here provides a far better description of nonlinear effects than diffraction.

It was demonstrated in analyzing the one-dimensional case that the characteristic times to obtain maximum amplitudes $t_2 \gtrsim 1$. Hence system (6.30) will yield reliable results on the wave form of the resonant curve only for small values of the parameter D_f. We recall that we use the term resonant curve to refer to the dependence of the stationary wave amplitude v_a on source velocity (detuning Δ). It is now clear that an exact plot of the phase portrait is not advisable; it is more convenient from the outset to utilize the condition $D_f t \ll 1$ or even the stronger conditions $D_f \ll 1$, $t \lesssim 1$.

It follows from characteristic system (6.42) that the time dependence of the coordinate of the liquid element within the heating zone is described by the equation

$$\frac{d^3\xi}{dt^3} = -D_f(1-\xi).$$

After determining the solution $\xi = \xi(C_1,C_2,C_3,t)$ of this equation we employ Eq. (6.42) to find $w_0(C_1,C_2,C_3,t)$ and $W_2(C_1,C_2,C_3,t)$.

At the instant of source activation ($t=0$) the representative points will lie on the initial profile:

$$\xi(t=0) = \xi_0 \ (|\xi_0| \leqslant 1), \quad w_0(t=0) = -\Delta, \quad W_2(t=0) = 0.$$

This condition makes it possible to find the constants C_1, C_2, C_3. We then have relations describing the motion of the points of the representative profile in three-dimensional phase space with any values of the parameter D_f. It is sufficient to utilize the first two relations to analyze the evolution of the acoustic wave profile in the plane of symmetry ($\eta = 0$). For $D \ll 1$ they take the form

$$\begin{aligned} \xi &= \xi_0 - t\Delta + (t^2/2!) + (t^3/3!)D_f(\xi_0 - 1) - (t^4/4!)D_f\Delta + (t^5/5!)D_f, \\ w_0 &= -\Delta + t + (t^2/2!)D_f(\xi_0 - 1) - (t^3/2!)D_f\Delta + (t^4/4!)D_f, \end{aligned} \tag{6.43}$$

where terms through first order in the parameter D_f are retained. In fact, representation (6.43) is a parametric description of the phase trajectory $w_0 = w_0(\xi,\xi_0,\Delta)$ (t is the parameter). We show how parametrically represented phase portrait (6.43) can be used to investigate the processes occurring within the distributed system under analysis.

We examine in greater detail a description of the dependence of the excited wave amplitude on the mismatch Δ. We note that in the analysis of sound generation in a relaxation medium, the details of such a calculation were dropped. We will illustrate the progression of the argument by qualitative formulations.

With rather large, positive detunings $\Delta \geqslant \Delta_{cr} > 0$ (totally supersonic regime) the stationary wave profile will be identical to the trajectory of the point ($\xi = 1$, $w_0 = -\Delta$) of the initial profile [Figs. 6.14(a) and 6.14(b)]. The amplitude of the stationary wave v_a is equal to the sum of the detuning and the maximum phase variable w_0 for the given regime. For $\Delta \geqslant \Delta_{cr}$ it is sufficient to set $\xi = -1$, $\xi_0 = 1$, $w_0 = w_0^{max}$ in Eq. (6.43) to determine $w_0^{max}(\Delta,D_f)$. In the case of large detunings

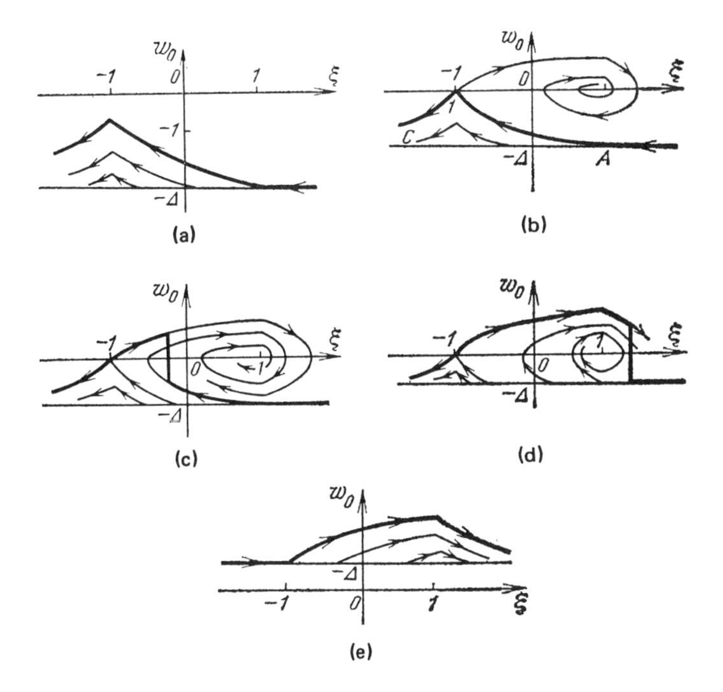

FIG. 6.14. Formation of a stationary acoustic wave profile in different quasi-one-dimensional source motion regimes: $a - \Delta > \Delta_{cr}$; $b - \Delta = \Delta_{cr}$; $c - \Delta_{max} \leqslant \Delta < \Delta_{cr}$; $d - 0 \leqslant \Delta < \Delta_{max}$; $e - \Delta < 0$.

$(\Delta \gg 1) w_0^{max}(D_f) - w_0^{max}(D_f = 0) \sim D_f$. When $\Delta - 2 \sim D_f$ the resonant curves will differ significantly depending on the presence or absence of diffraction

$$w_0^{max}(\Delta, D_f) - w_0^{max}(\Delta, D_f = 0) \sim (\Delta - 2 + 8D_f/15)^{1/2}.$$

The detuning corresponding to a crossover from the supersonic regime to the mixed regime is determined with the additional condition $w_0^{max} = 0$: $\Delta_{cr} = 2 - 8D_f/15$. Therefore the velocity range in which the mixed nonlinear wave excitation regime is implemented $(0 < \Delta < \Delta_{cr})$ diminishes with increasing value of the "diffraction/nonlinearity" parameter.

Figure 6.14(b) shows the phase trajectories for $\Delta = \Delta_{cr}$. The point $(\xi = 1, w_0 = 0)$ is a stable focus. The physical meaning of this singularity is the same as in the case of frequency-independent damping: in the heating zone $(|\xi| \leqslant 1)$ the liquid element is accelerated and overtakes the region of constant-speed sources, although outside the thermal layer the velocity of the liquid element diminishes due to diffraction, the sources catch it, etc. Such an anomaly of the crossover trajectory eliminates the discontinuity on the resonant curve for $\Delta = \Delta_{cr}$, which occurs in the absence of diffraction (see Figs. 1.8 and 6.15). Formally, this is related to the fact that it is impossible to plot a stationary wave profile other than *ABC* for $\Delta = \Delta_{cr}$ [Fig. 6.14(b)], since it is not possible to plot a fixed wave front that will interconnect the segments of the trajectory lying in the upper and lower semiplanes.

For $\Delta_{max} \leqslant \Delta < \Delta_{cr}$ the stationary wave profile is determined by the behavior of two trajectories [Fig. 6.14(c)]. The first trajectory passes through the point $(\xi = 1,$

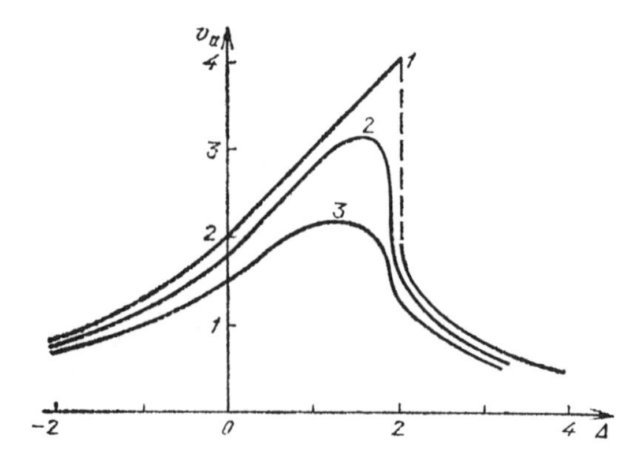

FIG. 6.15. Acoustic wave amplitude plotted as a function of source velocity for different values of the "diffraction/nonlinearity" parameters D_f of 0 (1), 0.1 (2), 0.2 (3).

$w_0 = -\Delta$), while the second passes through the point ($\xi = -1$, $w_0 = 0$). Their equations $\xi_1 = \xi_1(w_0,\Delta,D_f)$ and $\xi_2 = \xi_2(w_0,\Delta,D_f)$ can be found using Eq. (6.43). Then the edge amplitude v_e and its coordinate ξ_e are determined by the relation $\xi_e = \xi_1(-v_e/2, \Delta, D_f) = \xi_2(v_e/2, \Delta, D_f)$. The wave amplitude can be calculated by the formula $v_a = -\Delta + v_e/2$. Analysis reveals that the amplitude of the front is equal to zero for $\Delta = \Delta_{cr}$ and grows continuously with diminishing detuning through $\Delta = \Delta_{max}$ when the front lies on the right boundary of the heating zone. The wave amplitude is maximized for $\Delta = \Delta_{max} = 2 - 24D_f/5$: $v_a^{max} = 2\Delta_{max}$ (the last two relations therefore define the shift of the peak of the resonant curve for small values of the parameter D_f).

The stationary wave profile for $0 \leqslant \Delta < \Delta_{max}$ is determined solely by the trajectory passing through ($\xi = -1$, $w_0 = 0$) [Fig. 6.14(d)], and for $\Delta < 0$ is identical to the trajectory of the point ($\xi = -1$, $w_0 = -\Delta$) of the initial profile [Fig. 6.14(e)]. In both cases the maximum w_0^{max} which must be calculated to determine the wave amplitude is reached at $\xi = 1$. Specifically, with exact velocity matching ($\Delta = 0$) $v_a = 2 - 68D_f/15$.

Figure 6.15 shows the amplitude of the excited nonlinear wave plotted as a function of source velocity for different values of the "diffraction/nonlinearity" parameter ($D_f \ll 1$).

We utilize the method proposed in Ref. 22 to determine the dependence of the maximum acoustic wave amplitude on the "diffraction/nonlinearity" parameter for $D_f \gg 1$. We consider stationary motion of a narrow (in the y direction) heating zone. Then we employ Eq. (6.37) for large values of the parameter D_f in the range ($|y| \leqslant b$) to determine the component $v_\perp$ of the dimensionless velocity disturbance:

$$H(\xi)\varphi(\eta) = D_f \frac{\partial v_\perp}{\partial \eta}, \quad v_\perp = D_f^{-1} H(\xi) \int_0^\eta \varphi(\eta')d\eta'. \tag{6.44}$$

The longitudinal component of the velocity disturbance in the range $|y| \leqslant b$ is determined by the "joining" conditions to the solution of the problem in an external domain $|y| \geqslant b$. In the external domain the velocity disturbance is described by the homogeneous equation

$$\frac{\partial}{\partial \xi}\left(w\frac{\partial w}{\partial \xi}\right) = -D_f\frac{\partial^2 w}{\partial \eta^2}, \tag{6.45}$$

which must be analyzed with a boundary condition for $|\eta| \sim 1$ [see Eq. (6.36)]:

$$\frac{\partial w}{\partial \eta} = \frac{\partial v_\perp}{\partial \xi}.$$

Consistent with the method of introducing two spatial scales[22] this boundary condition will be represented using Eq. (6.44) and $\varphi(\eta) = \exp(-\eta^2)$ as

$$\left.\frac{\partial w}{\partial \eta}\right|_{\eta=0} = \left.\frac{\partial v_\perp}{\partial \xi}\right|_{\eta\to\infty} = \frac{\sqrt{\pi}}{2D_f}\frac{\partial H(\xi)}{\partial \xi}. \tag{6.46}$$

An exact solution of Eq. (6.45) for supersonic velocities ($w<0$) was found in Ref. 28:

$$w = \mathscr{F}[\xi + D_f^{-1/2}(-w)^{1/2}|\eta|].$$

After determining function $\mathscr{F}$ by means of boundary condition (6.46) we obtain

$$w = -\{\Delta^{3/2} - (3/4)(\pi/D_f)^{1/2}H[\xi + (-w/D_f)^{1/2}|\eta|]\}^{2/3}. \tag{6.47}$$

The crossover detuning Δ_{cr} is determined by Eq. (6.47) for $w(\xi,\eta=0)=0$. If the source distribution $H(\xi)$ is normalized to unity, the points of the representative profile may for the first time fall on the semiplane $w>0$ for

$$\Delta_{cr} = (3\pi^{1/2}/4)^{2/3}D_f^{-1/3}.$$

It follows from the analysis carried out above for $D_f \ll 1$ that the maximum wave amplitude will differ only insignificantly from the wave amplitude in the crossover regime even for $D_f \simeq 0.2$ (see Fig. 6.15). This makes it possible to use the last relation of those derived here as an estimate of the peak of the resonant curve: $v_a^{max} \propto D_f^{-1/3} (D_f \gg 1)$.

In conclusion, we note that the rise of the acoustic signal under quasisynchronous excitation amounting to several orders of magnitude compared to the case of fixed sources[14–16] can be used in optoacoustic spectroscopy.

6.2. Coherent Acoustic Resistance to an Electron-Hole Plasma Front Moving at Near Sound Speed

In accordance with the concept of deformation potential, the charge carrier energy in an acoustic field changes (Sec. 3.4). The change ΔF in free energy F of the electron-hole pair in the field of a plane longitudinal acoustic wave propagating along the z axis can be represented in the isotropic crystal model as $\Delta F = d(\partial u/\partial z)$, where u is mechanical displacement. Consequently, a force $f_a = -\partial(\Delta F)/\partial\delta = -d(\partial^2 u/\partial z^2)$ will be acting on each EH pair from the coherent acoustic field.

So far we have not considered the reverse effect of sound pulses excited from interband light absorption on the expansion of the photogenerated EH plasma. At the same time hydrodynamic expansion of an EH plasma at transonic velocities may substantially amplify acoustic waves (Sec. 4.2). In such cases it is necessary to account for the inverse effect of the crystal lattice deformation on EH pair motion.

We note in support of this statement that all research carried out to date aimed at accelerating EH droplets in an inhomogeneous strain field to supersonic velocities have not yielded the desired result.[63–65] The most probable reason is the effective interaction of droplets moving at transonic velocities with the coherent acoustic fields that initially decelerate,[66,67] deform,[68] and, in the final stage, destroy[69] the droplets. We note that although the theory focuses substantial attention on the effect of longitudinal acoustic waves on EH droplets,[66,68] of primary importance is analyzing droplet interaction with transverse acoustic waves,[67] since under actual experimental conditions[63–65] the EH droplet velocity did not exceed the velocity of the slow transverse acoustic mode c_{ST}.

An analysis of quasisynchronous EH plasma interaction with longitudinal acoustic waves becomes predominant in analyzing the transonic expansion of surface layers of photoexcited EH plasma. A quasi-one-dimensional initial plasma expansion regime[64,65,70,71] is realized in the majority of experiments and coherent transverse acoustic wave generation becomes inefficient on the basis of symmetry. Such experiments have recorded plasma fronts traveling away from a photoexcited surface at velocities exceeding that of the longitudinal acoustic mode.[65,70–72] However, theoretical models developed to describe fast plasma drift[73–75] have neglected the deceleration of the plasma front by coherent acoustic fields as the front crosses the sound barrier ($v_e \approx c_L$).

In order to describe coherent acoustic resistance to EH plasma, it is only necessary to introduce the force $f_a = -d(\partial^2 u/\partial z^2)$ in the equation of motion of the EH pairs (5.34):

$$M\left(\frac{\partial v}{\partial t} + v\frac{\partial v}{\partial z} + \frac{v}{\tau_d}\right) = n^{-1}\frac{\partial p}{\partial z} - d\frac{\partial^2 u}{\partial z^2}. \tag{6.48}$$

In this section we will drop the index on the hydrodynamic plasma velocity ($v_d \equiv v$). According to Eq. (6.48) the action of a coherent acoustic field on a plasma can be characterized by its pressure $\mathcal{P}_a$ on the entire region occupied by the plasma:

$$\mathcal{P}_a = -\int_{-\infty}^{\infty} nd\left(\frac{\partial^2 u}{\partial z^2}\right)dz. \tag{6.49}$$

This is particularly convenient when the EH plasma can be treated as incompressible and, consequently, capable of responding to external actions as an integrated whole.

In order to develop a convenient representation of the physical processes that may occur in such a system, we first analyze the constant transonic velocity motion of the incompressible EH plasma front. We describe acoustic wave excitation by the concentration-deformation mechanism by the equation[76,77]

$$\frac{\partial^2 u}{\partial t^2} - c_L^2\frac{\partial^2 u}{\partial z^2} + \varepsilon c_L^2\frac{\partial}{\partial z}\left(\frac{\partial u}{\partial z}\right)^2 = \frac{d}{\rho_0}\frac{\partial n}{\partial z}. \tag{6.50}$$

Equation (6.50) accounts for the acoustic nonlinearity [see Eq. (3.62)]. We assume that the plasma is distributed in space as $n = n_e N[(z - v_e t)/a]$, where $0 \leqslant N \leqslant 1$

is a smoothed step function. The quantity a represents the width of the EH plasma edge $[N(a\to 0)\to 1-\theta(z-v_e t),\ N(z\to\infty)=0,\ N(z\to-\infty)=1]$; n_e is the plasma concentration at the edge.

The comoving acoustic waves will be efficiently amplified only at the transonic plasma front velocities $[|\Delta_0| = |(v_e - c_L)/c_L| \ll 1]$ of interest to us. Equation (6.50) can be simplified by the slowly varying profile method (Sec. 6.1) for use in describing quasisynchronous excitation of sound waves that copropagate with the EH plasma front in the positive direction of the z axis. In the accompanying coordinate system $\xi = z - v_e t$ the equation for crystal deformation $D = \partial u/\partial z = \partial u/\partial \varepsilon$ can be presented as

$$\partial\mathscr{D}/\partial t - c_L\Delta_0\,\partial\mathscr{D}/\partial\xi - \varepsilon c_L\mathscr{D}\,\partial\mathscr{D}/\partial\xi = -(d\eta_e/2\rho_0 c_L)dN/\partial\xi.$$

In dimensionless variables and functions $\xi=\xi/a$, $t=t/t_0$, $\mathscr{D}=\mathscr{D}/\mathscr{D}_0$, $\mathscr{P}_a=\mathscr{P}_a/\mathscr{P}_0$, $\Delta=\Delta_0/\varepsilon\mathscr{D}_0$, where $t_0=a/c_L\varepsilon\mathscr{D}_0$, $\mathscr{P}_0=|d|n_e\mathscr{D}_0$, $\mathscr{D}_0=(|d|n_e/2\varepsilon\rho_0c_L^2)^{1/2}$, it is transformed to

$$\frac{\partial\mathscr{D}}{\partial t} - \Delta\frac{\partial\mathscr{D}}{\partial\xi} - \mathscr{D}\frac{\partial\mathscr{D}}{\partial\xi} = -(\operatorname{sgn} d)\frac{\partial N}{\partial\xi}. \tag{6.51}$$

Using the same designations for calculating acoustic field pressure (6.49) on the plasma, we obtain the integral representation

$$\mathscr{P}_a = -\operatorname{sgn} d\int_{-\infty}^{\infty} N\left(\frac{\partial\mathscr{D}}{\partial\xi}\right)d\xi. \tag{6.52}$$

In order to find the stationary acoustic wave profile accompanying the plasma front it is necessary to solve the pulse rise problem with the initial condition $\mathscr{D}(\xi, t=0)=0$.

Neglecting nonlinear acoustic effects ($\varepsilon = 0$) the stationary solution of Eq. (6.51) takes the form

$$\begin{aligned}\mathscr{D}_\infty(\xi,\Delta<0) &= (\operatorname{sgn} d/\Delta)[N(\xi)-1],\\ \mathscr{D}_\infty(\xi,\Delta>0) &= (\operatorname{sgn} d/\Delta)N(\xi).\end{aligned} \tag{6.53}$$

This solution satisfies the following physical conditions: at subsonic velocities of the front ($\Delta<0$) all excited acoustic waves overtake the sources, and hence $\mathscr{D}(\xi\to-\infty)\to 0$. For $\Delta>0$, all acoustic waves trail the sources: $\mathscr{D}(\xi\to+\infty)\to 0$.

Substituting deformation profiles (6.53) into relation (6.52), we obtain for the coherent acoustic field pressure on the EH plasma: $\mathscr{P}_a(\varepsilon=0)=(2\Delta)^{-1}$. Therefore, the acoustic field decelerates the plasma drift at subsonic front velocities ($\Delta<0$) independent of the sign of the deformation potential, while the acoustic wave pressure grows without limit as the sound barrier is approached ($\Delta\to 0$). Unlike the case of accelerated sound barrier crossing (Sec. 4.2) incorporating the finite width of the EH plasma front ($a\neq 0$) is not sufficient to explain the limit on the amplitude of the excited deformation waves or the pressure on the plasma under linear resonance conditions ($\Delta = 0$). Consequently, it is fundamentally important to account for the acoustic nonlinearity.

It is convenient to analyze inhomogeneous quasilinear equation (6.51) on the phase plane using the methods described in Sec. 6.1. When the constant of the

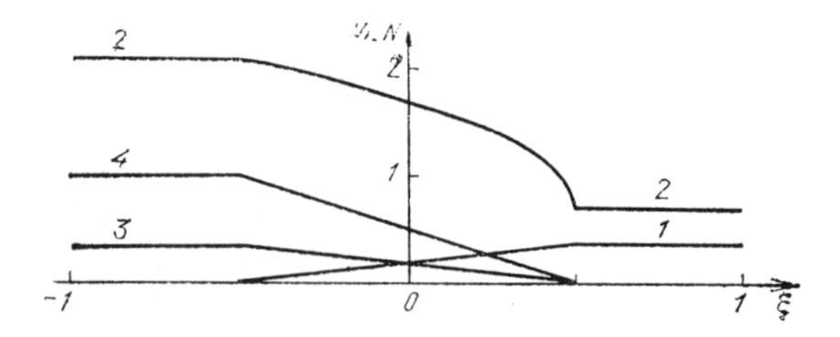

FIG. 6.16. Strain distribution in an acoustic wave near the EH plasma front (4) at different plasma velocities Δ of $-2\sqrt{2}$ (1); $-1/\sqrt{2}$ (2); $2\sqrt{2}$ (3).

deformation potential is positive ($d>0$) the phase portrait of the system on the plane ($\xi, \mathscr{D}' = \mathscr{D} + \Delta$) is analogous to that shown in Fig. 6.2. These two phase portraits overlap for the case of specular reflection relative to the vertical axis. The stationary solutions take the form

$$\mathscr{D}_\infty(d>0) = \begin{cases} -\Delta - \{\Delta^2 + 2[N(\xi) - 1]\}^{1/2}, & \Delta \leqslant -\sqrt{2}, \\ -\Delta + [2N(\xi)]^{1/2}, & -\sqrt{2} \leqslant \Delta \leqslant 0, \\ -\Delta + [\Delta^2 + 2N(\xi)]^{1/2}, & \Delta \geqslant 0. \end{cases} \quad (6.54)$$

The totally subsonic acoustic wave generation regime is realized in accordance with Eq. (6.54) for $\Delta < -\sqrt{2}$ when all excited waves overtake the EH plasma front. In Fig. 6.16 deformation profile 1 corresponds to a dimensionless detuning $\Delta < \Delta_{cr} = -\sqrt{2}$, while profile 4 describes the distribution of EH plasma concentration near the plasma front.

The crystal lattice compresses in semiconductors in which $d>0$ upon photogeneration of EH pairs, which serves to dilate the underlying regions to which the dilatational waves travel. The diminished pressure regions in an acoustic wave of finite amplitude ($\varepsilon \neq 0$) propagate at subsonic velocities (if $\varepsilon > 0$). Hence even when $v_e < c_L$, a portion of the excited waves may lag the sources. It is for this reason that the so-called mixed regime of deformation wave generation (Sec. 6.1) where acoustic disturbances may reach $\xi = +\infty$ and $\xi = -\infty$ (Fig. 6.16, profile 2) is realized for $-\sqrt{2} \leqslant \Delta \leqslant 0$. Finally, for $\Delta > 0$ the totally supersonic regime is realized: sound lags the moving front (Fig. 6.16, profile 3).

In order to employ Eq. (6.54) to calculate the acoustic pressure on the plasma it is convenient, using the stationary limit ($\partial \mathscr{D}/\partial t = 0$) of Eq. (6.51), to transform integral (6.52) into

$$-(\operatorname{sgn} d) \int N\left(\frac{\partial \mathscr{D}_\infty}{\partial \xi}\right) d\xi = -(\operatorname{sgn} d) N \mathscr{D}_\infty + \frac{\Delta \mathscr{D}_\infty^2}{2} + \frac{\mathscr{D}_\infty^3}{3} + \text{const.} \quad (6.55)$$

Using Eqs. (6.52) and (6.55), we obtain

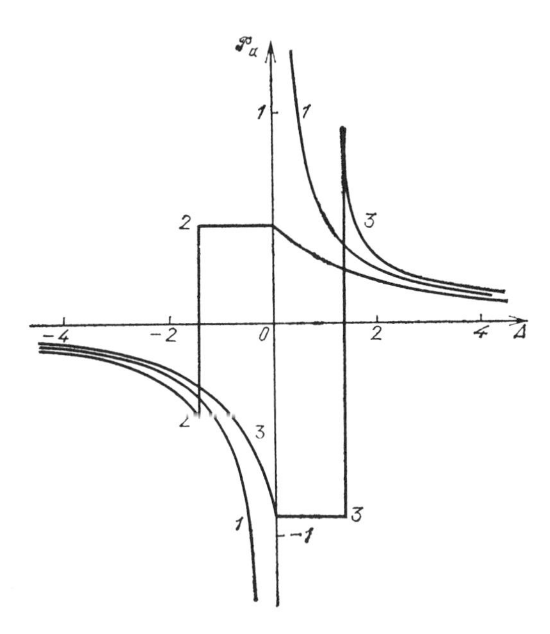

FIG. 6.17. Coherent acoustic wave pressure on the plasma plotted as a function of plasma velocity: $1 - \varepsilon = 0$, $2 - \varepsilon > 0$, $d > 0$, $3 - \varepsilon > 0$, $d < 0$.

$$\mathscr{P}_a(d>0) = \begin{cases} \Delta - (1/3)[(\Delta^2-2)^{3/2} + \Delta^3], & \Delta \leqslant -\sqrt{2}, \\ \sqrt{2}/3, & -\sqrt{2} \leqslant \Delta \leqslant 0, \\ (\Delta+2)^{1/2} - (1/3)[(\Delta^2+2)^{3/2} - \Delta^3], & \Delta \geqslant 0. \end{cases} \tag{6.56}$$

This dependence of wave pressure on the EH plasma velocity is represented in Fig. 6.17 (curve 2). Also shown in this figure (curve 3) is an analogous relation in the case $d<0$, when the phase portrait of the system on the plane (ξ, $\mathscr{D}' = \mathscr{D} + \Delta$) is identical to that shown in Fig. 6.2 with the single change in the direction of point motion along the trajectory.

An analytic representation of the solution for $d<0$ takes the form

$$\mathscr{D}_\infty(d<0) = \begin{cases} -\Delta - \{\Delta^2 - 2[N(\xi)-1]\}^{1/2}, & \Delta \leqslant 0, \\ -\Delta - \{2[1-N(\xi)]\}^{1/2}, & 0 \leqslant \Delta \leqslant \sqrt{2}, \\ -\Delta + [\Delta^2 - 2N(\xi)]^{1/2}, & \Delta \geqslant \sqrt{2}, \end{cases} \tag{6.57}$$

$$\mathscr{P}_a(d<0) = \begin{cases} -\Delta - (1/3)[\Delta^3 + (\Delta^2+2)^{3/2}], & \Delta \leqslant 0, \\ -2\sqrt{2}/3, & 0 \leqslant \Delta \leqslant \sqrt{2}, \\ -(\Delta^2-2)^{1/2} + (1/3)[\Delta^3 - (\Delta^2-2)^{3/2}], & \Delta \geqslant \sqrt{2}. \end{cases} \tag{6.58}$$

The primary physical difference between Eqs. (6.57) and (6.58) on the one hand and Eqs. (6.54) and (6.56) on the other is the shift of the acoustic wave generation mixed regime towards $\Delta > 0$. This is directly related to the fact that compressional waves are emitted from the plasma localization region in semiconductors with $d < 0$; such waves propagate at supersonic velocities for $\varepsilon > 0$.

These results demonstrate that the acoustic nonlinearity limits the sound barrier in the case of fast EH plasma expansion (Fig. 6.17). Far from velocity matching ($|\Delta| \gg 1$) Eqs. (6.54) and (6.56)–(6.58) are consistent with the conclusions of linear theory (6.53), $\mathscr{P}_a(\varepsilon = 0) \sim \Delta^{-1}$ (Fig. 6.17, curve 1).

We note that the regions of sharp variation in the relation $\mathscr{P}_a(\Delta)$ naturally smooth out if the weak (at low temperatures) damping and (or) weak diffraction of the acoustic waves (Sec. 6.1) are accounted for. Unlike the results from an analysis of EH droplet motion at near sonic speed[66] in this case the resistance to plasma drift will arise as well at subsonic velocities. This is related to the assumption of a semibounded distribution of plasma concentration in space. It will be demonstrated in analyzing the motion of an incompressible plasma layer limited on the z axis that for $v_e \neq c_L$ in a linear approximation the acoustic field pressures on the leading and trailing edges of the layer surface are compensated.

Results (6.56) and (6.58) of this calculation are independent of the specific plasma front structure and hence are valid as $a \to 0$. In this limit, a coherent stationary acoustic field acts directly on the EH plasma surface. The acoustic field will only act on the plasma surface also in the case of nonstationary plasma motion in the subsonic regime, since all acoustic waves overtake the wave front. This fact is particularly clear from an analysis of analytic solutions of Eq. (6.51) with δ-localized sources: $\partial N/\partial\xi = -\delta(\xi)$, δ is the Dirac δ function.[79]

From the evenness of the δ function the solutions of equation

$$\frac{\partial \mathscr{D}}{\partial t} - \Delta \frac{\partial \mathscr{D}}{\partial \xi} - \mathscr{D}\frac{\partial \mathscr{D}}{\partial \xi} = (\operatorname{sgn} d)\delta(\xi) \tag{6.59}$$

satisfy the following symmetry property:

$$\mathscr{D}(\xi,\Delta,d) = -\mathscr{D}(-\xi,-\Delta,-d).$$

It is therefore sufficient to analyze Eq. (6.59) in, for example, the case $d < 0$. A phase plane analysis makes possible an explicit solution of Eq. (6.59). This solution derives from the description of acoustic pulse profiles provided in Sec. 6.1 for moving thermal sources, in the limiting case $a \to 0$.

Figure 6.18 shows the characteristic representation of the solutions of Eq. (6.59) with the initial condition $\mathscr{D}(t=0) = 0$. The integral relation $\int_{-\infty}^{\infty} \mathscr{D}\, d\xi = (\operatorname{sgn} d)t$ analogous to Eq. (6.22) was used to plot the shock front that overtakes the plasma in the subsonic [$\Delta < 0$, Fig. 6.18(a)] and mixed [$0 \leqslant \Delta \leqslant \sqrt{2}$, Fig. 6.18(b)] regimes.

In the subsonic regime, the acoustic waves do not penetrate the plasma localization region. Their pressure is applied directly to the EH plasma boundary. It is also important that the rise time of the stationary deformation wave profile in the plasma localization region diminishes without limit as $a \to 0$. Hence, if the EH plasma shock front begins to move in the subsonic regime, the pressure determined by solutions (6.56) (for $d > 0$, $\Delta < -\sqrt{2}$) or Eq. (6.58) (for $d < 0$, $\Delta < 0$) will act on this plasma instantaneously. This makes it possible in describing EH pair motion

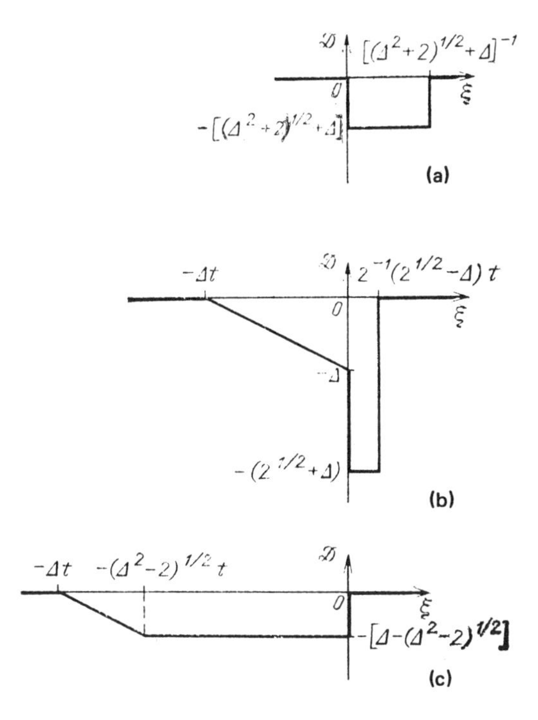

FIG. 6.18. Profile of strain pulses excited by traveling δ-localized sources in the subsonic (a), mixed (b), and supersonic (c) regimes.

to go from inhomogeneous equation (6.48) to a homogeneous equation with additional boundary conditions at the front of the expanding EH plasma.

Assume a laser source generates a step-like EH plasma distribution near the surface $z = 0$ of a semiconductor by the initial instant in time (Fig. 6.19, profile 1):

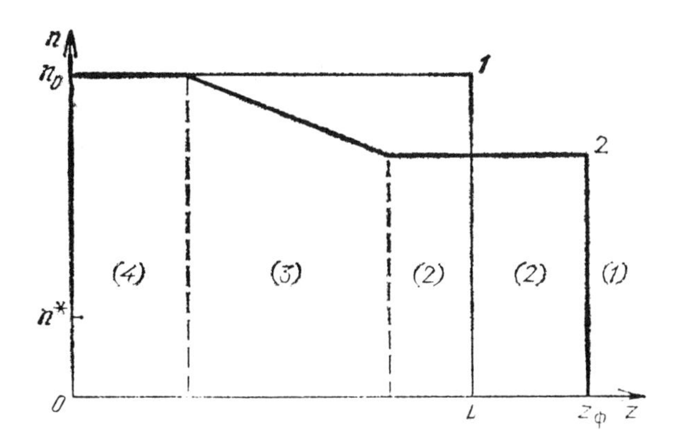

Fig. 6.19. Plasma distribution at the initial time (1) and the stage preceding surface reflection of the rarefactional wave (2).

$$n = n_0[1 - \theta(z - L)], \; z \geqslant 0. \tag{6.60}$$

Here it is assumed that the initial plasma concentration n_0 exceeds the equilibrium concentration of the EH liquid n^* (Sec. 5.3): $n_0 > n^*$. In order to describe expansion of the nonequilibrium carrier cloud we utilize the continuity equation and equation of motion:

$$\frac{\partial n}{\partial t} + \frac{\partial (nv)}{\partial z} = 0, \tag{6.61}$$

$$\frac{\partial v}{\partial t} + v\frac{\partial v}{\partial z} = -(c^2/n)\frac{\partial n}{\partial z}. \tag{6.62}$$

Here, $c = c(n)$ represents the velocity of the acoustic waves in the EH plasma [$Mc^2 \, \partial n/\partial z = \partial p/\partial z$, $p = n^2(\partial F/\partial n)_T$ is the internal plasma pressure, Sec. 5.3].

Compared to Eq. (6.48), Eq. (6.62) does not account for carrier scattering by phonons, crystal lattice defects, etc., since a dominant role of deceleration by the coherent sound field is assumed. Equation (6.61) does not account for EH pair recombination, since only the initial instants of the evolution of spatially inhomogeneous distribution (6.60) generated by the ultrashort optical irradiation are investigated. We, in fact, will analyze a classical hydrodynamic problem of the decay of a discontinuity in the initial conditions,[80] although we account for the resistance of coherent nonlinear acoustic waves.

Consistent with general theoretical concepts of the decay of an initial discontinuity,[80] it is possible to identify three regions of the spatial distribution in the first stage of the process [$0 \leqslant t \leqslant L/c(n_0)$] (Fig. 6.19, profile 2): region (2) directly behind the shock front of the EH liquid ($L_1 \leqslant z \leqslant z_e$); region (4), the undisturbed plasma near the crystal surface ($0 \leqslant z \leqslant L_0 = L - c[n_0)t]$; and region (3), in which the rarefactional wave traveling through the plasma couples regions (4) and (2) ($L_0 \leqslant z \leqslant L_1$). A solution of Eqs. (6.61) and (6.62) in region (3) is an autosimilar solution[80]:

$$z = L + [v(n) - c(n)]t,$$

$$v = \int_n^{n_0} \left(\frac{c(n')}{n'}\right) dn'. \tag{6.63}$$

Estimates indicate that the boundary velocities $z = L_0(v_{L0} = -c(n_0))$ and $z = L_1(v_{L1} = v(n_2) - c(n_2))$ are substantially different from the sound speed of the crystal c_L, which made it possible to neglect asynchronous sound generation in region (3) [to drop the second term on the right-hand side of Eq. (6.48)]. Solution (6.63) is "joined" to fixed region (4), since $v(n_0) = 0$. According to Eq. (6.63) plasma velocity in region (2)

$$v_2 = v(n_2) = \int_{n_2}^{n_0} \left(\frac{c(n)}{n}\right) dn, \tag{6.64}$$

where n_2 is the plasma concentration in this region.

The boundary conditions at the shock front ($z = z_e$) yield one relation to supplement Eq. (6.64) for determining v_2 and n_2[80]:

$$n_2 v_{20} = n_1 v_{10}, \tag{6.65}$$

$$n_2 v_{20}(E_2 + p_2/n_2 + v_{20}^2/2) = n_1 v_{10}(E_1 + p_1/n_1 + v_{10}^2/2), \tag{6.66}$$

$$p_2 + n_2 v_{20}^2 = p_1 + n_1 v_{10}^2, \tag{6.67}$$

where the subscript 1 applies to the crystal region in front of the shock wave $(z > z_e)$, $v_{i0} = v_i - v_e (i = 1,2)$ is the velocity of the plasma relative to the shock front; E is the internal energy of the EH plasma.

Since the plasma expands in a vacuum $(n_1 = 0)$ under the conditions of our problem, from the law of flow continuity of matter (6.65) $v_{20} = 0$, i.e., plasma velocity in region (2) is identical to the velocity of the shock front $v_2 = v_e$. Then Eq. (6.66) holds identically, while the law of conservation of momentum (6.67) reduces to equality between the internal plasma pressure in region 2 and the external pressure on the front from the coherent acoustic field:

$$p_2(n_2) = \mathscr{P}_0 |\mathscr{P}_a(d,\Delta)|. \tag{6.68}$$

Consistent with this analysis, resistance pressure (6.56), (6.58) will depend on the velocity of the shock front ($\Delta = (v_e - c_L)/c_L \varepsilon \mathscr{D}_0$) and, consequently, relations (6.64) and (6.68) determine the dependence of v_e on the initial plasma concentration since $v_e = v_2$. And, vice versa, we can calculate the concentration n_0 that is required for the plasma to expand at a predetermined subsonic velocity or to cross the sound barrier.

In order to illustrate these results we employ for the EH pair free energy F a representation that does not account for the contribution from correlations between carriers[74]:

$$F(n) \approx a n^{2/3} - b n^{1/3}. \tag{6.69}$$

The first term in (6.69) corresponds to kinetic energy, and the second term corresponds to the exchange energy; a and b are constants. Then

$$p = 2an^{5/3}/3 - bn^{4/3}/3, \quad n^* = (b/2a)^3,$$

$$c(n) = (10an^{2/3} - 4bn^{1/3})^{1/2}/3M^{1/2},$$

and relations (6.64) and (6.68) in the case $d < 0$, $\Delta \leqslant 0$ reduce to the form

$$\frac{v_e}{c_L} = 1 + \Delta\varepsilon\mathscr{D}_0$$

$$= \left(\frac{10a}{M}\right)^{1/2} \frac{(n^*)^{1/3}}{c_L} \left\{ n\left(n^2 - \frac{4}{5}\right)^{1/2} - \frac{4}{5}\ln\left| n + \left(n^2 - \frac{4}{5}\right)^{1/2} \right| \right\} \Bigg|_{N_e^{1/6}}^{N_0^{1/6}}, \tag{6.70}$$

$$N_e^{1/6} - N_e^{-1/6} = \left[\frac{3|d|^{3/2}}{2ac_L(2\varepsilon\rho_0)^{1/2}(n^*)^{1/6}}\right] \left\{ \Delta + \frac{1}{3}[\Delta^3 + (\Delta^2 + 2)^{3/2}] \right\}, \tag{6.71}$$

where $N_0 \equiv n_0/n^*$, $N_e \equiv n_2/n^*$, and from the previous normalization

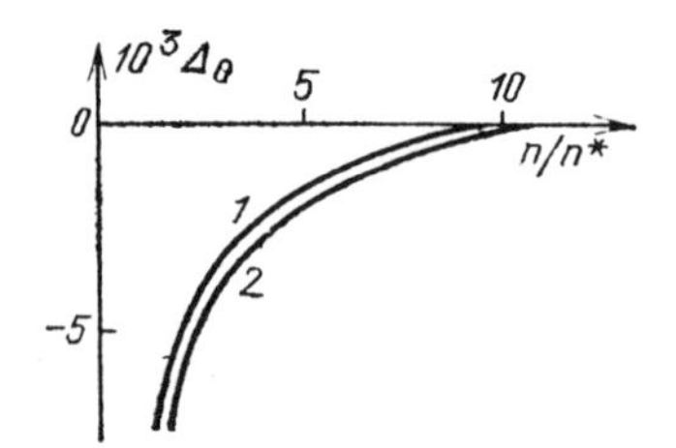

FIG. 6.20. Velocity of EH plasma shock front plotted as a function of concentration at the front (1) and the initial concentration (2).

$$\Delta = (v_e - c_L)/\varepsilon c_L \mathscr{D}_0 = (2\rho_0/\varepsilon |d| n^* N_e)^{1/2}(v_e - c_L). \qquad (6.72)$$

Equalities (6.71) and (6.72) permit finding $N_e = N_e(v_e)$ and then we determine $N_0(v_e)$ by means of Eq. (6.70).

The results of such an analysis for germanium ($d<0$) are shown in Fig. 6.20. The following parameter values were used in the calculation: $|d| \approx 7$ eV,[65] $M \approx 0.4 \times 10^{-27}$ g,[65] $\varepsilon \approx 14.6$, $a \approx 1.6 \times 10^{-26}$ erg cm^2, $b \approx 1.4 \times 10^{-20}$ erg cm. The numerical values of constants a and b correspond to calculations in the Hartree-Fock approximation (HF) accounting for the actual band structure of germanium.[81] We note, however, that neglecting the correlation energy leads roughly to a twofold reduction in the ground-state density within the framework of this model: n^*(HF, Ge) $\approx 0.8 \times 10^{17}$ cm^{-3}.

According to the calculations (Fig. 6.20) the counteraction of the coherent acoustic field to EH plasma front motion begins to be manifested in direct proximity to velocity matching: $|v_e - c_L|/c_L \leqslant 10^{-2}$. In order to cross the sound barrier, the plasma concentrations at the front (and the initial concentration) must be approximately an order of magnitude higher than the equilibrium concentration. Estimates demonstrate the validity of these conclusions for silicon as well: $d>0$, $\Delta \leqslant -\sqrt{2}$, n^*(HF,Si) $\approx 10^{18}$ cm^{-3}. Therefore, an acoustic barrier may lead to an increase in photogenerated plasma concentration near the semiconductor surface.

These calculations have demonstrated that when a plasma expands at transonic velocities a rather uniform distribution may arise along the z axis ($n_0 \approx n_2 = n_e \gg n^*$). This makes it possible to qualitatively estimate the optical radiation intensity I_0 required to sustain fast ($v_e \approx c_L$) EH plasma motion during light absorption: $n_e v_e \sim I_0/\hbar\omega_L$. The latter estimate is analogous to relation (4.52). Using this estimate we find that an intensity $I_0 \gtrsim 10^5$ W/cm^2 is required to cross the sound barrier in germanium under laser irradiation at $\lambda_1 \approx 1.06$ μm. This is in agreement with experimental results.[65]

We consider the features of resistance to a moving EH plasma layer by coherent acoustic fields. Reference 68 has demonstrated that as the EH droplet approaches the sound barrier, it flattens out in its direction of motion without limit. It was accounted for here, however, that the results are not applicable in direct proximity to linear resonance ($v_e = c_L$). This, specifically, is related to the fact that as a droplet becomes thinner (as $v_e \to c_L$, $v_e < c_L$) it transforms into a quasiplanar EH liquid layer. The role of diffraction, which limits the amplitude of quasisynchronous excited longitudinal acoustic waves, diminishes sharply in this case and the pressure of the acoustic fields on the EH droplet grows without limit. Under these conditions the acoustic nonlinearity may play a fundamental role in limiting the sound barrier.[78]

We utilize model (6.51), (6.52) with a Gaussian-like distribution of the nonequilibrium carrier concentration $[0 \leqslant N \leqslant 1, N(\xi) \to \pm\infty \to 0]$ to describe the coherent acoustic resistance to motion of the incompressible EH liquid layer. In this case, the characteristic spatial scale a represents the thickness of the EH plasma layer. In addition to $\mathscr{P}_a$ (6.52) it is useful to introduce the acoustic field pressure at the leading $\mathscr{P}_1$ and trailing $\mathscr{P}_2$ fronts of the layer:

$$\mathscr{P}_1 = -\operatorname{sgn} d \int_0^\infty N\left(\frac{\partial \mathscr{D}}{\partial \xi}\right) d\xi, \quad \mathscr{P}_2 = -\operatorname{sgn} d \int_{-\infty}^0 N\left(\frac{\partial \mathscr{D}}{\partial \xi}\right) d\xi. \tag{6.73}$$

For definiteness we treat $N(\xi)$ as an even function of the coordinate $N(\xi)$
$= N(-\xi)$, $N(0) = 1$, $\partial N/\partial \xi(0) = 0$.

The stationary solution of linearized Eq. (6.51) off exact velocity matching takes the form

$$\mathscr{D}_\infty(\Delta \neq 0, \varepsilon = 0) = (\operatorname{sgn} d/\Delta) N(\xi). \tag{6.74}$$

Using this solution, we find for Eqs. (6.52) and (6.73)

$$\mathscr{P}_1 = -\mathscr{P}_2 = (2\Delta)^{-1}, \quad \mathscr{P}_a = \mathscr{P}_1 + \mathscr{P}_2 = 0. \tag{6.75}$$

Therefore although the entire EH plasma layer does not experience resistance to its transonic motion ($\mathscr{P}_a = 0$), compressive forces on this layer grow without limit as $\Delta \to 0$ $(\Delta < 0)$ in accordance with Eq. (6.75).

When the velocity of the layer matches the sound speed ($\Delta = 0$) the linear acoustic field accompanying the EH layer grows without limit over time

$$\mathscr{D}(\Delta = 0, \varepsilon = 0) = -(\operatorname{sgn} d)\left(\frac{\partial N}{\partial \xi}\right) t.$$

Then,

$$\mathscr{P}_a = 2\mathscr{P}_1 = 2\mathscr{P}_2 = -\left[\int_{-\infty}^{\infty} \left(\frac{\partial N}{\partial \xi}\right)^2 d\xi\right] t. \tag{6.76}$$

The pressure decelerating layer motion grows without limit in accordance with Eq. (6.76) for $\Delta = 0$. The sound barrier for a quasistationary traveling EH plasma layer is insurmountable within the framework of linear acoustics.

The stationary solutions of the nonlinear equation are easily found on the phase plane $(\xi, \mathscr{D}' = \mathscr{D} + \Delta)$. The trajectory equation takes the form

$$(\mathscr{D}')^2/2 - (\operatorname{sgn} d) N(\xi) = C_1.$$

Consequently, the distribution $(-\operatorname{sgn} d) N(\xi)$ plays the role of the potential function $\mathscr{H}(\xi)$ (Sec. 6.1). Unlike the results of linear theory (6.74)–(6.76) the solution of the nonlinear generation problem (6.51) $[\mathscr{D}'(t=0,\xi) = \Delta]$ is highly dependent on the sign of the deformation potential.

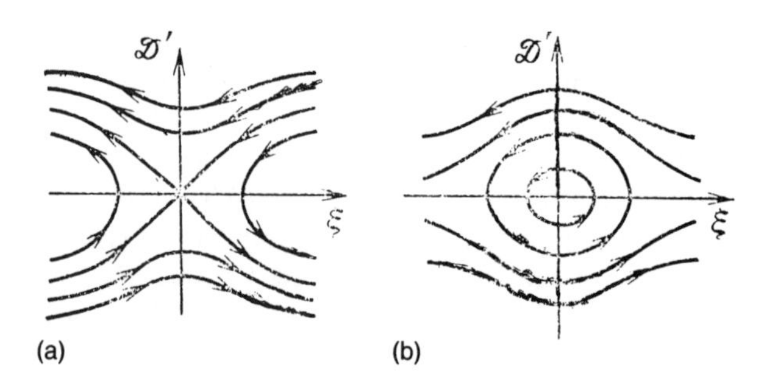

FIG. 6.21. Phase portrait of sound generation by a moving EH plasma layer in the cases $d<0$ (a) and $d>0$ (b).

The maximum of the potential function (the potential barrier) is realized in the EH plasma localization region in the case $d<0$. The phase portrait of the system [Fig. 6.21(a)] contains at the point (0, 0) a saddle singularity. The form of the stationary profiles changes radically depending on whether or not the initial profile $\mathscr{D}'=\Delta$ intersects the separatrix $\mathscr{D}'=\pm\operatorname{sgn}\xi\{2[1-N(\xi)]\}^{1/2}$. For $|\Delta|\geqslant\sqrt{2}$ the stationary profiles lie in the totally supersonic range (if $\Delta>0$); or in the totally subsonic range (if $\Delta<0$): $\mathscr{D}_\infty(|\Delta|\geqslant\sqrt{2})=-\Delta+\operatorname{sgn}\Delta[\Delta^2-2N(\xi)]^{1/2}$.

Near resonance ($|\Delta|\leqslant\sqrt{2}$) the profile of stationary distribution $\mathscr{D}'_\infty$ is identical to the separatrix $\mathscr{D}'=-\operatorname{sgn}\xi\{2[1-N(\xi)]\}^{1/2}$ independent of the initial velocity mismatch: $\mathscr{D}_\infty(|\Delta|\leqslant\sqrt{2})=-\Delta-\operatorname{sgn}\xi\{2[1-N(\xi)]\}^{1/2}$.

Using auxiliary function (6.55) to calculate the pressure on the plasma, we obtain

$$\mathscr{P}_1(\Delta)=-\mathscr{P}_1(-\Delta)=-\mathscr{P}_2(\Delta),\ \mathscr{P}_a=0,$$

$$\mathscr{P}_1(\Delta\geqslant\sqrt{2})=-(\Delta^2-2)^{1/2}+[\Delta^3-(\Delta^2-2)^{3/2}]/3,\ |\Delta|\geqslant\sqrt{2}, \tag{6.77}$$

$$\mathscr{P}_1=\mathscr{P}_2=-2\sqrt{2}/3,\ \mathscr{P}_a=-4\sqrt{2}/3,\ |\Delta|\leqslant\sqrt{2}.$$

In the case $d>0$ the potential function minimum (the potential well) is realized in the plasma localization region. The phase portrait of the system [Fig. 6.21(b)] contains a center-type singularity at the point (0,0). Off velocity matching $\Delta\neq0$ the stationary profile $\mathscr{D}'_\infty(\xi)$ fall either in the totally supersonic range (for $\Delta>0$) or in the totally subsonic range (for $\Delta<0$). Here $\mathscr{D}_\infty(\Delta\neq0)=-\Delta+(\operatorname{sgn}\Delta)[\Delta^2+2N(\xi)]^{1/2}$ is valid for the steady-state deformation profile.

In the case of exact velocity matching ($\Delta=0$) the stationary profile consists of sections of the separatrix $\mathscr{D}=\pm[2N(\xi)]^{1/2}$ and contains a discontinuity: $\mathscr{D}_\infty(\Delta=0)=\operatorname{sgn}\xi[2N(\xi)]^{1/2}$.

Calculations of the pressure of coherent acoustic fields on the EH plasma layer for $d>0$ lead to the following results:

$$\mathscr{P}_1(\Delta)=-\mathscr{P}_1(-\Delta)=-\mathscr{P}_2(\Delta),\ \mathscr{P}_a=0,$$

$$\mathscr{P}_1(\Delta>0)=(\Delta^2+2)^{1/2}+[\Delta^3-(\Delta^2+2)^{3/2}]/3,\ \Delta\neq0, \tag{6.78}$$

$$\mathscr{P}_1=\mathscr{P}_2=\mathscr{P}_a=0, \Delta=0.$$

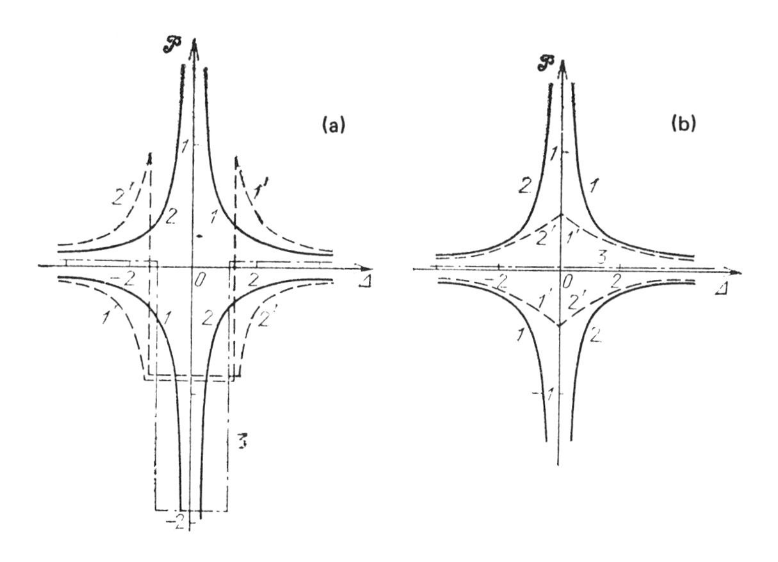

FIG. 6.22. The compressive and decelerating pressures plotted as a function of EH plasma layer velocity in the cases $d<0$ (a) and $d>0$ (b).

According to these relations, the nonlinear acoustic effects limit the amplitudes of quasisynchronously excited deformation waves that compress plasma, of pressure $\mathscr{P}_1$ and $\mathscr{P}_2$ as well as the sound barrier $\mathscr{P}_a$. The physical cause of this limiting process is the change in the propagation velocity of the nonlinear waves with increasing wave amplitude such that they exit the localization region of the sources moving with constant velocity.

Relations (6.75)–(6.78) of the pressures acting on the plasma layer are presented in Fig. 6.22 in the case of a negative (a) and positive (b) deformation potential constant. In this figure, 1, 2 and 1′, 2′ represent the compressive forces in the linear ($\varepsilon=0$) and nonlinear ($\varepsilon>0$) regimes; 3 is the resistance pressure $\mathscr{P}_a$ in the nonlinear regime ($\varepsilon>0$).

We can easily see that the nonlinear interaction of acoustic waves expands the resistance region for the EH plasma layer if $d<0$ (for example, in Ge) and destroys the stationary sound barrier if $d>0$ (in, for example, Si). The following qualitative explanation can be provided. If $d<0$, at subsonic layer velocities ($\Delta<0$) the crystal lattice will expand from the interference of linear acoustic waves excited at the leading and trailing fronts of the EH layer in accordance with Eq. (6.74). Rarefactional waves of finite amplitude propagate at subsonic velocities for $\varepsilon>0$ and hence the EH plasma layer will begin to be effectively counteracted as early as $-\sqrt{2}\leqslant\Delta\leqslant 0$. Supersonic layer motion (if $d<0$) causes crystal compression in accordance with Eq. (6.74). Compressional waves of finite amplitude propagate at supersonic velocities (for $\varepsilon>0$). This leads to effective deceleration of the layer through shifts $\Delta\leqslant\sqrt{2}$ [Fig. 6.22(a)].

In the case $d>0$ acoustic waves propogating at supersonic velocities are excited for $\Delta < 0$ while propogating at subsonic velocities for $\Delta > 0$. The acoustic nonlinearity therefore inhibits quasisynchronous deceleration of the EH plasma layer [Fig. 6.22(b)].

For purposes of clarity we emphasize that these conclusions apply to the stationary acoustic fields accompanying the plasma ($t \to \infty$). A resistance to plasma motion associated with nonstationary processes may appear during the rise of the acoustic fields as in the case when such fields are altered due to the variation of source velocity.[66] We consider as an example $d>0$, $\Delta=0$, $N(\xi)=\theta(\xi+1) - \theta(\xi-1)$.

In this case an excited acoustic field results from the summing of nonlinear acoustic waves generated by two moving delta sources

$$\mathscr{D}(t\geqslant 0) = \sqrt{2}\{-[\theta(\xi+1)-\theta(\xi+1-t/\sqrt{2})] + [\theta(\xi-1+t/\sqrt{2})-\theta(\xi-1)]\}.$$

A calculation from Eqs. (6.51) and (6.52) of the plasma-counteracting pressure yields

$$\mathscr{P}_a = -\frac{\partial}{\partial t}\int_{-\infty}^{\infty}\frac{D^2}{2}d\xi = -\sqrt{2}[\theta(t)-\theta(t-\sqrt{2})].$$

The coherent acoustic field will therefore counteract the EH layer to the time $t=\sqrt{2}$ when a stationary deformation distribution is established.

In concluding this section we estimate the characteristic acoustic field pressure inhibiting sound barrier crossing in Ge(Si) by EH plasma layers[79]: $|\mathscr{P}_a| \sim \mathscr{P}_0 \sim (|d|n^*)^{3/2}(2\varepsilon\rho_0 c_L^2)^{-1/2} \sim 5\times 10^2 (5\times 10^4)$ dyn/cm^2.

6.3. Excitation of Acoustic Waves of Finite Amplitude by Quasisynchronous Monochromatic Sources

Sections 6.1 and 6.2 provide a detailed description of nonlinear sound generation by moving broadband sources. Laser radiation makes it possible to generate moving harmonic sources of acoustic waves as well. For example, a traveling interference pattern generated on the surface of the absorbing medium by crossed laser beams at different frequencies can serve as a quasisynchronous monochromatic SAW source. The possibility of forming transonic electrostriction sources of longitudinal acoustic waves in the field of two crossed optical waves was established in Sec. 3.4. When phase-matching conditions hold in a linear approximation the amplitude of the difference frequency wave grows without limit with increasing light-crystal interaction length.

In order to describe the limiting of sound amplitude excited by moving monochromatic sources we introduced into wave equation (3.66) high-frequency absorption of acoustic waves (Sec. 2.3) and account for the acoustic nonlinearity (3.62):

$$\frac{\partial^2 u}{\partial t^2} - c_L^2\frac{\partial^3 u}{\partial z^2} - \frac{b}{\rho_0}\frac{\partial^2 u}{\partial t \partial z} + c_L^2\varepsilon\frac{\partial}{\partial z}\left(\frac{\partial u}{\partial z}\right)^2$$

$$= -\frac{YE_1E_2}{8\pi\rho_0}\frac{\partial}{\partial z}[\cos(\omega t - k_a z - \Delta kz - \varphi_0)]. \qquad (6.79)$$

In this section we are interested in quasisynchronous sound generation, i.e., the case where the velocity of the striction sources $V_0 = \omega/(k_a + \Delta k)$ is nearly identical to the velocity of the acoustic wave $|V_0 - c_L|/c_L \equiv |\Delta_0| \sim \mu \ll 1$. Under this condition, the wave accompanying the sources (the comoving wave) is amplified preferentially. In order to describe the slow variation in the comoving wave profile as it propogates from the boundary [assuming $u = \mu u(\mu z, \tau = t - z/V_0)$] it is possible to obtain from Eq. (6.79) an evolution equation[82,83]

$$\partial v/\partial z + (\Delta_0/c_L)\partial v/\partial \tau - (\varepsilon v/c_L^2)\partial v/\partial \tau - (b/2\rho_0 c_L^3)\partial^2 v/\partial \tau^2$$
$$= -\operatorname{sgn} Y(|Y|E_1E_2\omega/16\pi\rho_0 c_L^2)\sin(\omega\tau - \varphi_0).$$

Here $v = \partial u/\partial t = \partial u/\partial \tau$ is the vibrational velocity of particles in the acoustic wave.

Going over to dimensionless variables and functions

$$\tau = \omega\tau - \varphi_0 - \pi(1 + \operatorname{sgn} Y)/2,\ z = z/L_{\text{NL}},\ \Delta = (V_0 - c_L)/\varepsilon v_0,$$

$$v = \frac{v}{v_0},\ v_0 = \left(\frac{N}{\varepsilon}\right)^{1/2} c_L,\ L_{\text{NL}} = \frac{c_L^2}{\varepsilon\omega v_0},\ N = \frac{|Y|E_1E_2}{16\pi\rho_0 c_L^2},$$

we obtain

$$\partial v/\partial z + \Delta\ \partial v/\partial \tau - v\partial v/\partial \tau - \Gamma\ \partial^2 v/\partial \tau^2 = \sin\tau. \qquad (6.80)$$

Here L_{NL} is the formation length of the discontinuity in a freely propagating harmonic wave of frequency ω and amplitude v_0. Parameter Γ characterizes the relative role of the dissipative and nonlinear effects: $\Gamma = L_{\text{NL}}/L_{\text{DS}} = (2\varepsilon\ \text{Re})^{-1}$.[33] The dimensionless parameter N dependent on optical source intensity determines the characteristic acoustic Mach number of the excited nonlinear wave: $M_a \sim |v|/c_L \sim v_0/c_L \propto N^{1/2}$ (in the absence of high-frequency absorption). Equation (6.80) is called an inhomogeneous Burgers' equation.[35,82]

In the case of exact synchronism $\Delta = 0$ neglecting dissipation and the acoustic nonlinearity, the solution of Eq. (6.80) is essentially identical to that found in Sec. 3.4: $v = z\sin\tau$. It satisfies the boundary condition $v(z=0) = 0$.

In order to describe the limit on the acoustic wave amplitude as it propogates from the boundary it is sufficient to account for the dissipative term in Eq. (6.80). For $\varepsilon = 0$, $\Delta = 0$ the solution of Eq. (6.80) takes the form

$$v(\Delta = 0, \Gamma \gg 1) = (1 - \exp(-\Gamma z))\Gamma^{-1}\sin\tau.$$

High-frequency ($L_{\text{DS}} \sim \omega^{-2}$) sound absorption therefore limits the amplitude of a synchronously excited wave at the level Γ^{-1}. In dimensionless variables $M_a \sim |v|/c_L \sim v_0/\Gamma c_L \propto N$ (in the absence of nonlinear effects).

In the case of large acoustic Reynolds' numbers ($\text{Re} \gg 1$) Eq. (6.80) becomes an inhomogeneous equation of simple waves with a harmonic right side:

$$\frac{\partial w}{\partial z} - w\frac{\partial w}{\partial t} = \sin\tau,\ w = v - \Delta. \qquad (6.81)$$

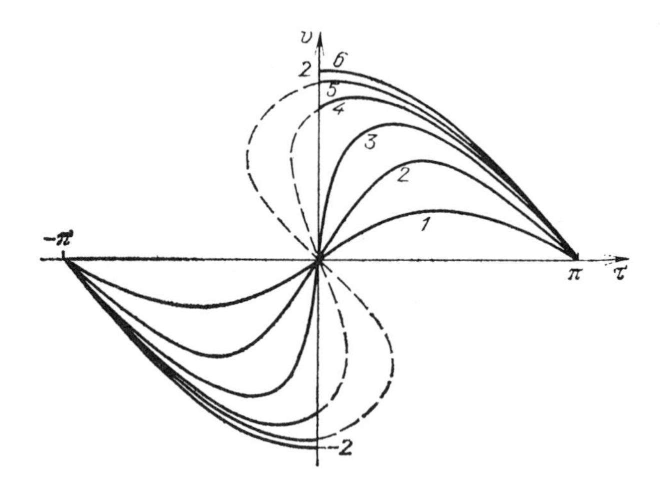

FIG. 6.23. Profiles of a wave synchronously excited by harmonic sources at different distances z from the boundary: 0.5 (1), 1 (2), 1.5 (3), 2(4), 2.5 (5), ∞ (6).

The problem of excitation of sea waves of finite amplitude by a traveling harmonic pressure system[84–86] and the problem of the stabilization and parametric interaction of nonlinear oscillations in acoustic resonators[87–89] can also be reduced to an analysis of Eq. (6.81).

If the boundary of the medium $z=0$ is rigid, the solution of Eq. (6.81) must satisfy the condition

$$w(z=0,\tau) = -\Delta. \tag{6.82}$$

One notable feature of problem (6.81), (6.82) is the existence of an exact solution by special functions.[39] This solution can be obtained using a scheme analogous to Eqs. (6.11)–(6.14),[39,35]

$$z = \pm\{F[\arcsin(\gamma^{-1}\sin(\tau_0/2)),\gamma] - F[\arcsin(\gamma^{-1}\sin(\tau/2)),\gamma]\},$$

$$\gamma^2 = w^2/4 + \sin^2(\tau/2) = \Delta^2/4 + \sin^2(\tau_0/2),$$

where F is Legendre's form of the elliptical integral of the first kind.[90] In the case of exact velocity matching ($\Delta=0$) the solution takes the form

$$z = F[\pi/2,\gamma] - F[\arcsin(\sin(\tau/2)\gamma^{-1}),\gamma]. \tag{6.83}$$

The results of a graphical analysis of relation (6.83) are given in Fig. 6.23. The vibrational velocity profiles of a periodic wave are plotted with increasing distance of the observation point from the boundary. Since the compressional regions ($v>0$) in an acoustic wave of finite amplitude propagate at supersonic velocities, while the rarefactional regions ($v<0$) move at subsonic velocities, the accumulating nonlinear distortions will cause the overlaps in the profile $v(t)$ (the dashed curves in Fig. 6.23). The front obeys the rule of equivalent intersected areas, although in this specific case ($\Delta=0$) its position is also determined by symmetry considerations.

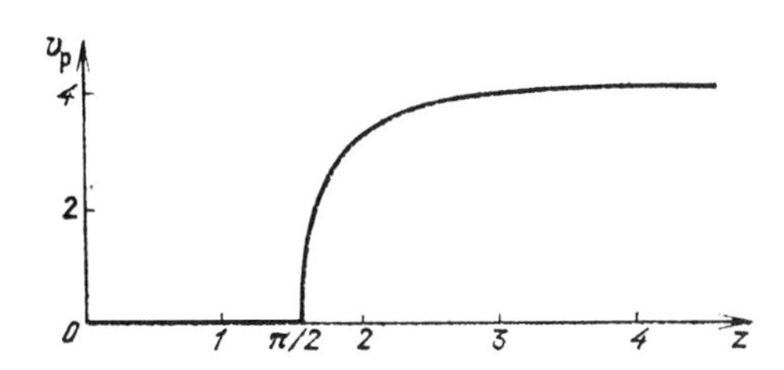

FIG. 6.24. Discontinuity amplitude as a function of distance covered by the excited wave.

Assuming $\tau = 0$ in (6.83), we obtain for the amplitude of the discontinuity $v_d = v(\tau + 0) - v(\tau - 0)$: $z = F[\pi/2, v_d/4] \equiv \mathscr{K}(v_d/4)$, where $\mathscr{K}$ is the complete normal elliptical integral of the first kind.[90] Figure 6.24 shows the discontinuity amplitude plotted as a function of the distance traveled by the wave. The dimensionless length of formation of the discontinuity z_{NL} is given by $z_{NL} = \pi/2$. Therefore it is roughly one and a half times the formation length of the discontinuity in a freely propagating wave of amplitude v_0.

Although it is possible to formulate an exact solution for any mismatch of the source velocity and the velocity of linear sound, it is advisable to use the phase plane method[87–89] to obtain a convenient physical representation of wave evolution in the system. The system of characteristic equations corresponding to Eq. (6.81) can be given as

$$\frac{dw}{dz} = \sin \tau,$$
$$\frac{d\tau}{dz} = -w. \tag{6.84}$$

This system defines the motion of each of the points on the excited wave profile in the periodic potential $\mathscr{H}(\tau) = -2\cos^2(\tau/2)$.

The phase trajectories of system (6.84) are described by

$$w = \pm[C_1 + 4\cos^2(\tau/2)]^{1/2}, \quad C \geqslant -4. \tag{6.85}$$

Phase portrait (6.85) [Fig. 6.25(a)] takes the characteristic form of Kelvin's "cat's eyes."[47] The arrows indicate the direction of the representative points motion along the trajectories, C_1 is the trajectory parameter. For $C_1 = 0$ relation (6.85) is a separatrix equation [curves 1 and 2 in Fig. 6.25(a)]. With negative values of the parameter $C_1(-4 \leqslant C_1 < 0)$, Eq. (6.85) describes on the phase plane closed trajectories of the liquid elements trapped by the potential wells. The infinite motion region corresponds to positive values of the parameter C_1 on the phase plane (τ, w).

The concepts of trapped and transit particles were apparently introduced for the first time in analyzing the interaction of particles with longitudinal waves in a plasma.[45–47] The "cat's eyes" appear in the phase space of the particles in the longitudinal wave field if the particle and wave velocities are similar. An analogous mathematical apparatus can be used in the theory of traveling wave tubes.[91,92] However, Refs. 45–47 in fact analyze a system of noninteracting particles with no

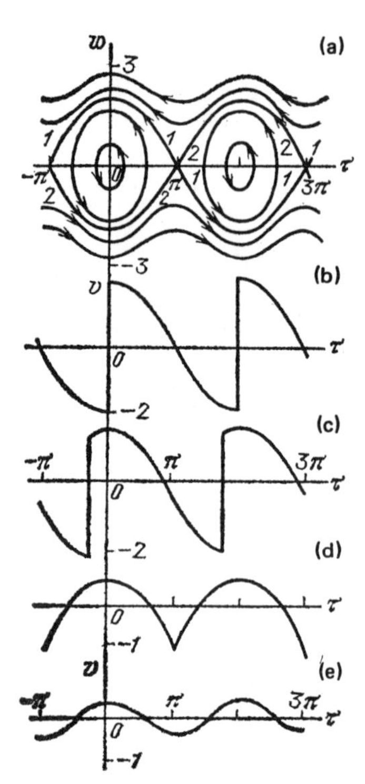

FIG. 6.25. Phase portrait (a) and stationary traveling wave profiles for different detunings Δ: 0 (b), -0.5 (c), $-4/\pi$ (d), -2.45 (e).

constraints imposed on their collective motion. Specifically, several particles with different velocities can simultaneously occupy a given point in space. This corresponds to a reversal of the profile of the particle velocity versus coordinate relation on the phase plane.[45] In contrast with this, the formation of the overlapped profile of the acoustic wave does not correspond to physical reality.[33]

The specific nature of our problem lies in the fundamental need to account for collective interactions of elements of a distributed system. Eliminating ambiguity of the wave profile by establishing the front, we in fact account for the infinitely weak energy dissipation of the high-frequency acoustic waves [the discarded term $\Gamma\partial^2 v/\partial\tau^2$ in Eq. (6.80)].

We first determine the wave form of the stationary stimulated nonlinear waves. The profile of the disturbance at the initial instant in time is represented on the phase plane (τ,w) by horizontal line (6.82), and it is transformed during the generation process in such a manner that on the average there is no deviation from the initial level over the period:

$$(2\pi)^{-1}\int_{\tau-\pi}^{\tau+\pi} w(z,\tau')d\tau' = -\Delta = \text{const.} \tag{6.86}$$

Integral relation (6.86) derives directly from Eq. (6.81) when (6.82) holds and is an analog of law (6.22) for monochromatic sources. It demonstrates that in these approximations the effect of moving harmonic sources will not result in directional motion of the medium: $\int_{\tau-\pi}^{\tau+\pi} v(z,\tau')d\tau' \equiv 0$.

Since relation (6.85) is an integral of (6.81) in the wave profile stabilization region ($\partial v/\partial z = 0$), the wave profile established at large distances ($z \gg 1$) will either be identical to one of the phase trajectories of (6.85) or will consist of parts of different phase trajectories and weak discontinuities. The primary criterion for selecting the profile in the stabilization region is satisfaction of condition (6.86) for each fixed detuning Δ.

Problem (6.81), (6.82) is not altered by the simultaneous substitution $\Delta \to -\Delta$, $w \to -w$, $\tau \to -\tau$ and hence it is sufficient, for example, to consider the case of source motion at subsonic velocities ($\Delta \leqslant 0$). The symmetry transform provided above can be used to determine the behavior of the test system for the case of supersonic source motion.

Analysis suggests that the stationary wave profile w for $|\Delta| < \pi/4$ consists of separatrix sections and contains a discontinuity whose position is determined by condition (6.86) [Figs. 6.25(b) and 6.25(c)]:

$$w_\infty = v_\infty - \Delta = 2|\cos(\tau/2)|\,\mathrm{sgn}[\tau - 2\pi m - 2\arcsin(\pi\Delta/4)],$$
$$(2m-1)\pi \leqslant \tau \leqslant (2m+1)\pi, \tag{6.87}$$
$$m = 0, \pm 1, \pm 2, \pm \ldots$$

Here the function in square brackets vanishes at the discontinuities of the acoustic wave profile: $\tau_d = 2\pi m + 2\arcsin(\pi\Delta/4)$.

In this range of source velocities ($|\Delta| < 4/\pi$) the stationary (as $z \to \infty$) wave amplitude v_a and the discontinuity amplitude v_d are given by

$$v_a = v_{\max} - v_{\min} = 2 + 2[1 - (\pi\Delta/4)^2]^{1/2},$$
$$v_d = |w_\infty(\tau_d + 0) - w_\infty(\tau_d - 0)| = 4[1 - (\pi\Delta/4)^2]^{1/2}.$$

The wave profile in the region $z \gg 1$ for $|\Delta| = 4\pi$ consists of sections of separatrixes, while the shock front vanishes [Fig. 6.25(d)]. For mismatches $|\Delta| \geqslant 4/\pi$ the stationary wave contains no discontinuities and is identical to one of the phase trajectories of (6.85) [Figs. 6.25(d) and 6.25(e)]. The wave polarity on the plane (τ, w) is dependent on the detuning sign ($\mathrm{sgn}\, w_\infty = -\mathrm{sgn}\,\Delta$), while the parameter C_1 is determined by the relation

$$(C_1 + 4)^{1/2} E[2(C_1 + 4)^{-1/2}] = \pi|\Delta|/2,$$

where E is the complete normal elliptical integral of the second kind.[90] The disturbance amplitude is calculated from $v_a = (C_1 + 4)^{1/2} - C^{1/2}$. Far from velocity matching ($|\Delta| \gg 1$) this amplitude decays in inverse proportion to the detuning ($v_a \propto |\Delta|^{-1}$). A plot of the wave amplitude in the stabilization region and of the discontinuity amplitude as a function of harmonic source velocity is shown in Fig. 6.26.

Neglecting the nonlinear acoustic effects, the amplitude of the harmonic wave oscillates in space (Sec. 3.4). The acoustic nonlinearity stabilizes the wave profile. There exists a trapping band $|\Delta| \leqslant 4/\pi$ in which wave amplification by quasisynchronous sources will lead to a nonlinear alteration of the wave propagation velocity in a manner that compensates the difference between the source velocity and the

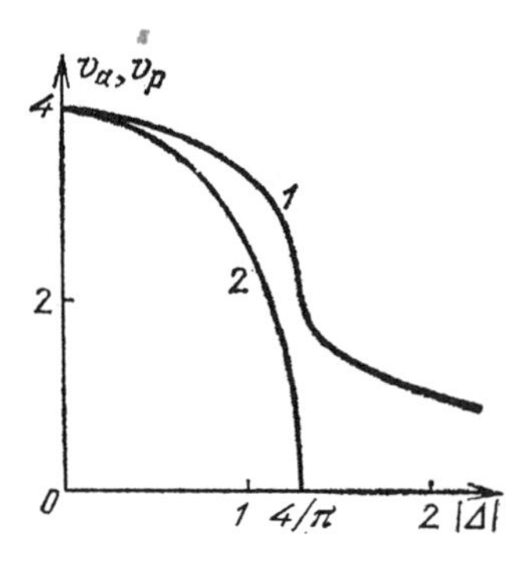

FIG. 6.26. Wave amplitude v_a (1) and discontinuity amplitude v_d (2) plotted as a function of harmonic source velocity.

linear sound speed. Finally, the acoustic nonlinearity limits the acoustic wave amplitude near synchronism (Figs. 6.23 and 6.26).

Wave profiles (6.87) are predicted by theory and have been observed experimentally from vibrations of cuvettes containing "shallow water"[93,94] and of acoustic tubes.[95–97] Experimental results[94,96,97] can be treated as confirmation of theoretical conclusions of nonlinear wave generation in dispersionless media by moving harmonic sources.

Profile steepening and formation of a weak shock front (Fig. 6.23) is characterized in the language of frequency as generation of harmonics of the acoustic fundamental frequency and a cascade energy transfer up the spectrum. We carry out a harmonic analysis[90] of periodic discontinuous profiles (6.87):

$$v_\infty(\tau) = 2 \sum_{m=1}^{\infty} |\tilde{v}_m| \cos(m\tau + \arg \tilde{v}_m),$$

$$\tilde{v}_m = \tilde{v}_{-m} = (2\pi)^{-1} \int_{-\pi}^{\pi} v_\infty(\tau) e^{-im\tau}\, d\tau.$$

The calculation result is as follows:

$$|\tilde{v}_m| = \frac{4}{\pi(4m^2-1)^{1/2}} \left[\frac{4m^2}{4m^2-1} - \left(\frac{\pi\Delta}{4} \right)^2 \right]^{1/2}.$$

The last relation indicates that both the harmonic amplitudes and the amplitude of the fundamental wave ($m = 1$) grow with increasing proximity to velocity matching. The relative harmonic value $|\tilde{v}_m|/|\tilde{v}_1|$ grows with diminishing absolute mismatch $|\Delta|$.

Reference 98 reported the experimental observation of hypersound second harmonic generation by stimulated Brillouin scattering (SBS) in a liquid. Unlike the case examined in Sec. 3.4, under SBS conditions the counterpropagating optical wave E_2 is not introduced to the medium from outside, but rather is formed from the scattering of incident electromagnetic field E_1 by the acoustic wave.[83,99]

We now analyze the dynamics of the wave profile stabilization process. In order to obtain the wave form at an arbitrary distance from the boundary it is necessary to plot on the plane (τ, w) boundary profile (6.82) and allow the representative

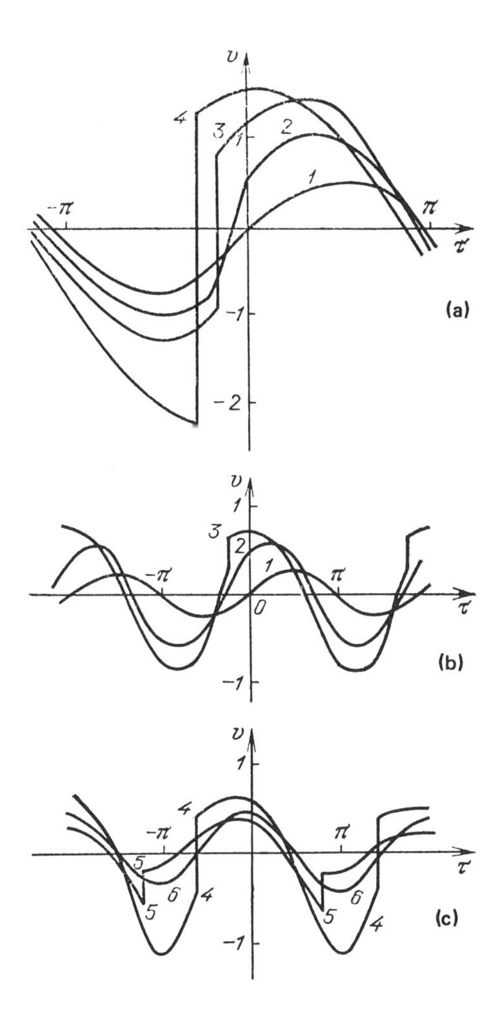

FIG. 6.27. Dynamics of linear wave profile stabilization: $a - \Delta = -0.5$; $z = 0.6$ (1); 1.2 (2); 1.8 (3); ∞ (4); $b - \Delta = -2.35$; $z = 0.2$ (1); 0.6 (2); 1 (3); 1.6 (4); 2.6 (5); ∞ (6).

points to follow trajectories (6.85) in accordance with the law of motion (6.84). The result of this plot for the two characteristic source velocities is shown in Fig. 6.27.

Figure 6.27(a) shows the evolution of the wave profile as it travels away from the boundary in the case $|\Delta| < 4/\pi$ where the sources travel at a subsonic velocity ($\Delta < 0$). The physical cause of the asymmetry of the positive and negative phases of the acoustic wave can be explained on a qualitative level as follows. The rarefactional regions of the acoustic wave $v < 0$) travels at velocities below c_L. Hence, with small negative detuning, the conditions of their synchronous amplification by sources are enhanced. For the compressional regions ($v > 0$) degraded generation efficiency is observed with detuning of the same sign. As a result, the amplitude of the rarefactional phase exceeds the amplitude of the compressional phase.

In the case illustrated by Fig. 6.27(b) the velocity of the sources is significantly different from linear sound speed ($|\Delta| > 4/\pi$). With such mismatches the stationary waves contain no profile discontinuities. An analysis of stabilization [Fig. 6.27(b)] shows that nonlinear interactions form shock waves at a certain distance from the boundary in this case as well (profile 3). However, subsequently the significant difference between wave-front velocity and source velocity drives the wave front towards the region of the antiphase sources [Fig. 6.27(b), profile 5] where it effectively "disperses" due to the simultaneous effect of the sources and nonlinear absorption.

Quasisynchronous sound generation by traveling monochromatic sources in the limit of infinitely weak dissipation ($\mathrm{Re} = \infty$) was described above. Since the inhomogeneous Burgers' equation (6.80) is linearized by the Hopf-Cole substitution ($v = 2\Gamma\partial \ln V/\partial\tau$, V is a new function), an analytic description of the simultaneous manifestation of nonlinear and dissipative effects is also possible.[83,35] However, both these solutions and their analysis are quite cumbersome. Overall, both analytic and numerical calculations within the framework of general model (6.80) confirm the physical anomalies of the process that are manifested in an analysis of the linearized problem as well as the phase plane of Eq. (6.81).

It is, however, useful to recall a simple method of deriving approximate solutions of Eq. (6.80) for the case of large, yet finite acoustic Reynolds' numbers Re. If the local Reynolds' number is significant ($v_d \mathrm{Re} \gg 1$, v_d is the dimensionless amplitude of the front) at the discontinuity of the solution of inhomogeneous equation of simple waves (6.81), the discontinuity will remain fixed in the transition to an inhomogeneous Burgers' equation. The wave form of the front is described by a hyperbolic tangent function.[33] For example, in order to obtain an approximate analytical description of the profiles of stationary waves it is sufficient to carry out the substitution $\mathrm{sgn}(\tau - \tau_d) \to \mathrm{th}[(\tau - \tau_d) v_d \mathrm{Re}/2]$ in solution (6.87).

A detailed analysis of the limits on the applicability of various approximate solutions of Eq. (6.80) as a function of the values of Δ and Γ is carried out in Ref. 35.

In conclusion, we note that these methods of describing quasisynchronous sound generation in the field of counterpropagating light waves[35] have been used successfully in analyzing the effect of an acoustic nonlinearity on the SBS process at low temperatures.[83,100]

6.4. Interaction of Coherent Nonlinear Acoustic Pulses with Thermal Pulses

Absorption of optical radiation by low-temperature crystals may cause substantially nonequilibrium heating, with the concept of temperature useless for analysis (Sec. 3.5). Nonstationary phonon heat conduction has been attracting the interest of researchers in recent years.[101] Traditionally kinetic equations in a form that does not account for the possibility of regular deformation wave propagation in the crystal are employed to describe the space-time evolution of nonequilibrium phonon fields. However, in many cases it is necessary to consider the interaction of thermal pulses with deformation waves.

On the one hand, under laser irradiation nonequilibrium phonons and coherent acoustic pulses may be simultaneous excited in the crystals. Such a case is characteristic of semiconductors that absorb light at energies exceeding the band gap. Indeed, when the EH pair concentration is altered the crystal lattice is deformed throughout the photoexcitation region (regular deformation) (Secs. 3.4 and 4.1); nonequilibrium phonons are excited from the nonequilibrium carrier relaxation to the bottom of the band, from nonradiative electron-hole recombination and also from indirect optical transition (Sec. 3.5). On the other hand, creation and interaction of nonequilibrium phonons will in turn result in the generation of coherent sound (Sec. 3.6). Moreover, deformation waves are synchronously excited from ballistic propagation of thermal pulses (Sec. 3.6).

Estimates reveal that acoustic pulses with characteristic acoustic Mach numbers of the order of 10^{-3} are excited in crystals in standard low-temperature experiments.[104] The possibility of soliton formation from nonlinear self-action of such intense acoustic waves was discussed in Ref. 105 under conditions of weak dispersion of the sound speed. However, their interaction with the nonequilibrium phonon field was neglected, although the acoustic pulse steepening process and the interaction between the regular wave and random lattice vibrations is due to the same effect: the quadratic nonlinearity in the elastic wave dynamics.

Reference 106 predicts the possibility of generating a series of thermal pulses in an intense coherent harmonic wave field, although it is noted that these solutions are not valid if the pump wave can generate harmonics. The problem here is that the formation length of a shock front in a regular wave is of the same order as the space-time phonon bunching length.

We consider a simple model[107] for describing the initial stage of interaction of regular nonlinear acoustic waves with spatially and temporally modulated phonon fields (thermal pulses). This model permits analysis of the formation of a coherent acoustic wave of finite amplitude from thermal pulse propagation; self-action of an acoustic pulse in a phonon flux, and space-time bunching of phonons in a regular wave field.

In its most general formulation the problem involves analyzing a system of equations: a wave equation for the atomic displacement u in the coherent wave (Sec. 3.4) and a kinetic equation for the phonon distribution function $N_{\text{ph}}(\mathbf{k},\mathbf{r},t)$ (Sec. 3.5). These equations are related, since the deformation waves shift the phonon frequencies (3.86), while the phonons make a contribution to elastic stress tensor (3.87).

We analyze the simplest case. As in Sec. 3.6, we neglect dispersion and anisotropy of acoustic wave propagation, and only account for phonons of longitudinal polarization. Then the system of equations in plane geometry (Secs. 3.5 and 3.6) can be given as

$$\frac{\partial^2 u}{\partial t^2} - c_L^2 \frac{\partial^2 u}{\partial z^2} + c_L^2 \varepsilon \frac{\partial}{\partial z}\left(\frac{\partial u}{\partial z}\right)^2 = -\frac{1}{\rho_0}\frac{\partial}{\partial z}\left[(2\pi)^{-3}\int \hbar c_L k(\varepsilon_2 + \varepsilon_1 \xi^2) N_k \, d\mathbf{k}\right], \tag{6.88}$$

$$\frac{\partial N_k}{\partial t} + \frac{\partial \omega}{\partial k_z}\frac{\partial N_k}{\partial z} - \frac{\partial \omega}{\partial z}\frac{\partial N_k}{\partial k_z} = St_{\text{ph-ph}} \,. \tag{6.89}$$

Here k_z is the projection of the phonon wave vector onto the z axis: $k_z = k\xi$, $k = |\mathbf{k}|$, while strain dependence of the phonon frequency (3.86) is simplified,

$$\omega = c_L k\left[1 - (\varepsilon_2 + \varepsilon_1 \xi^2)\frac{\partial u}{\partial z}\right]. \tag{6.90}$$

The remaining designations are the same as in Chap. 3. Equation (6.88), in fact, coincides with Eq. (3.91), while Eq. (6.89) follows from Eqs. (3.72) and (3.74) when only phonon-phonon interactions are taken into account for the case of the geometry selected here.

We employ equation set (6.88)–(6.90) to analyze synchronous interaction of deformation waves with phonons propagating in the same direction (along the z axis). Interaction with such ($\xi = 1$) phonons is most efficient according to Eqs. (6.88) and (6.90). We recall that phonons propagating quasicollinearly to the z axis compose a thermal pulse leading front where the most efficient generation of coherent sound occurs (Sec. 3.6).

We note that such a formulation of the problem has a real physical basis. First, selected phonon emission directions from the scattering of photoexcited electrons exist due to the anisotropy of the electron-phonon deformation potential (in, for example, semiconductors),[108] while the elastic anisotropy of the crystal gives rise to preferred directions of phonon field energy propagation (phonon focusing).[109] Second, unidirectional random high-frequency acoustic waves may be generated. Such a situation is realized in the surface piezoeffect (Sec. 3.4) if damage to the crystal structure is irregular in the sound excitation region.[110]

Assuming $N_k \propto \delta(k_x)\delta(k_y)$ in equation system (6.88)–(6.90) (where δ is the Dirac δ function), multiplying Eq. (6.89) by the phonon energy and integrating over the wave vectors, we obtain a closed system of equations for the displacement u in a regular wave and the free-energy density of the phonon field $F = (2\pi)^{-3}\int \hbar c_L \times k N_k \, d\mathbf{k}$:

$$\begin{gathered}\frac{\partial^2 u}{\partial t^2} - c_L^2 \frac{\partial^2 u}{\partial z^2} + c_L^2 \varepsilon \frac{\partial}{\partial z}\left(\frac{\partial u}{\partial z}\right)^2 + \frac{\varepsilon}{\rho_0}\frac{\partial F}{\partial z} = 0, \\ \frac{\partial F}{\partial t} + c_L\left(1 - \varepsilon \frac{\partial u}{\partial z}\right)\frac{\partial F}{\partial z} + 2c_L \varepsilon \frac{\partial^2 u}{\partial z^2} F = 0.\end{gathered} \tag{6.91}$$

In going over to Eqs. (6.91) we additionally neglect umklapp processes. In this case the collision integral $St_{\text{ph-ph}}$ conserves energy of phonons propagating in the given direction.

Simplifying Eqs. (6.91) by the slowly varying profile method [$u = \mu u(\mu z, \tau = t - z/c_L)$, $F = \mu^2 F(\mu z, \tau = z/c_L)$, $\mu \ll 1$], we arrive[107] at a system of quasilinear equations for the vibrational velocity in a coherent acoustic wave $v = \partial u/\partial t = \partial u/\partial \tau$ and the quantity $V = (F/\rho_0)^{1/2}$:

$$\begin{gathered}\frac{\partial v}{\partial z} - \left(\frac{\varepsilon}{c_L^2}\right) v \frac{\partial v}{\partial \tau} - \left(\frac{\varepsilon}{c_L^2}\right) V \frac{\partial V}{\partial \tau} = 0, \\ \frac{\partial V}{\partial z} - \left(\frac{\varepsilon}{c_L^2}\right) v \frac{\partial V}{\partial \tau} - \left(\frac{\varepsilon}{c_L^2}\right) V \frac{\partial v}{\partial \tau} = 0.\end{gathered}$$

If we draw a parallel between the equilibrium incoherent phonon field and acoustic noise (Sec. 3.6), physically the quantity V will be the dispersion of the vibrational velocity of the random acoustic wave field. Following this analogy, we will refer to V as the dispersion of the phonon field.

It is convenient to use dimensionless variables and functions in the analysis below:

$$v = v/v_0,\ V = V/v_0,\ \tau = \tau/\tau_0,\ z = z/L_{NL},\ L_{NL} = c_L^2\tau_0/\varepsilon v_0. \quad (6.92)$$

Then the system of equations takes the form

$$\frac{\partial v}{\partial z} - v\frac{\partial v}{\partial \tau} - V\frac{\partial V}{\partial \tau} = 0, \quad (6.93)$$

$$\frac{\partial V}{\partial z} - v\frac{\partial V}{\partial \tau} - V\frac{\partial v}{\partial \tau} = 0. \quad (6.94)$$

System of equations (6.93), (6.94) is apparently the simplest possible model for describing the interaction of regular longitudinal waves of finite amplitude with a nonequilibrium phonon flux. The second term in Eq. (6.93) describes the self-action of a regular wave (Sec. 2.3). The third term in this equation describes the excitation of the regular component in regions containing energy density gradients of the phonon field ($\partial V^2/\partial\tau \neq 0$) (Sec. 3.6). Equation (6.94) accounts for the effect of the regular wave on phonon motion. This can also be obtained formally by the slowly varying profile method proceeding from a wave equation with a parametric source.[106] For a given coherent acoustic field [$v = v(\tau)$] Eq. (6.94) describes all processes identified in the analysis.[106]

From Eqs. (6.93) and (6.94) the total energy of the regular and phonon components of the acoustic field is conserved:

$$\frac{\partial}{\partial z}\int_{-\infty}^{\infty}(v^2 + V^2)d\tau = 0.$$

This is valid through the point where ambiguities appear in the solutions that are characteristic of quasilinear systems of hyperbolic equations.[33] We analyze the evolution of regular wave profiles and the phonon field dispersion only at distances comparable to the possible formation length of weak shock waves (the initial interaction stage).

If a modulated phonon flux is represented at the boundary $z = 0$, while no coherent wave is present, i.e.,

$$V(z=0,t) = f_N(t),\ v(z=0,t) = 0, \quad (6.95)$$

in describing the generation of a regular wave it is advisable to initially neglect its reverse effect on the phonon field as well as self-action. Then system of equations (6.93), (6.94) defines the linear growth of the coherent component with distance:

$$v(z,t) = \frac{1}{2}\frac{\partial}{\partial\tau}[f_N^2(\tau)]z.$$

The linear rise of amplitude of a synchronously excited acoustic wave [unlike the logarithmic growth described in Eq. (3.98)] is related to the fact that in this case the thermal pulse is generated by a narrow phonon distribution [$N \propto \delta(k_x)\delta(k_y)$]. As a result, neglecting the reverse effect of sound on phonon heat conduction, the phonon flux and the coherent acoustic wave sources will not decay with distance from the boundary $z=0$. If $f_N(\tau)$ describes a spatially limited phonon field, a regular compressional wave will be excited in the vicinity of its leading edge $\partial f_N/\partial\tau>0$, while a rarefactional wave will be excited in the vicinity of its trailing edge $\delta f_N/d\tau<0$ (the cases are reversed when $\varepsilon<0$). It is necessary to find an exact solution of system (6.98), (6.94) with boundary conditions (6.95) to describe the nonlinear limiting of a synchronously excited wave.

If a weak regular wave (signal) is given at the boundary $z=0$, while the phonon flux is not modulated, i.e.,

$$v(z=0,t)=f_S(t),|\,f_S|\ll 1,\;\; V(z=0,t)=V_0=\text{const}, \qquad (6.96)$$

it is natural in describing the effect of the regular wave on the phonon field to linearize the system of equations with respect to the small signal and the variation of the phonon field dispersion:

$$V=V_0+V',\;|V'|\ll V_0,\;|v|\ll V_0, \qquad (6.97)$$

$$\frac{\partial v}{\partial z}-V_0\frac{\partial V'}{\partial\tau}=0,\;\;\frac{\partial V'}{\partial z}-V_0\frac{\partial v}{\partial\tau}=0. \qquad (6.98)$$

System (6.98) makes it possible to obtain a closed equation for the regular wave: $\partial^2 v/\partial z^2-V_0^2\,\partial^2 v/\partial\tau^2=0$. Its solution, which satisfies boundary condition (6.96), takes the form

$$v(z,\tau)=(1/2)[\,f_S(\tau+V_0z)+f_S(\tau-V_0z)\,]. \qquad (6.99)$$

Using this solution as well as Eqs. (6.96) and (6.98), we find the representation for the modulated component of the phonon field dispersion:

$$V'(z,\tau)=(1/2)[\,f_S(\tau+V_0z)-f_S(\tau-V_0z)\,]. \qquad (6.100)$$

Solution (6.99) describes the decay of an initially fixed (in the comoving coordinate system) regular wave into two waves propagating in opposite directions.

We explain the physical mechanism of this process in the case of a regular compressional pulse ($f_s\geqslant 0$) for example. Assume for definiteness that the pulse is a bell-shaped pulse. The local sound speed rises in the acoustic pulse localization region. This will cause the phonons in this region to begin to overtake the regular pulse. A bipolar disturbance profile in phonon field dispersion (6.100) is formed. Here, interaction of the variation V' with the principal phonon field at the boundaries of the disturbed region according to the first equation of (6.98) excites compressional pulses. In the centrum at the disturbed region a rarefactional pulse is excited that gradually compensates the peak of the initial signal. This process step-by-step leads to the splitting of the compressional pulse into two pulses.

Solution (6.100) describes the space-time redistribution of noise within the regular wave field. In the case of harmonic signals $f_s(t)=\sin t$ it is similar to the solutions obtained in Ref. 106, and determines the periodic addition to the unmodulated phonon field $V'(z,\tau)\;=\;\sin(V_0z)\cos\tau$.

In this case phonon bunching is realized at a distance $z_b\propto V_0^{-1}$. The form of boundary condition (6.96) indicates that (6.92) was normalized to the characteristic amplitude of the regular wave. Hence, in dimensionless variables, the formation length of the discontinuity in the regular wave $z_d \sim 1$, and from Eq. (6.97) $V_0^{-1} \ll 1$. Therefore, in this situation, $z_b \ll z_d$ and unlike Ref. 106 solutions (6.99), (6.100) provide a correct physical description of weak phonon bunching.

One advantage of model (6.93), (6.94) proposed in Ref. 107 is that it can be used to obtain an analytic description of the interaction of regular acoustic waves with phonons when the linearization of this system of equations is not physically justified. Indeed, adding and subtracting Eqs. (6.94) and (6.93) we arrive at two independent quasilinear equations

$$\frac{\partial w_+}{\partial z} - w_+ \frac{\partial w_+}{\partial \tau} = 0, \quad \frac{\partial w_-}{\partial z} + w_- \frac{\partial w_-}{\partial \tau} = 0$$

for the functions $w_\pm = V \pm v$. According to these equations the profiles of functions $w_\pm$ are distorted as simple waves (Sec. 6.1).

An exact analytic solution of these equations in implicit form can be obtained for arbitrary boundary conditions.[33] However, a graphical analysis of the solutions of these equations is more convenient and accessible in the majority of practical cases (see Fig. 6.1). The regular wave profiles and dispersion of the phonon field at a certain distance z from the boundary $z=0$ are normally found in practice in three stages: in the first stage, the profiles $w_+(z=0) = f_N(\tau) + f_S(\tau)$ and $w_-(z=0) = f_N(\tau) - f_S(\tau)$ are plotted on the planes (w_+,τ) and (w_-,τ) for the initial boundary conditions $V(z=0,t) = f_N(t)$, $v(z=0,t) = f_S(t)$, respectively. The profiles $w_+(z,\tau)$ and $w_-(z,\tau)$ are determined graphically in the second stage and, finally, we find the spatial distribution of the phonon field dispersion and

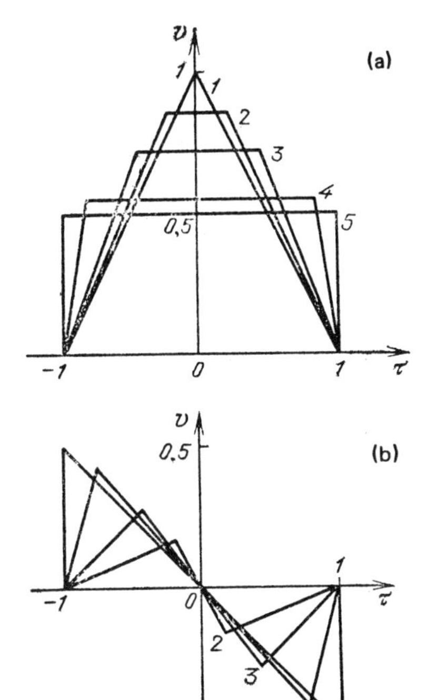

FIG. 6.28. Thermal pulse (a) and excited nonlinear regular wave (b) profiles at different distances z from the boundary of 0 (1), 0.2 (2), 0.4 (3), 0.8 (4), 1 (5).

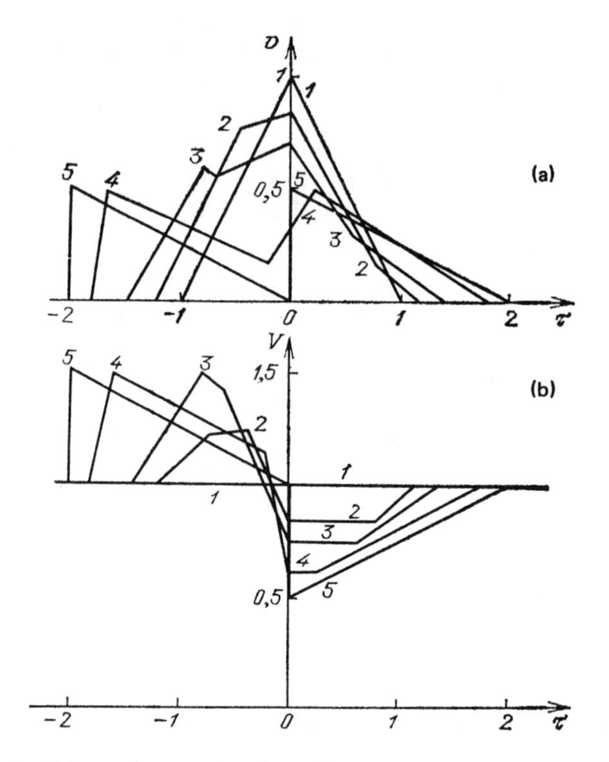

FIG. 6.29. Evolution of the regular acoustic pulse profile (a) and the phonon field dispersion profile (b) with a distance z from the boundary of the nonlinear medium of 0 (1), 0.2 (2), 0.4 (3), 0.8 (4), 1 (5).

the profile of the regular component of the acoustic field from the equations $V=(w_+ + w_-)/2$, $v=(w_+ - w_-)/2$ for the distance z from the boundary of interest to us.

Figure 6.28 shows the generation $[v(z=0)=0]$ of regular acoustic pulses from thermal pulse propagation. For simplicity the graphical profile of the thermal pulse is approximated by the linear sections [Fig. 6.28(a)]. Self-action limits the amplitude of the synchronously excited pulse [Fig. 6.28(b)]. The reverse effect of the regular waves on the phonon field leads to broadening of the pulse and steepening of its fronts [Fig. 6.28(a)]. A compressional wave is excited near the leading front of the thermal pulse, the local sound speed rises and the leading front is accelerated. The effect on the phonons of the regular rarefactional pulse is manifested as a slowing of the trailing front. The thermal pulse is broadened overall. The solutions provide a qualitative description of the process through $z \sim z_d \sim 1$, where in this case upon normalization of Eq. (6.92) τ_0 is the characteristic duration of the thermal pulse, and v_0 is the characteristic dispersion of the vibrational velocity in the phonon field.

Figure 6.29 shows the decay of the regular acoustic pulse into two pulses in the initially unmodulated phonon flux for the case where the formation distance of the discontinuity in the regular wave is equal to the phonon bunching distance $(v_0=1)$. Thus, the trend described by solution (6.99) is conserved when self-

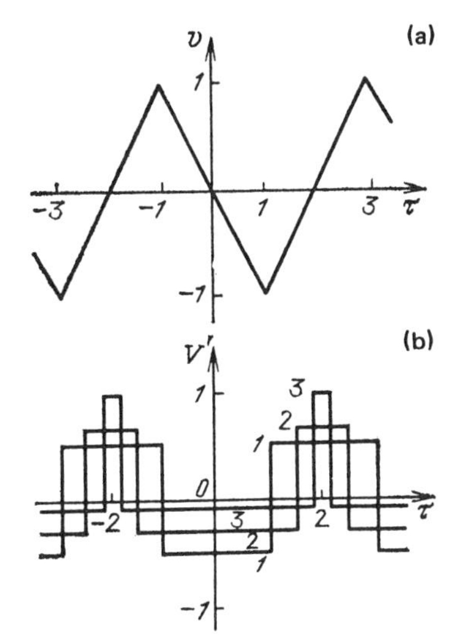

FIG. 6.30. Regular periodic wave at the boundary $z = 0$ (a) and the resulting sequence of thermal pulses generated in its field at a distance $z = z_d = 1$ (b) for different stationary phonon fluxes V_0 at the boundary of 1 (1), 0.6 (2), 0.2 (3).

action of the regular wave is taken into account. If the amplitude of the coherent pulse exceeds the dispersion of the phonon field, there will be no complete spatial separation of the acoustic pulses into which the initial pulse decays by the beginning of front formation.

Figure 6.30 shows the result of the graphical analysis of space-time phonon rebunching in the field of a regular periodic wave of finite amplitude. In the initial stage $z \leqslant 1$ nonlinear distortion to the regular wave profile does not inhibit periodic modulation of the phonon field. In the absence of phonon modulation at the boundary the derived description [Figs. 6.29 and 6.30] is valid through $z \sim z_d \sim 1$, where in normalization (6.92) τ_0 is the characteristic time scale of the regular wave, v_0 is its amplitude.

With diminishing phonon source intensity at the boundary the phonons bunch in the region $z \sim z_d$ into a sequence of increasingly short pulses [Fig. 6.30(b)]. The duration of such pulses τ_N can be estimated as $\tau_N \sim \tau_0 V_0 v^{-1}$ in this model. Therefore, formally $\tau_N \rightarrow 0$ for $V_0 \rightarrow 0$. In the actual case the minimum duration of the generated thermal pulses is determined by the same processes neglected in model (6.93), (6.94) which determine the finite width of the resulting wave fronts.

The quasicollinear interactions of acoustic waves ought to be primarily considered. In describing the structure of the wave fields for $z \geqslant z_d$ it is therefore necessary to either account for the dispersion of the acoustic wave velocities (which is possible even within the framework of the one-dimensional problem[105,111]) or go to an three-dimensional description of at least noise [see Eqs. (6.89) and (6.90)]. The latter approach makes it possible to supplement Eqs. (6.93) and (6.94) with terms that describe the preferential damping of the high-frequency components of the plane acoustic wave as energy is gradually contributed to noncollinear phonon field components.

In conclusion, we note that the applicability conditions of the concept of phonon wave packets may break down as the spatial scales of acoustic field variations diminish (for example, near the front of the regular wave) (Sec. 3.5). In this connection it is promising to use the mathematical apparatus of statistical nonlinear acoustics to analyze this problem.[33,112,113]

REFERENCES

1. L. M. Lyamshev and L. V. Sedov, Akust. Zh. **27**, 5 (1981) [Sov. Phys. Acoust. **27**, 4 (1981)].
2. V. N. Lugovoy and V. N. Strel'tsov, Zh. Eksp. Teor. Fiz. **65**, 1407 (1973) [Sov. Phys. JETP **38**, 701 (1973)].
3. J. N. Hayes, Appl. Opt. **13**, 2072 (1974).
4. J. W. Ellinwood and H. Mirels, Appl. Opt. **14**, 2238 (1975).
5. J. Wallace and J. Pasciak, Appl. Opt. **15**, 218 (1976).
6. A. I. Bozhkov, F. V. Bunkin, and Al. A. Kolomenskiy, Kvantovaya Elektron. **4**, 942 (1977) [Sov. J. Quantum Electron. **7**, 536 (1977)].
7. L. M. Lyamshev and L. V. Sedov, Akust. Zh. **25**, 906 (1979) [Sov. Phys. Acoust. **25**, 510 (1979)].
8. V. A. Belokon', O. V. Rudenko, and R. V. Khokhlov, Akust. Zh. **23**, 632 (1977) [Sov. Phys. Acoust. **23**, 361 (1977)].
9. A. I. Bozhkov and Al. A. Kolomenskiy, Kvantovaya Elektron. **5**, 2577 (1978) [Sov. J. Quantum Electron. **8**, 1449 (1978)].
10. Al. A. Kolomenskiy, Akust. Zh. **25**, 547 (1979) [Sov. Phys. Acoust. **25**, 312 (1979)].
11. L. M. Lyamshev and L. V. Sedov, Pis'ma Zh. Tekh. Fiz. **5**, 970 (1979) [Sov. Tech. Phys. Lett. **5**, 403 (1979)].
12. A. I. Bozhkov, F. V. Bunkin, and Al. A. Kolomenskiy, Akust. Zh. **25**, 786 (1979) [Sov. Phys. Acoust. **25**, 443 (1979)].
13. A. I. Bozhkov, F. V. Bunkin, and Al. A. Kolomenskiy, Akust. Zh. **26**, 35 (1980) [Sov. Phys. Acoust. **26**, 18 (1980)].
14. F. V. Bunkin, A. I. Malyarovskiy *et al.*, Kvantovaya Elektron. **5**, 457 (1978) [Sov. J. Quantum Electron. **8**, 270 (1978)].
15. F. V. Bunkin, A. I. Malyarovskiy, and V. G. Mikhalevich, Akust. Zh. **27**, 179 (1981) [Sov. Phys. Acoust. **27**, 98 (1981)].
16. Y. H. Berthelot and I. J. Busch-Vishniac, J. Acoust. Soc. Am. **81**, 317 (1987).
17. E. P. Velikhov, E. V. Dan'shchikov *et al.*, Pis'ma Zh. Eksp. Teor. Fiz. **38**, 483 (1983) [JETP Lett. **38**, 584 (1983)].
18. M. N. Kogan and A. N. Kucherov, Dokl. Akad. Nauk SSSR **241**, 48 (1978) [Sov. Phys. Dokl. **23**, 437 (1978)].
19. M. N. Kogan and A. N. Kucherov, Zh. Tekh. Fiz. **50**, 465 (1980) [Sov. Phys. Tech. Phys. **25**, 281 (1980)].
20. A. A. Karabutov, Pis'ma Zh. Tekh. Fiz. **5**, 429 (1979) [Sov. Tech. Phys. Lett. **5**, 174 (1979)].
21. A. A. Karabutov and O. V. Rudenko, Akust. Zh. **25**, 536 (1979) [Sov. Phys. Acoust. **25**, 306 (1979)].
22. M. N. Kogan, A. N. Kucherov *et al.*, Izv. Akad. Nauk SSSR Mekh. Zhidk. Gaza **5**, 95 (1978).
23. S. A. Akhmanov, O. V. Rudenko, and A. T. Fedorchenko, Pis'ma Zh. Tekh. Fiz. **5**, 934 (1979) [Sov. Tech. Phys. Lett. **5**, 387 (1979)].
24. V. E. Gusev and A. A. Karabutov, Akust. Zh. **27**, 293 (1981) [Sov. Phys. Acoust. **27**, 117 (1984)].
25. A. T. Fedorchenko, Akust. Zh. **27**, 595 (1981) [Sov. Phys. Acoust. **27**, 330 (1981)].
26. V. E. Gusev and A. A. Karabutov, Akust. Zh. **28**, 38 (1982) [Sov. Phys. Acoust. **28**, 22 (1982)].
27. V. E. Gusev and A. A. Karabutov, Akust. Zh. **28**, 178 (1982) [Sov. Phys. Acoust. **28**, 108 (1982)].
28. A. A. Karabutov, O. V. Rudenko, Dokl. Akad. Nauk SSSR **248**, 1082 (1979) [Sov. Phys. Dokl. (1979)].
29. R. V. Khokhlov, Radiotekh. Elektron. **6**, 917 (1961).
30. S. I. Soluyan and R. V. Khokhlov, Vestn. Moscow State Univ. Phys. Astron. Ser. **3**, 52 (1961).
31. E. A. Zabolotskaya and R. V. Khokhlov, Akust. Zh. **15**, 40 (1969) [Sov. Phys. Acoust. **15**, 35 (1969)].
32. L. A. Ostrovskiy and E. N. Pelinovskiy, Prikl. Mat. Mekh. **32**, 122 (1974).
33. O. V. Rudenko and S. I. Soluyan, *Teoreticheskie osnovy nelineynoy akustiki* [*Theoretical Foundations of Nonlinear Acoustics*] (Nauka, Moscow, 1975).

34. S. A. Akhmanov, V. E. Gusev *et al.*, in, *Proceedings of the International Symposium on Nonlinear Acoustics,* Leeds, 1981, p. 2.
35. O. A. Vasil'eva, A. A. Karabutov, E. A. Lapshin, and O. V. Rudenko, *Vzanmodeystvie odnomernykh voln v sredakh bez dispersii* [*Interaction of One-Dimensional Waves in Nondispersive Media*] (Moscow State University, Moscow, 1983).
36. R. Courant, *Uravneniya s chastnymi proizvodnymi* [*Partial Differential Equations*] (Mir, Moscow, 1964) (English translation).
37. V. E. Gusev and O. V. Rudenko, Akust. Zh. **25**, 875 (1979) [Sov. Phys. Acoust. **25**, 493 (1979)].
38. V. E. Gusev and O. V. Rudenko, Akust. Zh. **27**, 869 (1981) [Sov. Phys. Acoust. **27**, 481 (1981)].
39. V. E. Gusev, Vestn. Moscow State Univ. Phys. Astron. Ser. **22**, 7 (1981) [Moscow University Physics Bulletin, **22**(4), 2, (1981)].
40. A. L. Hoffman, J. Plasma Phys. **1**, 193 (1967).
41. G. Withem, *Linear and Nonlinear Waves* (Mir, Moscow, 1977) (English translation).
42. A. A. Karabutov, Vestn. Moscow State Univ. Phys. Astron. Ser. **23**, 26 (1982).
43. V. V. Migulin, V. I. Medvedev, E. R. Mustel', and V. N. Parygin, *Osnovy teorii kolebaniy* [*The Principles of Oscillation Theory*] (Nauka, Moscow, 1978).
44. T. Hayasi, *Nelineynye kolebaniya v fizicheskikh sistemakh* [*Nonlinear Oscillations in Physical Systems*] (Mir, Moscow, 1968).
45. V. D. Shapiro and V. I. Shevchenko, Izv. Vyssh. Uchelon. Zaved. Radiofiz. **19**, 767 (1976).
46. V. I. Karpman, Izv. Vyssh. Uchen. Zaved. Radiofiz. **19**, 812 (1976).
47. A. A. Andronov and A. L. Fabrikant, *Nelineynye volny* [*Nonlinear Waves*] (Nauka, Moscow, 1979), p. 68.
48. L. D. Landau and E. M. Lifshitz, *Mekhanika sploshnykh sred* [*The Mechanics of Continuous Media*] (Gostekhizdat, 1954).
49. A. G. Bogdoev and G. G. Ozhinyan, Izv. Akad. Nauk SSSR Mekh. Zhidk. Gaza **1**, 133 (1980).
50. O. S. Ryzhov, *Problemy prikladnoy matematiki i mekhaniki* [*Problems of Applied Mathematics and Mechanics*] (Nauka, Moscow, 1971).
51. V. G. Gasenko, V. E. Nakoryakov, and I. R. Shreyber, Akust. Zh. **25**, 681 (1979) [Sov. Phys. Acoust. **25**, 385 (1979)].
52. I. S. Southern and N. H. Johannesen, J. Fluid Mech. **99**, 343 (1980).
53. C. C. Lin, E. Riessner, and H. S. Tsien, J. Math. Phys. **27**, 220 (1948).
54. O. S. Ryzhov, *Issledovanie transzvukovykh techeniy v soplakh Lavalya* [*Investigation of Transonic Flows in Laval Nozzles*] (VTs AN SSSR, Moscow, 1985).
55. S. Morioka and T. Yoshinaga, Phys. Fluids **23**, 689 (1980).
56. A. A. Karabutov, Vestn. Moscow State Univ. Phys. Astron. Ser. **23**, 26 (1982).
57. A. A. Karabutov and O. A. Sapozhnikov, Akust. Zh. **34**, 865 (1988) [Sov. Phys. Acoust. **34**, 501 (1988)].
58. E. A. Zabolotskaya and R. V. Khokhlov, Akust. Zh. **15**, 40 (1969) [Sov. Phys. Acoust. **15**, 35 (1969)].
59. O. V. Rudenko, Akust. Zh. **21**, 311 (1975) [Sov. Phys. Acoust. **21**, 196 (1975)].
60. S. A. Akhmanov, V. E. Gusev *et al.*, *Tez. dokl. V Vses. soteshch. po nerezonansnomu vzaimodeystviyu opticheskс. izlucheniya s veshchestvom* [*Conferenece Proceedings of the Fifth All-Union Conference on the Nonresonant Interaction of Optical Radiation with Matter*] (State Optics Institute Press, Leningrad, 1981), p. 371.
61. O. V. Rudenko, S. I. Soluyan, and R. V. Khokhlov, Dokl. Akad. Nauk SSSR **225**, 1053 (1975) [Sov. Phys. Dokl. **20**, 836 (1975)].
62. Henri Poincaré, *O krivykh opredelyaemykh differentsial'nymi uravneniyami* [*Curves Defined by Differential Equations*] (Gostekhizdat, Moscow, Leningrad, 1947).
63. M. A. Tamor and J. P. Wolfe, Phys. Rev. B **26**, 5743 (1982).
64. J. P. Wolfe, J. Lumin. **30**, 82 (1985).
65. M. A. Tamor, M. Greenstein, and J. P. Wolfe, Phys. Rev. B **27**, 7353 (1983).
66. M. I. D'yakonov and A. V. Subashtsev, Zh. Eksp. Teor. Fiz. **75**, 1943 (1978) [Sov. Phys. JETP **48**, 980 (1978)].
67. A. V. Subashiev, Fiz. Tverd. Tela **22**, 738 (1980) [Sov. Phys. Solid State **22**, 431 (1980)].
68. S. G. Tikhodeev, Pis'ma Zh. Eksp. Teor. Fiz. **29**, 392 (1979) [JETP Lett. **29**, 355 (1979)].

69. I. V. Kukushkin and V. D. Kulakovskiy, Fiz. Tverd. Tela **25**, 2360 (1983) [Sov. Phys. Solid State **25**, 1355 (1983)].
70. A. Forchel, H. Schweizer, and G. Mahler, Phys. Rev. Lett. **51**, 501 (1983).
71. B. Laurich, A. Forchel *et al.* **31-32**, 681 (1984).
72. C. L. Collins and P. Y. Yu, Solid State Commun. **51**, 123 (1984).
73. M. Combescot and J. Bok, J. Lumin. **30**, 1 (1985).
74. M. Combescot, Solid State Commun. **30**, 81 (1979).
75. G. Mahler and A. Fourikis, J. Lumin. **30**, 18 (1985).
76. V. E. Gusev, Pis'ma Zh. Eksp. Teor. Fiz. **45**, 288 (1987) [JETP Lett. **45**, 362 (1987)].
77. V. E. Gusev, Fiz. Tverd. Tela **29**, 2316 (1987) [Sov. Phys. Solid State **29**, 1335 (1987)].
78. V. E. Gusev, Akust. Zh. **33**, 624 (1987) [Sov. Phys. Acoust. **33**, 364 (1987)].
79. V. E. Gusev, Vestn. Mosk. Univ. Fiz. Asronomiya **28**, 75 (1987) [Moscow Unov. Phys. Bulletin, **28**(6), 84 (1987)].
80. L. D. Landau and E. M. Lifshitz, *Gidrodinamika. Teoreticheskaya fizika* [*Hydrodynamics. Theoretical Physics*] (Nauka, Moscow, 1986).
81. T. Rice, G. Hensel, T. Phillips, and G. Thomas, *Elektronno-dyrochnaya zhidkost' v poluprovodnikakh* [*Electron-Hole Liquid in Semiconductors*], edited by T. I. Galkina and B. G. Zhurkin (Mir, Moscow, 1980), p. 352.
82. O. V. Rudenko, Pis'ma Zh. Eksp. Teor. Fiz. **20**, 445 (1974) [JETP Lett. **20**, 203 (1974)].
83. A. A. Karabutov, E. A. Lapshin, and O. V. Rudenko, Zh. Eksp. Teor. Fiz. **71**, 111 (1976) [Sov. Phys. JETP **44**, 58 (1976)].
84. S. V. Nesterov, Izv. Akad. Nauk SSSR Fiz. Atmos. Okeana **4**, 1123 (1968).
85. A. I. Leonov and Yu. Z. Miropol'skiy, Izv. Akad. Nauk SSSR Fiz. Atmos. Okeana, 851 (1973).
86. V. E. Gusev and A. A. Karabutov, *Tez. dokl. VIII Vsesoyuzn. simp. po difr. i raspr. voln.* [*Conference Proceedings of the Eighth All-Union Symposium on Wave Diffraction and Propagation*] (IRE AN SSSR, Moscow,1981), Vol. 2, p. 122.
87. V. E. Gusev, Akust. Zh. **30**, 204 (1984) [Sov. Phys. Acoust. **30**, 121 (1984)].
88. V. E. Gusev, Akust. Zh. **30**, 298 (1984) [Sov. Phys. Acoust. **30**, 176 (1984)].
89. V. E. Gusev, Vestn. Moscow State Univ. Phys. Astron. Ser. **25**, 29 (1984) [Moscow Univ. Phys. Bulletin, N4, 33 (1984)].
90. G. Korn and T. Korn, *Spravochnik po matematike* [*Mathematics Handbook*] (Nauka, Moscow, 1977).
91. E. Scott, *Volny v aktivnykh i nelineynykh sredakh v prilozhenii k elektronike* [*Waves in active and nonlinear media with application to electronics*] (Sov. radio, Moscow, 1977).
92. N. S. Erokhin and R. K. Mazitov, PMTF **5**, 11 (1968).
93. W. Chester, Proc. R. Soc. London Ser. A **306**, 5 (1968).
94. W. Chester and J. A. Bones, Proc. R. Soc. London Ser. A **306**, 22 (1968).
95. W. Chester, J. Fluid Mech. **18**, 44 (1964).
96. Sh. U. Galiev, M. A. Il'gamov, and A. V. Sadykov, Mekh. Zhidk. Gaza **2**, 58 (1970).
97. D. B. Cruikshank, J. Acoust. Soc. Am. **52**, 1024 (1972).
98. R. G. Brever, Appl. Phys. Lett. **6**, 165 (1965).
99. G. Tacker, *V. Rampton. Giperzvuk v fizike tverdogo tela* [*Hypersound in Solid State Physics*] (Mir, Moscow, 1975) (English translation).
100. N. A. Bez"yazychnyy, A. A. Karabutov, and O. V. Rudenko, *Tez. dokl. XI Vsesoyuzn. konf. po kogerentn. i nelin. optike* [*Conference Proceedings of the Eleventh All-Union Conference on Coherent and Nonlinear Optics*] (Yerevan State University, Yerevan, 1982), Vol. 2, p. 551.
101. *Phonon Scattering in Condensed Matter*, edited by W. Eisenmenger, K. Lassmann, and S. Dottinger (Springer, New York, 1984).
102. P. S. Kowk, Phys. Rev. **175**, 1208 (1968).
103. N. M. Guseynov and I. B. Levinson, Zh. Eksp. Teor. Fiz. **85**, 779 (1983) [Sov. Phys. JETP **58**, 452 (1983)].
104. V. S. Bagaev, M. M. Bonch-Osmolovskiy *et al.*, Pis'ma Zh. Eksp. Teor. Fiz. **32**, 356 (1980) [JETP Lett. **32**, 332 (1980)].
105. V. E. Gusev and A. A. Karabutov, *Dokl. X Vsesoyuzn. akust. konf., sektsiya I* [*Proceedings of the Tenth All-Union Acoustical Conference Section I*] (Acoustics Institute, Moscow, 1983), p. 20.

106. N. Sharron, *Fizika fononov bolyshikh energiy* [*The Physics of High Energy Phonons*], edited by I. B. Levinson (Mir, Moscow, 1976) (English translation), p. 178.
107. V. E. Gusev, Akust. Zh. **32**, 322 (1986) [Sov. Phys. Acoust. **32**, 196 (1986)].
108. E. Conwell, *Kineticheskie svoystva poluprovodnikov v sil'nykh elektricheskikh polyakh* [*The Kinetic Properties of Semiconductors in Strong Electrical Fields*] (Mir, Moscow, 1970).
109. F. Rosch and O. Weis, Z. Phys. B **25**, 101 (1976).
110. W. E. Bron, M. Rossinelli *et al.*, Phys. Rev. B **27**, 1370 (1983).
111. M. B. Vinogradova, O. V. Rudenko, and A. P. Sukhorukhov, *Teoriya voln.* [*Wave Theory*] (Nauka, Moscow, 1979), pp. 150–153, 209–216.
112. O. V. Rudenko, Usp. Fiz. Nauk **149**, 413 (1986) [Sov. Phys. Usp. **29**, 620 (1986)].
113. S. N. Gurbatov, A. I. Saichev, and I. G. Yakushkin, Usp. Fiz. Nauk **141**, 221 (1983) [Sov. Phys. Usp. **26**, 857 (1983)].

Chapter 7

Experimental Methods of Pulsed Optoacoustics

The optoacoustic effect discovered by Bell as early as 1880 was used solely in the IR spectroscopy of gases prior to the development of lasers. The "laser era" of optoacoustics began with the pioneer studies of White[1,2] and an experimental work.[3] The experimental and theoretical research of the early 1960s identified the fundamental routes for the development of pulsed optoacoustics: laser excitation of acoustic wide-band pulses in liquids and solids,[4–10] semiconductors,[11] laser excitation of hypersound waves,[12–18] and Rayleigh waves.[19] The primary achievements of this period were reviewed in a survey study.[20]

The "renaissance" of optoacoustics occurred in the 1970s along with the simultaneous development of both traditional photoacoustic spectroscopy techniques as well as its pulsed version. The achievements of photoacoustic spectroscopy have been discussed in numerous surveys, volumes, and monographs (we cite only the most recent and accessible studies[21–27]). The bibliography on photoacoustics compiled by Gedrovits is also worth noting.[28] This chapter will be devoted to a review of experimental methods of pulsed optoacoustics which remains neglected in the monograph literature and can be found primarily in the journals (see Refs. 27 and 29–33).

As discussed in the preceding chapters, the OA-signal spectrum is the product of the laser radiation intensity spectrum and the frequency transfer function determined by the optical, thermal, and mechanical properties of the medium. The purpose of any version of OA spectroscopy is therefore to determine the transfer function at different wavelengths of incident radiation. In photoacoustic spectroscopy the spectral relation of the light absorptivity is most commonly measured; hence, the recording scheme is selected so that the transfer function is strongly dependent on α within the modulation frequency band chosen. In this case, as a rule, it is not necessary to analyze the entire acoustic spectrum, but rather it is sufficient to utilize a fixed modulation frequency. It is also largely irrelevant which laser operating principle—continuous (externally switched) or pulsed-periodic—is used; the most important element is that the laser intensity has a line spectrum. Only the noise suppression methods are fundamentally different. These include synchronous detection or strobing, although a combination in the pulse periodic regime is also possible.

Adequate recording of the temporal wave form of the OA signal is fundamentally important in pulsed OA spectroscopy, i.e., the capacity to record the signal over a broad modulation frequency band. This makes it possible to measure not

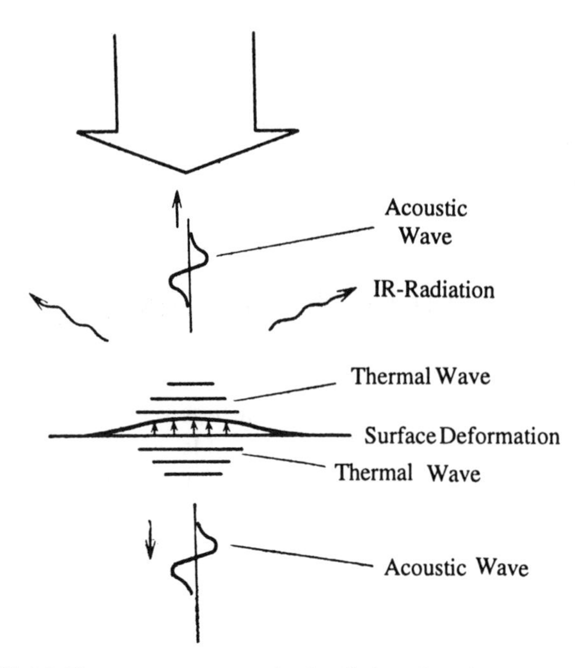

FIG. 7.1. Phenomena accompanying irradiation of an absorbing medium.

only the light absorptivity and other characteristics of the medium but also their depthwise distribution. Given the high power of pulsed lasers, OA signals, as a rule, are sufficiently strong in this case. Photoacoustic spectroscopy employing pseudo-random radiation modulation is closely related to pulsed techniques. However, the modulation frequency band is rather limited (ordinarily no higher than 1 kHz). Therefore, pulsed optoacoustics is primarily characterized by a broad frequency band. It is this element that we will focus on in analyzing different versions of OA spectroscopy.

7.1. Types of Optoacoustic Spectroscopy

Absorption of a laser pulse results in a nonstationary elevation of the temperature of the surface layer of both absorbing and—due to heat conduction—transparent media. Acoustic waves are excited in this case in both transparent and absorbing media. These phenomena were discussed in Chaps. 2–4. Spectroscopy techniques are arbitrarily divided into thermal and acoustic classes. In the first case the temperature field is registered and in the second case the acoustic field is recorded. If the measurements are taken in a transparent medium the method is called an indirect method, and if an absorbing medium is used it is called a direct method. The phenomena accompanying the OA effect are shown schematically in Fig. 7.1.

We first consider thermal methods. Measurement of the surface layer temperature seems to be the most natural process. Thermocouples, thermistors, pyroelectric films, and IR radiometers have been used for this process. Contact methods are applicable only for good thermally conducting media and at very low modulation

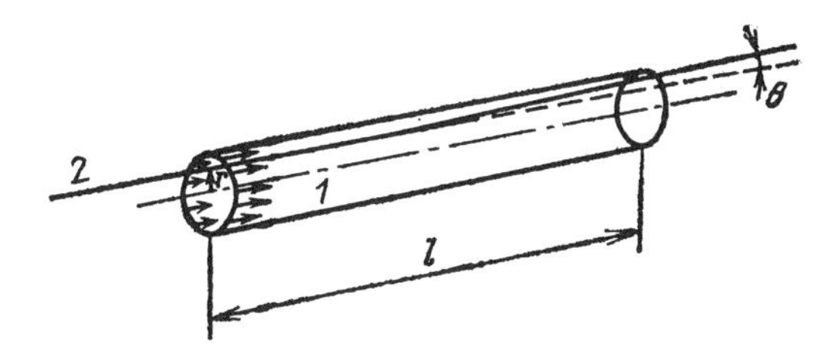

FIG. 7.2. Geometry of the exciting (1) and probe (2) rays in the thermal lens method.

frequencies. One exception is the analysis of metallic films deposited on pyroelectric films.[34] The advantage of this method is absolute calibration to absorption, although the limited bandwidth does not make it possible in practice to use this method for pulsed measurements.

Methods of OA-signal recording based on the change in the refractive index of media with temperature represent another group of thermal methods. Analogous variations can be induced by an acoustic wave, although the latter will travel at the sound speed and its contribution can be isolated from the total signal. These methods are fundamentally based on the deflection or distortion to a test light beam transmitted through the region containing the inhomogeneous thermal field.

The thermal lens method was proposed in Ref. 35 and a scheme employing a probe beam was employed in Ref. 36 (Fig. 7.2). In this case the probe beam is focused or defocused by the thermal lens produced by the main beam-induced inhomogeneous heating of the medium. The angle of deflection of the probe beam θ can be expressed by the formula[37,38]

$$\theta(r) \approx \frac{dn}{dT} \frac{\alpha \mathscr{E}_0 l}{\rho_0 c_p} \frac{2r}{a^2} \exp\left(-\frac{r^2}{a^2} \right),$$

where the profile of the main beam is assumed to have a Gaussian radius a and l is the length of the probe beam in the heating region. The most significant effect will be observed when the probe beam travels in the range of maximum temperature gradients $r \sim a$.

The thermal lens method is most convenient for analyzing transparent media and makes it possible to measure absorptivities through 10^{-7}–10^{-8} cm^{-1}. It can be employed both in the direct and indirect versions for measuring the temperature distribution, the thermal diffusivities, the velocity of gas flows, etc. (see Ref. 27, and the references cited therein). With an orthogonal configuration of the main and probe beams, the deflection of the beam by the thermal lens is often called the "mirage effect."

The frequency band of this method is largely limited by the noise of the probe radiation source and the photodetector as well as (in indirect recording) by beam diameter. Diagnostics in the heating zone also encounter the difficulties of isolating the temperature and acoustic contributions. Nonetheless, this method has become widely used (see, for example, Refs. 39–46) specifically in optoacoustic spectros-

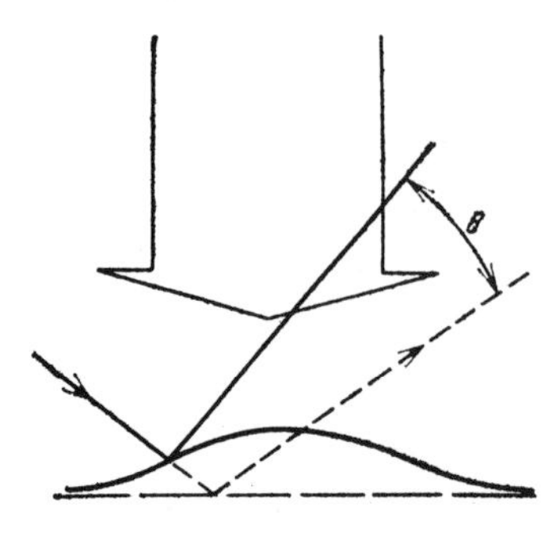

FIG. 7.3. Ray geometry in the photodeflection spectroscopy method.

copy (see Sec. 7.2). Changes in the refractive index can also be determined by other methods including interferometric, heterodyne, etc., which we will just mention.

The so-called photodeflection spectroscopy method[47,48] (see also Amer's article in Ref. 22) is closely related to the thermal methods. The essence of this method lies in scattering of the probe radiation by the deformed surface of the absorbing medium attributable to inhomogeneous laser heating (an analog of the transfer function $K_0(\omega)$ was examined in this case in Chap. 3 [see Eq. (3.24)]). Ordinarily, an effort is made to focus the probe radiation on the "swell" in the vicinity of the most severe surface slope (Fig. 7.3) to obtain the maximum possible signal. The frequency characteristics of this method have not been analyzed; it has been used nearly always in cw operation. The convenience of the method derives from the complete noncontact nature of the signal excitation and recording design, which is used in OA microscopy. There are certain difficulties associated with stabilizing the probe beam and its coincidence with the irradiated spot.

IR radiometry is based on measuring variations in the intensity of the thermal IR radiation emitted by the medium heated by the laser radiation. The total radiated power W is given in accordance with the Stephan-Boltzmann law by the equation

$$W = \varepsilon\sigma T^4,$$

where σ is the Stephan-Boltzmann constant, ε is the "grayness" coefficient of the body, i.e., the emissivity. The change in radiated power is, in turn, related to the change in surface temperature by the relation

$$\delta W = 4\varepsilon\sigma T_0^3 T'.$$

It is therefore possible to measure the heating of either the face surface irradiated by the laser or the rare surface of the absorbing medium (Fig. 7.4). The temperature modulation spectrum $T(\omega)$ according to Eqs. (3.16) is proportional to the laser radiation intensity spectrum.

In IR radiometry of the back side of the absorbing medium the temperature modulation frequency ω must be sufficiently small so that the thermal wave at this depth will have a detectable amplitude, ordinarily $(\omega/\chi)^{1/2}h \lesssim 3$–$5$ (h is the thickness of the test medium). In this case the most efficient method involves measuring the thermal diffusivity of the medium based on the phase of thermal wave[49,50] (see also the corresponding section in Ref. 22).

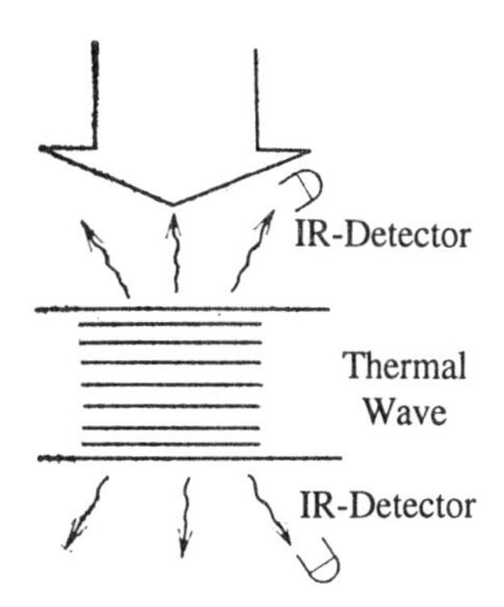

FIG. 7.4. Recording of OA signal by means of IR detectors.

IR radiometry of the front surface, like photodeflection spectroscopy, is a very convenient version of photoacoustic microscopy. Generally, all thermal versions of OA spectroscopy are usable with "difficult" samples: powders, light-scattering, radioactive, high-temperature, etc., media (see Ref. 27, and the references cited therein).

IR radiometry of the front surface, unlike its other version, makes it possible, in principle, to record temperature oscillations over a broad modulation frequency band. The frequency band is determined by the response of the recording section of the radiometer. Naturally, the depthwise resolution of the method improves with expanding bandwidth (the depth of penetration of the thermal wave is proportional to $\sqrt{\chi/\omega}$). Hence, IR radiometry is one of the promising OA-microscopy techniques for thermal waves.

A change in surface temperature will not only alter the refractive index, which is used in photorefraction methods, but also will cause variations in the light absorptivity and scattering coefficient. A version of photoacoustic spectroscopy of thermal waves employing this effect was proposed in Ref. 51. Its advantage over the "mirage effect" is that the signal maximum is observed when the probe beam is focused on the center of the main spot.

All thermal methods are promising for use in OA microscopy. However, there have apparently been no experiments to measure the transfer function of the temperature of the medium or the corresponding instrument spectral characteristics. Hence, the problem of pulsed broadband spectroscopy employing thermal waves remains open.

We now analyze methods of recording acoustic waves excited by laser radiation. Like thermal waves, acoustic waves alter the refractive index and absorptivity of media (these effects are analyzed by acousto-optics); as is the case for thermal waves, both direct and indirect recording is possible for acoustic waves. These effects make it possible to implement a purely optical—optical excitation and optical recording—version of OA spectroscopy.

An analog of IR radiometry in OA spectroscopy is "contact" recording of sound waves employing microphones, piezoelectric or capacitive transducers, etc. A frequency band of up to hundreds of megahertz is accessible by direct piezorecording. From the viewpoint of pulsed optoacoustics, acoustic wave recording is, generally speaking, preferred. We will therefore consider the application of transfer functions to OA spectroscopy in greater detail.

The transfer functions for the vibrational velocities in transparent $[K^{tr}(\omega)]$ and absorbing $[K(\omega)]$ media as well as surface vibrations $[K_0(\omega)]$ are given in Chap. 3. When the boundary of the absorbing medium is an impedance boundary, the information capabilities of all three spectroscopy versions are similar, since the waves excited in the transparent and absorbing media are comparatively poorly reflected by the boundary. Hence the selection of the type of recording (direct, indirect, or boundary vibrations) in the case of comparable impedances of the transparent and absorbing media is determined exclusively on the basis of convenience. Direct recording is most commonly used in this case.

In the case of a free boundary of the absorbing medium the acoustic waves excited in the transparent and absorbing media will be generated by different processes: the wave in the absorbing medium will be excited from the expansion of the heated volume, while in the transparent medium it will be excited from the expansion of a thin surface layer heated by conduction. Hence, when a gas is used as the transparent medium, and condensed matter is used as the absorbing medium (as is the case in gas microphone photoacoustic spectroscopy) indirect recording of the acoustic wave (like recording of boundary vibrations) yields the same information as thermal wave spectroscopy. This can easily be verified by comparing the temperature transfer functions with $K^{tr}(\omega)$, $K_0(\omega)$.

A potential advantage of indirect pulsed recording of an acoustic wave compared to spectrophones is the capacity to expand the detection band above the acoustic resonances of the chamber by means of temporal signal selection. This, in turn, makes it possible to increase the depthwise resolution of the measurements. However, as a consequence of the strong absorption of sound in gases, the frequency band in this case is most commonly limited to a range of up to 1 MHz. Bragg diffraction of the probe beam can be used to significantly expand the frequency band.[52] Therefore, indirect photoacoustic spectroscopy has much in common with OA spectroscopy of thermal waves, since the acoustic wave in the gas is excited from the diffusion of heat from the bulk of the absorbing medium towards its surface. Due to the strong attenuation of the thermal wave, sources in the surface layer of the absorbing medium of thicknesses of the order of $(\omega/\chi)^{-1/2}$ make a contribution to $K^{tr}(\omega)$. Hence, indirect photoacoustic spectroscopy represents, to the same degree as temperature wave spectroscopy, a convenient method for analyzing the surface layers of various media. We examine these methods separately below (Sec. 7.2).

The specific nature of pulsed optoacoustics is most clearly manifested when registering sound in an absorbing medium. We analyze methods of recording OA signals from the viewpoint of their broadband nature and locality. A similar analysis was carried out in Ref. 53.

We begin with optical methods. Recording of acoustical waves based on the deflection of a probe beam by wave-induced inhomogeneities in the refractive index was employed as early as 1932.[54] In optoacoustics this method was first employed in Ref. 55. Here, by using several probe beams it is possible to register disturbances simultaneously at several points and thereby measure the wave velocity and attenuation.[56] In the probe-beam deflection method, the signal $s(t)$ is proportional to the temporal derivative of the pressure: $s(t) \propto n^{-1}(\partial n/\partial r) \propto (\partial p/\partial t)$. In this case the receiving frequency band is largely limited by probe beam diameter d to a frequency c_0/d (c_0 is the sound speed in the medium under test); ordinarily this band is of the order of 10 MHz. Moreover, this is not a local method, since beam deflection is determined by the refractive index gradient at the intersection of the probe beam and the acoustic field. Therefore additional assumptions regarding the spatial structure of the acoustic field (such as a Gaussian transverse distribution) are ordinarily made in comparing the signal to the OA pulse wave form. This method was successfully used for acoustic spectroscopy of gases and liquids in the research of Tam *et al.*[57,58]

Changes in the refractive index of an acoustic wave can also be detected by other optical techniques, including interferometric, heterodyne, and schlieren photography techniques, etc. (see, for example, Refs. 59–62). Interferometric techniques are the most commonly used methods (such as a Mach-Zehnder interferometer). However, the band limits and the locality of measurements involved in using these techniques are the same as listed above. Since optical methods are rather traditional and have been described in detail in the literature we will not discuss these. We simply note that these methods are applicable in media transparent to probe radiation.

Another major group of OA-signal recording techniques is based on the detection of vibrations of the rare target surface. The following techniques are used: interferometers (primarily a Michelson interferometer)[62,63] (see also the survey by Hutchins and Tam in Ref. 25, pp. 429–449), dynamic holography,[64] capacitive detectors,[65,66] and various piezoelectrics, including both film and bulk detectors. In these cases there are rather weak requirements on the aperture of the detection system, since the plane of the sensitive element coincides with the wave phase front. The frequency band of sound in the interferometric recording method is determined by the photodetector characteristics and ordinarily amounts to several tens of megahertz. Piezoelectric methods make it possible to record acoustic signals over a broader frequency band: up to the gigahertz range[67,68] (as with capacitive techniques[66]). The maximum bandwidth of individual electrical single pulses has therefore been reached in practice. An even broader band and even shorter acoustic pulses are recorded by means of acoustooptic techniques.[69]

The specific applications of the OA-spectroscopic methods discussed above will be examined in the sections below.

7.2. Pulsed Optoacoustic Microscopy by Thermal Waves

Optoacoustic microscopy, like any other microscopy, is a method of acquiring rather high-resolution images of surface inhomogeneities. Unlike traditional reflection or transmission microscopy, OA microscopy makes it possible to detect surface defects in optically opaque samples. It is therefore possible to use introscopy methods while conserving the extraordinarily high—ideally micron level—resolution characteristic of optical methods. Since the measurements are carried out point by point it is necessary to utilize computer technology for data storage and processing.

Depending on the recording method (direct or indirect, acoustic or thermal) OA microscopy for nondestructive testing applications can employ thermal or acoustic waves. The former are preferred for detecting surface defects given the short wavelength of the thermal wave (compared to an acoustic wave of the same frequency). Since the temperature wave decays exponentially in the bulk of a medium with a decrement $(\omega/2\chi)^{1/2}$ defects at depths of $z > 3(2\chi/\omega)^{1/2}$ will have no effect on the amplitude of the variation in surface temperature. On the other hand, acoustic waves do not decay in the bulk of the medium and can be used to detect defects in the sample bulk.

Expressions for the spectral components of the temperature fields were derived in Chap. 3 [see Eq. (3.17)]. Here their amplitude, which is proportional to light intensity, is determined by the light absorptivity and thermal conductivity of the medium. Consequently, OA microscopy can be used to detect inhomogeneities in the absorption and heat conduction properties of the medium. However, the greatest difficulty lies in separating the contributions of the different inhomogeneities, since both types affect the image contrast.

The primary advantage of broadband microscopy, i.e., microscopy using a broad modulation frequency band is the capability to identify the distribution of absorption or heat conduction inhomogeneities throughout the material bulk. At the same time, since a signal at frequency ω is generated by heat sources in a layer of thickness of approximately $3(2\chi/\omega)^{1/2}$, sources in the bulk of the medium will cease to affect the signal with increasing frequency. Therefore, it is possible to analyze a three-dimensional rather than a two-dimensional image of the inhomogeneities. Naturally, analysis of the signal over a broad modulation band requires a longer measurement period in the case of continuous radiation modulation. The advantages of pulsed broadband spectroscopy are obvious from this viewpoint.

We examine these methods in greater detail. Since the thermal field in an absorbing medium can only be recorded at low frequencies (high frequencies do not penetrate through to the back side of the sample), indirect (in a transparent medium) signal recording is the primary method used for microscopy by thermal waves. The spectral temperature density in the transparent medium $\tilde{T}^{\text{tr}}(\omega,\mathbf{k}_\perp)$ can be represented as

$$\tilde{T}^{\text{tr}} = -\frac{\beta^* I_0 \tilde{f}(\omega)\tilde{H}(\mathbf{k}_\perp)\hat{g}[(-i\omega/\chi + \mathbf{k}_\perp^2)^{1/2}]}{\varkappa^{\text{tr}}(-i\omega/\chi^{\text{tr}} + \mathbf{k}_\perp^2)^{1/2} + \varkappa(-i\omega/\chi + \mathbf{k}_\perp^2)^{1/2}} \exp\left[\left(-\frac{i\omega}{\chi^{\text{tr}}} + \mathbf{k}_\perp^2\right)^{1/2} z\right] \tag{7.1}$$

[compare to Eq. (3.17)]. The designations of Chap. 3 are used in Eq. (7.1) and the equation is generalized to the case of inhomogeneously absorbing media:

$$\hat{g}(p) = \int_0^\infty g(z)\exp(-pz)dz,$$

$\varkappa$ is thermal conductivity.

It is clear from Eq. (7.1) that the spectral temperature density is proportional to the laser radiation intensity spectrum $\tilde{f}(\omega)$. The proportionality factor is determined by the parameters of the medium: the temperature coefficient of volume expansion; the light absorptivity; the thermal diffusivity, and is dependent on frequency.

If the medium has homogeneous thermophysical parameters, i.e., if the parameters are independent of the coordinates, in laser spot scanning the microscope

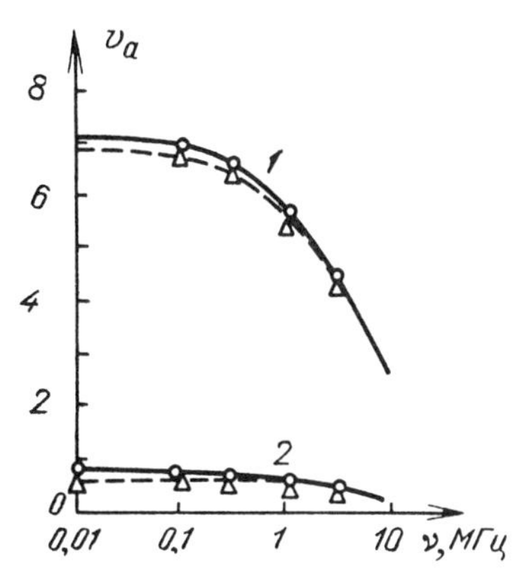

FIG. 7.5. Frequency dependence of an OA signal from an aluminum (1) and silicon (2) surface in a vacuum (solid curves) and in air (dashed curves) (Ref. 75).

pattern will be generated by the irregularity of light absorption. In the simplest version (when the OA signal is recorded at a single fixed frequency) the integral light absorption to depths of $z \leqslant (\chi/\omega)^{1/2}$ is determined and the microscope image is generated by inhomogeneities of light absorption on the sample surface. A typical application of this version is microscopic analysis of metallic coatings on nontransparent substrates such as semiconductors. The possibilities for using this method in various fields of physics, chemistry, and biology, as well as engineering, are discussed in surveys and textbooks.[22,23,25,70–74]

A fundamental advantage of the pulsed version of OA microscopy is that a broad frequency spectrum of temperature waves is excited and can be used from Eq. (7.1) to determine the spectrum of the depthwise absorption distribution $\hat{g}(p)$. This makes it possible to find the distribution $g(z)$. In order to implement such a "three-dimensional" version of microscopy it is necessary to achieve high-resolution recordings of temperature variations. The broadest frequency band can be achieved in photodeflection spectroscopy.[75] With use of a collinear geometry of the excitation and probe beams, and sharp focusing of both rays with weak (within the focal spot) beam separation, it is possible to achieve a band of up to 10 MHz. Further broadening of the band in thermal methods is hardly possible given the decay in OA-signal amplitude with increasing frequency. Since in this method the surface deformation is registered, then, in accordance with Eq. (3.24) the signal will depend on frequency (for a free boundary $N \to \infty$) and the transfer function

$$K_0(\omega) = \frac{I_0\beta^*}{\rho_0 c_p}(1+R_T)^{-1}\left(1+\frac{i\omega}{\omega_T}\right)^{-1}. \tag{7.2}$$

In the case of strong absorption $\omega_T = \alpha^2\chi \gg \omega$ the transfer function is constant. Hence measurements of the OA signal from a metal surface shown in Fig. 7.5 (curve 1),[75] in fact, yield the instrument function of the method. Its capabilities to determine the thickness d of aluminum films on silicon are illustrated by Fig. 7.6.[75] Waves were excited in a continuous regime in these experiments.

Pulsed photodeflection spectroscopy was realized in Ref. 76. Its experimental setup is shown in Fig. 7.7. A fundamental difference between the pulsed mode is the need for high-speed analog-to-digital converters that make it possible to record the system response to laser pulse-initiated shock excitation. Such responses are shown

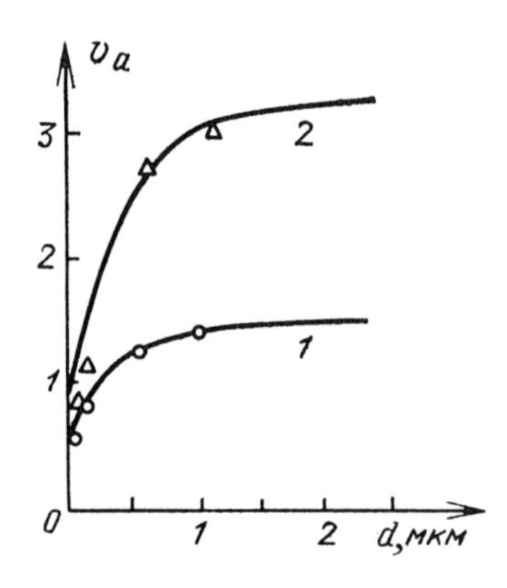

FIG. 7.6. OA signal at $\nu = 1$ MHz plotted as a function of the thickness of an aluminum film on a silicon substrate (1) and a silicon substrate coated by a silicon oxide layer 1000 Å thick (2) (Ref. 75).

in Fig. 7.8.[76] Since the surface of strongly absorbing media were analyzed, this signal is purely thermal in nature ($\alpha^2\chi \gg \omega$). The cooling rate of the surface

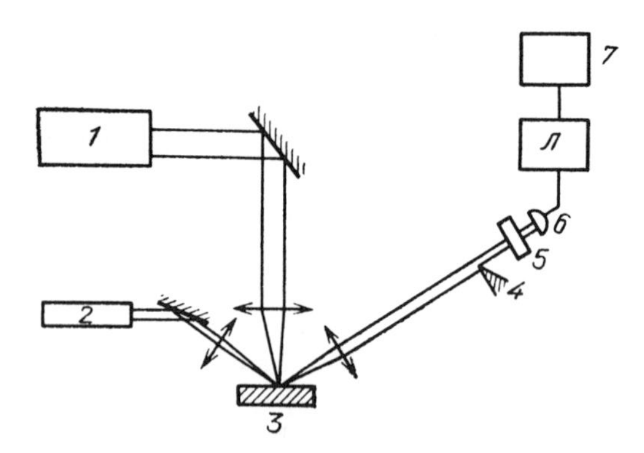

FIG. 7.7. Pulsed photodetection spectroscopy setup: 1—pulsed Nd-YAG laser; 2—He-Ne probe radiation laser; 3—test sample; 4—knife; 5—interference filter; 6—photodiode; 7—analog-to-digital converter.

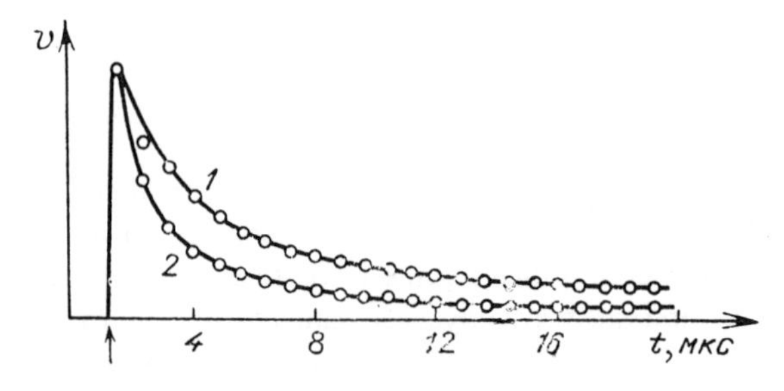

FIG. 7.8. Temporal profile of photodeflection signal from a brass (1) and aluminum (2) surface. The arrow indicates the instant of laser irradiation.

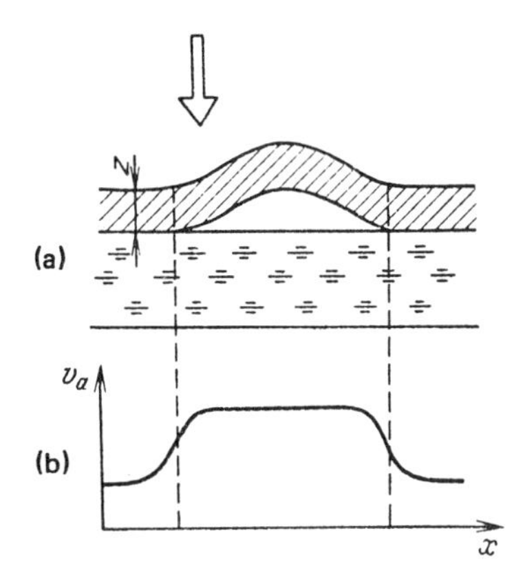

FIG. 7.9. Formation of OA signal for the case of irregularities in the heat conductivity of the medium.

correlates with the thermal diffusivity of the material. The frequency band can be determined from the rate of signal rise. In this setup it is limited to 1–2 MHz.

The OA microscopy of thermal waves therefore makes it possible to obtain images of irregularities in light absorption with a plane resolution of the order of a few micrometers[75] and a depthwise resolution of the order of $(\chi/\omega)^{1/2}$ of 0.1–1 μm.

Another important application of OA microscopy employing thermal waves is noncontact flaw detection of surface irregularities. We consider as an example a strongly absorptive thin layer on a substrate [Fig. 7.9(a)]. In this case the layer-separated regions can be treated as irregularities in the thermal diffusivity of the medium, since the thermal wave will be reflected differently by the layer-substrate boundary and the layer-vacuum boundary. In OA microscopy these regions will appear as bright regions, since heat will not penetrate into the substrate and the layer is heated more effectively [Fig. 7.9(b)]. Obviously, by analyzing the frequency characteristic of the OA signal it is also possible to determine the layer thickness, since at high frequencies $(\omega/\chi)^{1/2}z > 3$ the thermal wave does not "sense" the substrate, while at low frequencies $(\omega/\chi)^{1/2}z \ll 1$ the situation is reversed: the surface layer has virtually no effect on the thermal wave. A substantial number of studies have been devoted to these problems; we only cite some of these studies.[77–81]

It is, however, necessary to note that the primary difficulty of OA spectroscopy is separating the contributions of the irregularities of light absorptivity and the thermal diffusivity when both types of inhomogeneities are present. In this case it is important to acquire not only an amplitude image but also a phase image. As a rule, the phase contrast of an optoacoustic signal is more severe than the amplitude contrast (see Sec. 3.2).

OA microscopy employing gas-microphone recording is closely related to thermal wave OA microscopy. As demonstrated by analysis (Sec. 3.2) the transfer functions of indirect OA spectroscopy $K^{\mathrm{tr}}(\omega)$ and the transfer functions of the surface vibrations $K_0(\omega)$ are nearly identical (for a free boundary $N \to \infty$), since in gas-microphone recording the modulation frequencies do not exceed a few kilohertz. This limits the depthwise resolution to $(\chi/\omega)^{1/2}$. However, given its comparative simplicity and advancement, OA microscopy employing indirect acoustic recording has become widely used (see, for example, Refs. 82–89, 43, as well as Refs. 25 and 22).

7.3. Optoacoustic Spectroscopy of Rayleigh Waves

OA spectroscopy of Rayleigh waves is used for diagnostics of surface layers of solids. As suggested by the theory developed in Chap. 3, the spectral density of a thermo-optically excited Rayleigh wave is, like thermal waves, determined by the Laplace transform of the spatial distribution of acoustic sources [see, for example, Eq. (3.50)]. Therefore sound sources at depths $z > 3\omega^{-1}(c_R^{-2} - c_L^{-2})^{-1/2}$, in the bulk of the medium make no contribution to excitation of a Rayleigh wave of frequency ω. This form of OA spectroscopy, like thermal wave spectroscopy, is suitable specifically for surface diagnostics.

There are, at the same time, significant differences between these methods. Above all, since the wave vector of the thermal wave $(\omega/\chi)^{1/2}$ is much greater than the wave vector of the Rayleigh wave ω/c_R across the entire ultrasonic frequency band through the gigahertz range (see Table 3.2), in order to achieve the same resolution as in thermal wave microscopy, it is necessary to use higher light modulation frequencies in the OA spectroscopy of Rayleigh waves. Unlike thermal wave microscopy, in the OA spectroscopy of Rayleigh waves, the wave phase does not contain that much information. This is due to the fact that the Laplace component of the sources in the first case [see Eq. (7.2)] is taken from the complex argument $\hat{g}[(-i\omega/\chi + k_\perp^2)^{1/2}]$, while in the second case it is taken from the real argument $\hat{g}[(|\omega|/c_R)(1 - c_R^2/c_L^2)^{1/2}]$ [see Eqs. (3.50)–(3.52)]. Hence, in the case of OA microscopy of thermal waves, the wave phase bears information on the distribution of the sources and the thermal inhomogeneities even in the absence of a strong amplitude contrast.

The advantages of OA spectroscopy by surface acoustic waves (SAW's) include the capability to record the signal outside the excitation zone and correspondingly less stringent requirements on the relative alignment of the excited and probe beams together with near-unambiguous interpretation of the detected signal (while the different mechanisms of probe-beam modulation such as the thermal lens, acoustooptic, deflection, and other mechanisms can be interpreted within the excitation zone). The flip side of this advantage is the requirement for higher modulation frequencies. Another important advantage of OA spectroscopy by SAW's is the higher sound excitation efficiency with frequency [see transfer functions (3.50)–(3.52)], which suggest the possibility of achieving a spatial resolution comparable to that obtained in thermal wave microscopy. The problems of broadband recording of Rayleigh waves, however, have not been completely resolved to date.

We estimate the resolution of the method. As in the case of thermal wave microscopy, surface resolution is determined by the radiation spot diameter and the absorptivity contrast can be determined with an optical resolution of the order of 0.5–2 μm. Therefore, high modulation frequencies are not required. Consequently, the capabilities of both types of microscopy are identical in these experiments.

In order to determine surface inhomogeneities in OA spectroscopy by surface acoustic waves it is necessary to find the Laplace transform $\hat{g}(p)$ over a sufficiently broad range of p. The Rayleigh wave spectrum is determined not only by $\hat{g}(p)$ but also by the distribution $\tilde{H}[p(1 - c_R^2 c_L^2)^{-1/2}]$ and modulation $\tilde{f}[c_R p(1 - c_R^2/c_L^2)^{-1/2}]$ spectra. It is necessary for these components not to be small within the range of values of p. Limits on the distribution spectrum of $\tilde{H}$ are

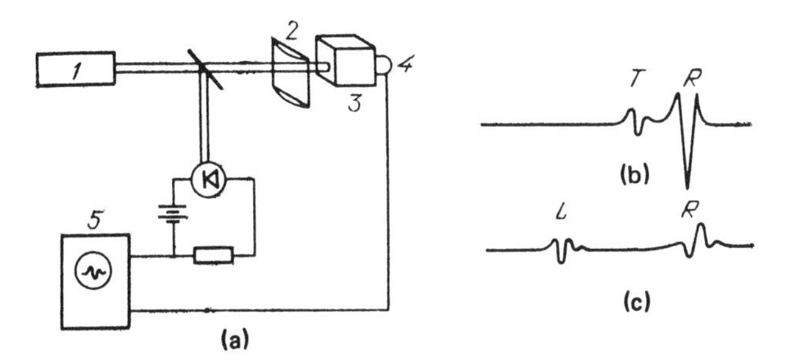

FIG. 7.10. (a) Setup of experiment: 1—pulsed laser; 2—cylindrical lens; 3—test sample; 4—piezotransducer; 5—oscilloscope; (b), (c): Wave forms of the transverse and longitudinal signals, respectively.

more severe. The light spot diameter may be of the order of the wavelength of light λ and consequently the spectral range p limited by $\widetilde{H}(\mathbf{k}_{\perp})$ extends to $p \sim [2(1-c_R^2/c_L^2)^{1/2}]\lambda^{-1}$.

Hence, the spatial resolution of SAW OA spectroscopy is limited to $[2(1-c_R^2/c_L^2)^{1/2}]^{-1}\lambda \sim 0.3-1\ \mu\text{m}$. This value is achieved at modulation frequencies $\omega/2\pi = c_R/\pi\lambda \sim (1 \div 3)$ GHz. The required modulation frequencies are attainable in mode-locked lasers (as harmonics of the pulse repetition rates). It is also quite possible to use other radiation types (such as electron beams) that permit sharper beam focusing, in order to increase resolution.

We consider experimental methods of SAW OA spectroscopy. The laser generation of surface acoustical waves was apparently first described in the study by Lee and White.[90] A Rayleigh wave was excited from the absorption of a single Q-switched laser pulse in an aluminum film deposited on the test surface. A variety of materials, including ceramic, crystal, and fused quartz, were used as the substrate. The laser radiation was focused onto a narrow rectilinear strip and hence a near-plane SAW wave front was achieved. However, the limited bandwidth of the Rayleigh wave detectors did not permit adequate resolution of the temporal wave form. Hence, the amplitude of the SAW spectral component at the fundamental resonance frequency of the detector was recorded.

The next study devoted to laser generation of SAW's only appeared 11 years later:[91] in 1979. Its authors employed noncontact laser excitation of acoustic waves (Q-switched laser radiation was focused by means of a cylindrical lens onto the end of a plane face within a narrow strip near the edge [Fig. 7.10(a)] in order to simultaneously measure the velocities of the longitudinal (L), transverse (T), and Rayleigh (R) waves. The velocities were determined from the delay of pulse arrival [Fig. 7.10 (b) and 7.10(c)] associated with wave propagation along the free face. These values were identical to those obtained by other methods. The velocity measurement errors from the optoacoustic method and traditional methods were within an order of magnitude. In proposing this method, the authors note its advantages: the possibility of conducting measurements on small samples of simple shape over a broad pressure and temperature range, the timeliness of the derived data, and the fact that the velocities of all three types of waves can be measured simultaneously.

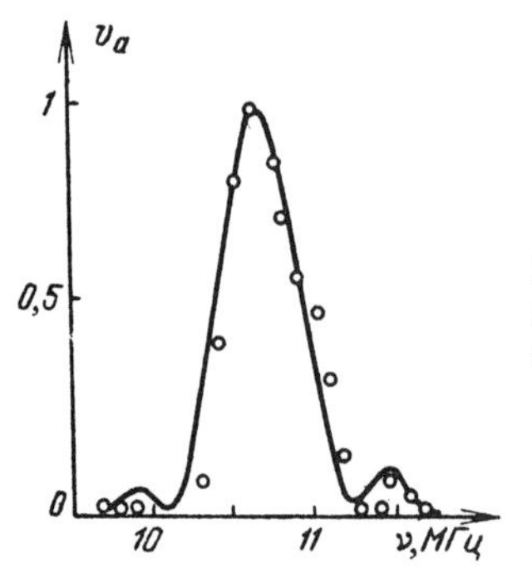

FIG. 7.11. Amplitude of the Rayleigh wave plotted as a function of radiation modulation frequency (circles) and the mask characteristic measured traditionally (solid curve).

Broadband pulses of Rayleigh waves were excited in Refs. 90 and 91. Laser generation of periodic SAW's was first proposed in Ref. 92. A harmonically modulated laser beam was guided to the test surface which absorbs radiation through a periodic mask. A sharp peak in SAW excitation efficiency is observed at the modulation frequency corresponding to the wavelength of the Rayleigh which is equal to the mask period (Fig. 7.11). The frequency dependence of the relative excitation efficiency was identical to the spatial spectrum of the mask measured by a SAW delay line. Measurements of the directional pattern of the SAW optoacoustic antenna yield results in good agreement with the theoretical results for a linear antenna (Fig. 7.12). Estimates of the SAW amplitude were within an order of magnitude of those measured experimentally.

It should be noted that a periodic mask is not necessary to excite periodic SAW's. Harmonic modulation of the intensity and beam focusing are sufficient (in this case the frequency response of the optoacoustic radiator will be a broadband response). This version was later implemented in Refs. 93 and 94.

A series of experimental and theoretical studies appeared in 1982–1986; such research analyzed different versions of laser generation of surfaced acoustical waves. References 93, 94, and 104 are devoted to the excitation of periodic Rayleigh waves, while pulsed SAW generation is discussed in Refs. 95, 98, and 103. SAW excitation by laser-beam scanning across the surface of the medium is proposed in Refs. 101 and 102.

A dye mode-locked laser was employed to excite periodic SAW's in Refs. 93 and 94. Radiation was absorbed by a dye deposited on the test surface. The laser beam

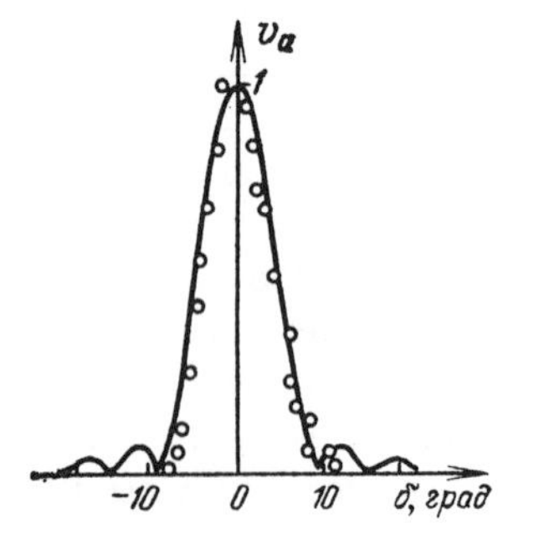

FIG. 7.12. Directivity characteristic of the Rayleigh-wave optoacoustic antenna: calculated (solid curve) and experiment (circles).

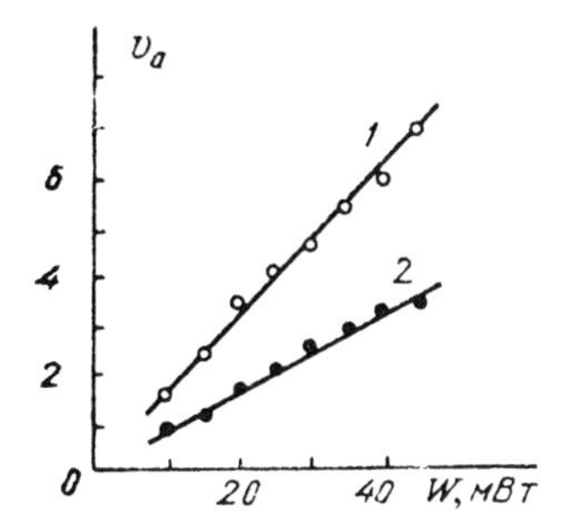

FIG. 7.13. SAW signal amplitude plotted as a function of laser fluence in cw operation for focusing by spherical (1) and cylindrical (2) lenses.

was focused by a spherical or cylindrical lens. The SAW's were recorded by an interdigital transducer tuned to the laser pulse repetition rate of 76.4 MHz. It was established in Ref. 93 that the acoustic signal was proportional to the fluence (Fig. 7.13).

This setup was used in Ref. 94 for SAW OA microscopy of light absorptivity. A laser spot of the order of 5 μm in diameter was scanned across the test surface, and the signal was synchronously output to a plotter. The sensitivity of the method was comparable to that of traditional OA microscopy.

In pulsed SAW excitation experiments[95,98,103] the primary difficulties are associated with recording the broadband Rayleigh wave. References 95 and 103 represent the development of Refs. 90 and 91, and the problem of resolving the wave form of the SAW pulse was not addressed. Reference 95 employed an experimental setup analogous to that described in Ref. 90: the Rayleigh wave was excited on the surface of a metallic block with the laser spot in the shape of a strip. A piezoelectric transducer with a 3 MHz resonant frequency was used for recording. The amplitude of surface vibrations at the resonant frequency of the transducer was recorded. The directional pattern of the acoustooptic Rayleigh wave antenna was investigated for different spot diameters. The Rayleigh wave velocity was recovered from the width of the directional pattern. The accuracy of these measurements does not appear to be significant due to the uncertainty of spot size.

Reference 103 investigated the dependence of the amplitude of surface vibrations on light spot diameter, pump power, and distance traveled by the wave. The SAW amplitude was proportional to the laser pulse energy through intensities of the order of 100 MW/cm^2. The Rayleigh wave amplitude decayed with distance roughly as $r^{-1/2}$ (Fig. 7.14), as expected for a cylindrical wave. Control experi-

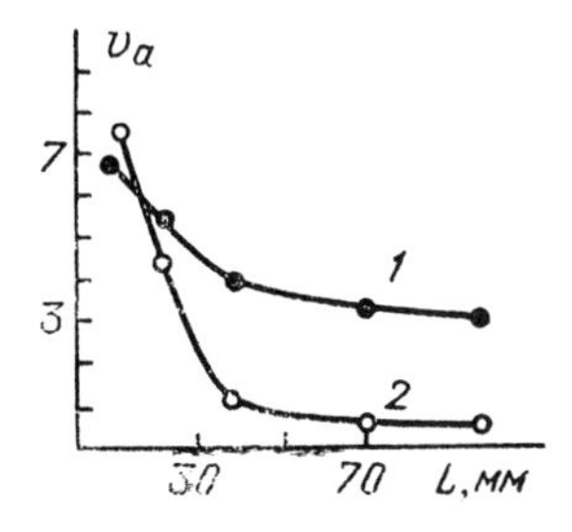

FIG. 7.14. SAW amplitude plotted as a function of distance between the excitation band and the detector for Rayleigh (1) and volume (2) waves.

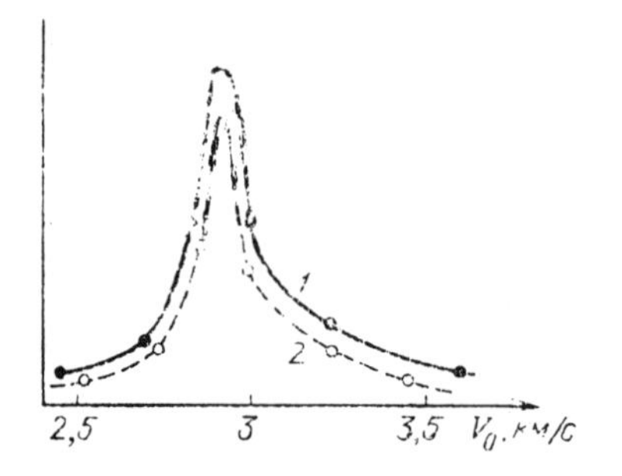

FIG. 7.15. SAW amplitude at the end of the scanning path plotted as a function of focus velocity for two path lengths $L = 16$ (1) and 12 (2) cm.

ments (the presence of a liquid layer, surface damage on the wave path) demonstrated that a surface wave was in fact observed. The SAW pulse duration was proportional to the laser spot diameter. Unfortunately, this relation was virtually neglected in this study.

Laser generation of Rayleigh waves as a new method of OA spectroscopy of surface layers was first discussed in Ref. 98. The SAW was excited by absorption of dye laser radiation by a quartz substrate surface (a lithium niobate substrate was also used). The radiation was focused by a cylindrical lens into a 30 μm $\times$ 2 cm spot. An R590 dye remaining on the substrate after washing was the absorption medium. The SAW pulses were recorded by an end-mounted piezoelectric transducer with a 50 MHz bandwidth. The light absorption spectra of the film were analyzed between 490 and 540 nm as well as their variation over time and under optical irradiation.

SAW OA spectroscopy has turned out to be much more sensitive to the nature of the bond between the absorbing film and the substrate compared to other methods (particularly the standard adsorption method). It was discovered that adding polystyrene to the dye increases by a factor of 4 the optical strength of the film (the rate of decay of the SAW signal with time diminishes with increasing polystyrene). The maximum sensitivity of the method was determined by the signal-to-noise ratio; it is roughly equivalent to absorption in a monomolecular dye layer.

The method used in Ref. 98 was more advanced than those described above. Specifically, in this study it was possible to resolve the wave form of thermooptically excited SAW's.

References 101 and 102 investigated SAW excitation from laser-beam scanning across the surface of a body near the Rayleigh wave velocity. The experimental[101] and theoretical[102] results are in good agreement. As in the case of volume waves, a sharp (resonant) increase in SAW excitation efficiency is observed as the beam velocity approaches the Rayleigh wave velocity (Fig. 7.15). The resonance width is determined by spot size a and the length of the scanning path L. In the event of a velocity shift δv from resonance c_R the amplitude will continue to rise until the SAW and ray diverge by a distance a. Therefore the resonance width is estimated as $\delta v \sim c_R a/L$, which corresponds to experimental data. The resonance curve narrows with increasing length of the scanning path (Fig. 7.16), while the wave amplitude grows. The deviation from linearity of the amplitude relation of the wave with distance is evidently due to SAW diffraction.

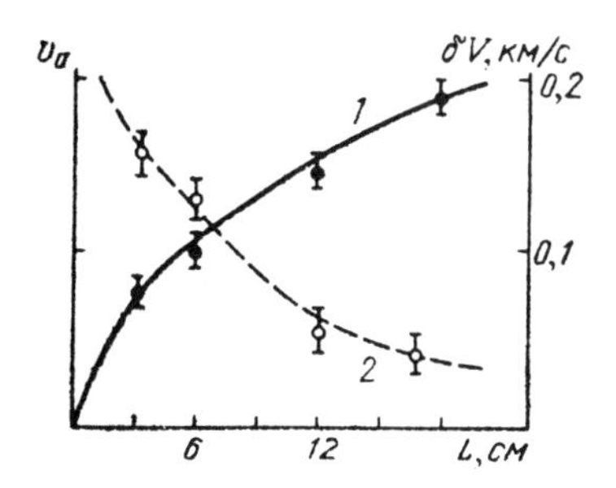

FIG. 7.16. Rayleigh-wave amplitude at resonance (1) and the resonance half-width (2) plotted as a function of the length of the scanning path.

The amplitude of the surface displacement in the scanning mode may be significant (in experiment[101] 0.1 μm) at a very weak surface heating level ($T \lesssim 3$ K). Nonetheless, it remains proportional to the laser radiation power. Estimates of the displacement[102] are in good agreement with measurement data.[101] Reference 102 obtained a solution for excitation of surface acoustical waves by a traveling strip-type laser beam. When the scanning velocity is exactly identical to the Rayleigh wave velocity, the latter grows linearly with time, and the sound velocity component normal to the surface reproduces the intensity distribution in the spot cross section. The relative change in volume under deformation for aluminum is of the order of $7 \times 10^{-12} I_0 L/a$ (I_0 is the light intensity measured in W/cm^2; a is the spot width; L is the scanning length). The advantages of the scanning mode include the capability for a purely mechanical influence on surface layers of the material without any significant heating.

Laser excitation of SAW's was analyzed theoretically based on a linear problem of thermoelasticity.[97,99,100,105–108] It is justified to limit the analysis to a linear acoustic approximation, as in the case of thermooptical generation of volume waves. It is sufficient to consider the excitation of a monochromatic wave given the linearity of the problem. All these studies are limited to the case of a plane monochromatic Rayleigh wave for which the theoretical analysis is significantly simplified.

A comparative analysis of the efficiency of laser excitation of acoustic waves in semiconductors by two different mechanisms—thermal and deformation—is reported in Ref. 96. The interference pattern generated by two beams incident on the surface at a certain angle to one another is considered. The light frequency exceeds the band gap. The deformation mechanism is related to the appearance of lattice deformations from concentration modulation of the electrons and holes generated by the laser radiation. The thermal and deformation mechanisms have a different frequency dependence of acoustic wave excitation efficiency.

Competition between the deformation and thermal mechanisms of sound excitation was observed in experiments.[109,110] As described in Chap. 4, the carrier recombination time diminishes with increasing intensity and the contribution of the thermal mechanism to OA-signal formation grows correspondingly. Since the thermal and deformation mechanisms act in antiphase, the SAW OA-signal profile becomes inverted with increasing intensity (see Fig. 5.6).

The primary difficulty of pulsed OA spectroscopy of Rayleigh waves lies in recording the broadband signals. If the test surface is piezoelectrically active relative to the surface acoustic waves, traditional interdigital converters can be used.[93,94] Broadband end-mounted transducers make it possible to record surface acoustic waves through 50 MHz.[98] Noncontact optical methods[111–115] utilizing an interferometric[111,112,115] or heterodyne[113,114] design have approximately the same frequency band. Expanding the band in interferometric designs primarily involves reducing spot size and sharp focusing.

The SAW OA-spectroscopy method is very promising as a means for real-time monitoring and control of industrial operations in microelectronics as a surface diagnostic technique. It is premature to discuss other possibilities for this technique such as local measurement of elastic constants in view of the limited experimental data. Moreover, the thermooptical method is extraordinarily convenient and flexible for exciting powerful high-frequency Rayleigh waves (Mach numbers on the order of 10^{-4} are attainable at a light intensity of the order of 10^8 W/cm^2). It may be useful in many acoustoelectronic applications.

On the whole, we again state that SAW OA spectroscopy, which has many of the advantages of OA spectroscopy by bulk waves (noncontact design, locality, high sensitivity to absorbed energy) is the most promising method for analyzing the surface layers of media.

7.4. Three-Dimensional Optoacoustic Spectroscopy by Bulk Waves

The dependence of signal amplitude on modulation frequency is of primary interest in broadband OA spectroscopy. As demonstrated in Chap. 3, the acoustic signal power is the product of the laser intensity spectrum and the transfer function $K(\omega)$ determined by the spatial distribution of light intensity and the acoustic impedance of the boundary. In the thermal and Rayleigh-wave OA-spectroscopy cases discussed above, the transfer function was expressed through the Laplace transform of the distribution of the acoustic sources $\bar{g}(z)$. The possibility for suppressing the contribution of the acoustic sources in the bulk of the medium and isolating this contribution from sources in the surface layer derives from this arrangement. Consequently, OA spectroscopy employing thermal and Rayleigh waves is more suitable for surface analysis. Unlike these versions of optoacoustics, the transfer function of the volume waves excited in the absorbing medium is determined by the Fourier transform of the spatial distribution of the sources. Hence, it is most advisable to utilize OA spectroscopy of volume waves in direct recording to analyze volume absorption.

Use of pulsed operation to measure low-light absorptivities makes it possible to significantly increase the signal level compared to the case of cw modulation. The harmonic amplitudes are identical at an identical average power of modulated cw and pulsed-periodic radiation, although given the large number of spectral components in the second case (the high peak power) the OA-signal amplitude is significantly higher. This fact dictates the signal detection method: strobing (unlike synchronous detection in the case of cw modulation). Ordinarily the strobe is delivered at the peak of the OA signal, and the result is averaged over a large number of shots. In this recording method the OA-signal wave form is neither established nor analyzed. Obviously this method can only be used to determine the light absorptivity (when it is sufficiently small).

Given the fact that the transfer function of thermooptical excitation of a volume

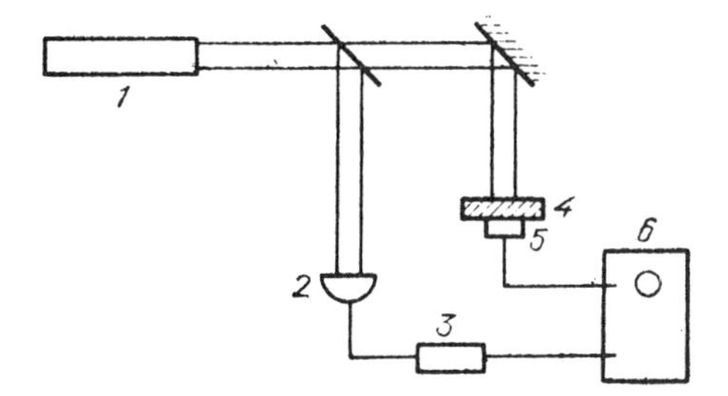

FIG. 7.17. Typical pulsed OA-spectroscopy experimental setup: 1—laser; 2—photodiode; 3—sync pulse delay unit; 4—test medium; 5—acoustic detector; 6—pulsed signal recording system.

acoustic wave is proportional to the Fourier spectrum of the intensity distribution it is most advisable to use this version of OA spectroscopy to detect anomalies in light absorption in the acoustic properties of the medium and the heating zone. Specifically, OA-signal recording may be extraordinarily useful for analyzing thermalization of energy in semiconductors through photoexcitation. Since the recombination and relaxation times in this case may vary over a significant range (from 10^{-6} to 10^{-10} s), the characteristic frequencies fall within the ultrasonic range (see Chap. 4) and the acoustic wave provides a more comprehensive reflection of the features of this process.

A typical experimental setup is shown in Fig. 7.17. Ordinarily, a *Q*-switched laser is used as the radiation source. Reliable recording of the OA signal requires rather strong and stable locking of the timing pulses to the laser pulse. Either a free surface or an absorbing surface applied to a transparent substrate is irradiated.

One of the primary difficulties is maintaining a rather broad band of acoustic wave frequencies. This requires thin-film piezoelectric transducers,[68,116–118] capacitive,[119] or interferometric detectors.[120] Acoustic pulses below 1 ns were registered in the first two cases. This suggests a frequency band of the order of 1 GHz. However, the frequency response of these detectors was not measured due to the specific difficulties of broadband acoustic measurements[121] and the lack of a standard methodology. Moreover, measurement of the transfer functions as a rule requires digital signal processing. It is therefore necessary to use high-speed analog-to-digital converters and memories. Digital stroboscopic systems (at a sufficiently high laser pulse repetition rate) can also be used.

A quantitative comparison of the experimental and theoretical wave forms (or spectra) of acoustic pulses has been carried out today for broad ($\alpha a \gg 1$) and narrow ($\alpha a \ll 1$) beams. References 122 and 123 have investigated the wave form of OA signals in the case of weak absorption. Reference 123 used a hydrophone with a 10 MHz bandwidth that substantially distorted the signal. Reference 122 succeeded in expanding the bandwidth by deflecting the probe beam. Since the bandwidth of the recorded frequencies in this case is largely determined by the diameter of the probe beam, while the excited frequency band is determined by the diameter of the heating radiation $2a$, it is sufficient for the first frequency to be below the second frequency for proper signal detection.

Figure 7.18 shows the experimental OA-signal profiles for different beam diameters and their comparison to the calculated profile. As we see, they are in good agreement, and we can conclude from Fig. 7.18(b) that the limited nature of the

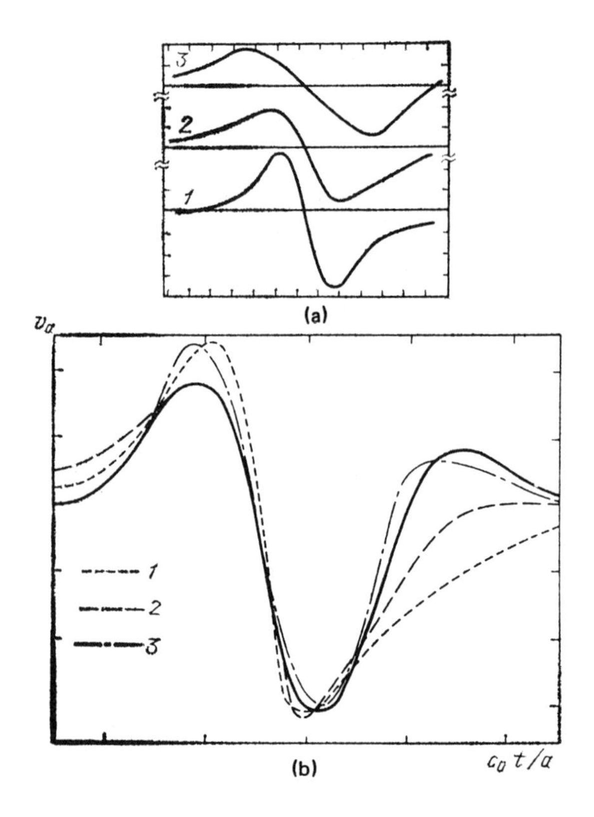

FIG. 7.18. (a) OA pulse profile (1–3) for the case of weak absorption and different beam diameters: division value, 10 ns/div; (b) a comparison of these profiles to the calculated profile (solid curve).

receive frequency band causes losses of minor profile details with diminishing diameter of the heating beam. Unfortunately, a spectral analysis was not carried out in these cases.

The first attempts at a spectral analysis of thermooptical sources were made in Ref. 124 for the case of strong absorption ($\alpha a \gg 1$). Sound was excited in a carbon black suspension in water with a free surface of the absorbing medium. A strongly damped piezoelectric transducer with a 30 MHz resonant frequency was employed. The experimentally derived spectrum was correlated accounting for acoustic wave diffraction. Figure 7.19 shows the experimental and calculated OA spectrum for $\alpha = 3.5 \times 10^7 \text{ cm}^{-1}$. The lack of details in Ref. 124 makes it impossible to determine the nature of such a high light absorptivity.

A quantitative comparison of the calculated and experimentally derived transfer functions was reported in Ref. 125. A *TE* CO_2 laser was used for sound excitation in transformer oil. OA signals were investigated for both a rigid and a free boundary. A damped lithium niobate piezoelectric element with a 100 MHz resonant frequency was employed for signal detection. The recorded frequency band was limited by the oscilloscope at 50 MHz. The optical and acoustic pulse profiles were recorded and digitized. The radiation absorptivity in the case of transformer oil $\alpha \approx 50 \text{ cm}^{-1}$ and the parameter $m_\chi \ll 1$. Hence the effect of heat conduction was not

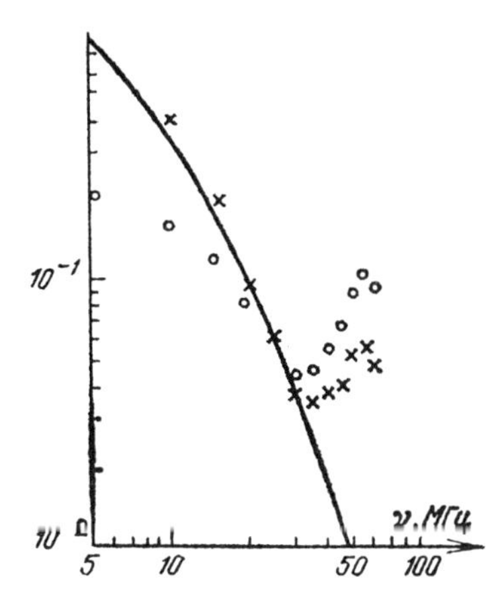

FIG. 7.19. OA-pulse spectrum (circles) and its correction accounting for diffraction (crosses) and the theoretical relation (solid curve).

significant across the entire recorded frequency range. Since the transfer functions K_r and K_f in Eqs. (2.24) and (2.25) can be written as

$$(\rho_0 c_p/\beta) K_r^{-1}(\omega) = 1 + \omega^2/(\alpha c_0)^2,$$

$$(\rho_0 c_p/\beta)\big(\omega/\alpha c_0 K_f(\omega)\big) = 1 + \omega^2/(\alpha c_0)^2,$$

on the coordinates $K_r^{-1} = \mathscr{K}(\omega^2)$ and $\omega K_f^{-1} = \mathscr{K}(\omega^2)$, respectively, these relations will be straight lines. Figure 2.8 shows the experimental relations that are accurately approximated by lines in the frequency range 1–4 MHz. Diffraction has no substantial effect on the wave amplitude in this frequency range (the diffraction parameter was of the order of 10^{-2}).

Reference 33 compared the theoretical and experimental wave form of the OA signals for the case of CO_2 laser irradiation of a water surface ($\alpha = 870$ cm^{-1}). However, inaccuracies in the laser pulse wave-form approximation made it impossible to obtain a complete correlation. The transfer functions were not calculated. This study focused on the manifestation of thermal nonlinearity in the thermooptical excitation of sound (see Sec. 5.1). Specifically, the fact that the peak of the signal excited in water lies at a water temperature below the density peak [where $\beta(T_0) = 0$] was noted. The temperature of the minimum diminished with increasing laser pulse energy (Fig. 7.20). This mechanism was confirmed theoretically in Ref. 126 within the framework of the thermal nonlinearity model (Sec. 5.1). Generally speaking, the thermal nonlinearity will be more strongly manifested in sound excitation in water (see Table 5.1). Figure 5.3 confirms the good agreement between experimental and theoretical data.

These data suggest that the theory developed here is applicable for describing thermooptical excitation of sound in the case of moderate absorption ($\alpha < 10^4$ cm^{-1}). Such an approach makes it possible to diagnose inhomogeneities in the surface layer. Short laser pulses ($\alpha c_0 \tau_L \ll 1$) are most suitable for this application, since the wave form of the leading edge of the acoustic wave is identical to the source distribution. Chapter 2 (Fig. 2.9) reports data from the diagnostics of irregularities in light absorptivity. If the thermal coefficient of volume expansion of

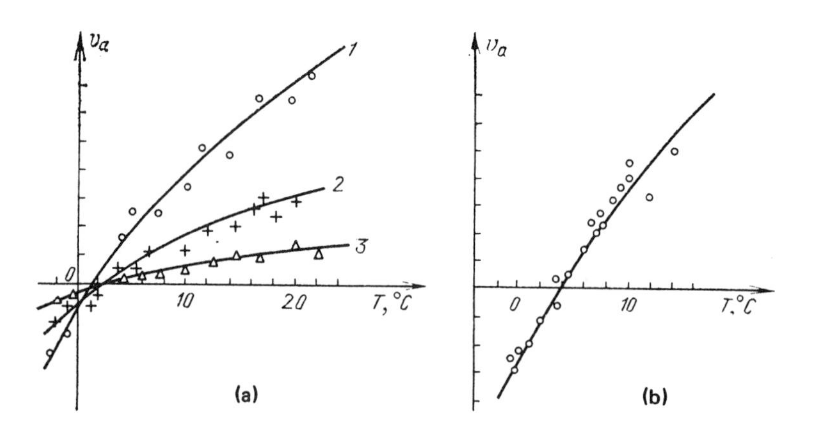

FIG. 7.20. OA-signal amplitude in water plotted against temperature: (a) at different radiation powers W_0 = 530 (1), 265 (2), 130 kW (3); (b) under weak pumping W_0 = 6.8 kW.

the medium is strongly temperature dependent (as is the case in water) it is also possible to investigate inhomogeneous temperature distributions. Figure 7.21 reports OA-signal profiles and calculated temperature distributions in an inhomogeneously heated surface layer.

Since the efficiency of sound excitation and its temporal wave form are essentially determined by the impedance of the boundary, we can naturally use the OA effect to analyze films on the surface of a medium. However, analysis of thin films requires rather short (for micrometer films: below the nanosecond range) light pulses and media with a high absorptivity. Ordinarily metals are used in such experiments as the absorbing medium. Reference 118 detected multiple reflections of acoustic pulses excited in a thin metallic film deposited on a sample under

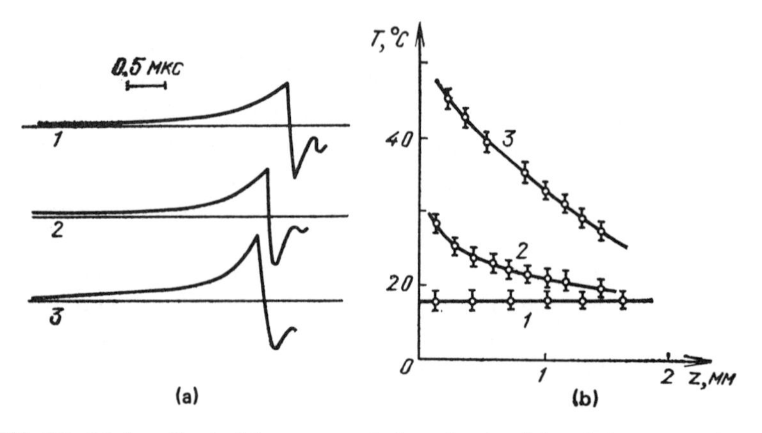

FIG. 7.21. OA-signal profiles in inhomogeneously heated water (a) and the recovered temperature distributions (b) before (1) and after 8 (2) and 15 (3) minutes heater initiation.

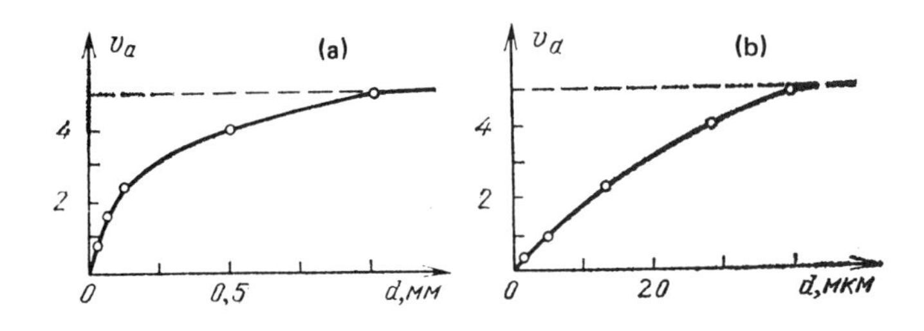

FIG. 7.22. Relative OA-signal amplitude in a metal plotted against the thickness of a glycerin (a) and F-23 fluorine lacquer (b) film on its surface.

irradiation by a mode-locked laser. The film thickness was determined from the pulse repetition period; the error was estimated at 1 % for a film thickness of 12 μm.

Such short acoustic pulses cannot be used in noncrystalline materials given their strong attenuation. It is therefore interesting to analyze the amplitude of sound at a rather low frequency as a function of the thickness of the transparent film on the metal surface. Such experiments are described in Ref. 127 and their results are presented in Fig. 7.22. The OA-signal amplitude (at a detector resonant frequency of 5 MHz) grows with increasing thickness of the transparent layer d and gradually goes to saturation. The presence of the transparent layer (even a thin layer) alters the conditions on the absorbing boundary of the metal and elevates excitation efficiency (see Sec. 3.2). In principle, this method makes it possible to detect and determine the thickness of transparent micron films on a metal surface.

A qualitative comparison of the wave forms of OA pulses excited in metals was reported in Ref. 128. Their wave form is universal due to strong light absorption ($m_\chi \sim 1$; see Sec. 3.2 and Fig. 3.4). Figure 7.23 gives the experimental and theoretically calculated OA-signal profiles for an impedance boundary and a free boundary (diffraction was accounted for in the calculations). Since the parameter $(\omega_\chi \tau_L)^{1/2} \approx 1.2 \times 10^2$ in these experiments, the OA-signal wave form with the impedance boundary is nearly identical to its wave form with a rigid boundary. The results indicate a good agreement between theoretical and experimental data.

The behavior of the shift of the back side of a plate in longitudinal and shear waves was investigated.[129–131] Optical excitation of sound in metals was proposed in these studies as a standard sound source for calibrating nondestructive testing systems.

In thermooptical excitation of sound in metals, changes in the thermal coefficient of volume expansion are not, as a rule, significant (see Table 5.1). In the case of metals, the light reflectance is most strongly dependent on temperature (due to growth of oxide film). However, given the surface character of light absorption ($m_\chi \sim 1$) this effect will principally alter the wave amplitude. We can assume that a change in absorptivity will cause a corresponding change in the optical pulse wave form (with fixed absorption). This effect was observed theoretically and experimentally in Ref. 120. The calculation relations provide good confirmation of the experimental relations over a light intensity range through 4×10^2 MW/cm^2.

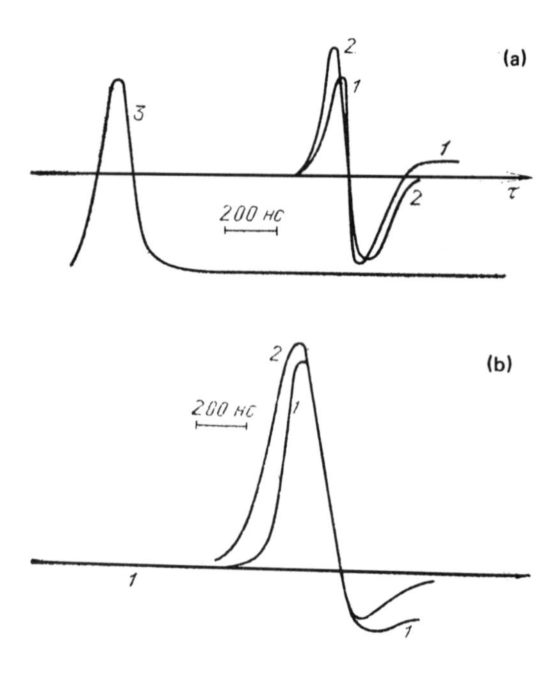

FIG. 7.23. Profiles of OA signals excited in a metal (copper) for the case of a free (a) and impedance (b) boundary: 1—experimental curve, 2—calculated curve; 3—laser pulse profile.

An analysis of OA signals makes it possible to restore the space-time image of light absorption and resulting changes in the state of the material. Sound excitation from phase transitions—evaporation, melting, structural phase transitions—have been widely investigated. The literature on these areas can be found in the bibliography of Ref. 28. However, the lack of analytic regularities makes it impossible to carry out a quantitative analysis of the experimental results.

7.5. Broadband Acoustic Spectroscopy With Laser Excitation

Optoacoustic spectroscopy makes it possible to obtain two types of spectra: optical spectra (dependence of the absorptivity on the wavelength of light) and acoustic spectra (dependence of sound speed, attenuation coefficient, etc., on the acoustic wave frequency). If the first type of spectra are obtained by changing the wavelength of light, the latter are obtained by changing its intensity modulation frequency. An investigation of the second type of spectra is a fundamental aspect of pulsed optoacoustics. In this case, OA excitation of sound is used as a source of powerful broadband waves with a known spectrum: a source of standard acoustic signals. By monitoring the change in spectrum (or profile) of the acoustic signal as it propagates in the test medium, it is possible to determine the dispersion of sound speed, the frequency dependence of the absorptivity, the acoustic nonlinearity coefficient, etc. The discussion therefore concerns new possibilities for broadband acoustic spectroscopy.

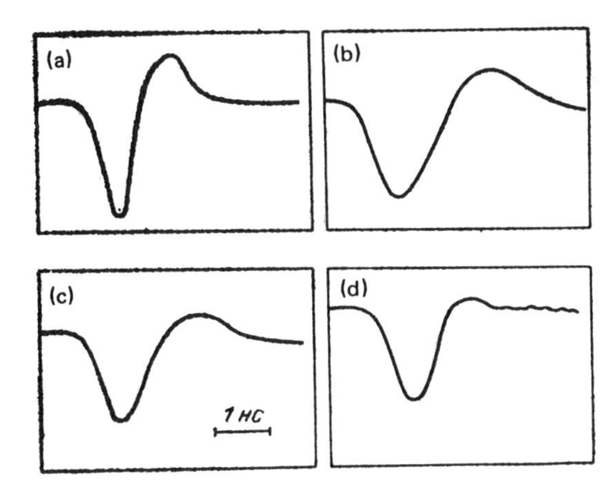

FIG. 7.24. Profiles of OA signals excited by ultrashort pulses and propagating in a variety of media: (a) quartz 190 μm thick; (b) polytrifluochloroethylene, 18 μm; (c) teflon, 25 μm; (d) polyethylene terephthalate, 25 μm.

One of the primary difficulties of broadband acoustic spectroscopy is maintaining a constant sound transmittance through the radiator-test medium and the test medium-detector interfaces.[121] It is likely that this is most easily achieved by broadband acoustic spectroscopy. Thus, acoustic pulses of subnanosecond duration were excited in a thin conducting film applied to a test sample in Ref. 134. A capacitive transducer recorded oscillations on the back side of the sample. Thin (of the order of 20 μm) quartz layers and polymer layers were analyzed.

The profiles of the recorded signals are given in Fig. 7.24. Sound absorption was estimated from the broadening of sound pulses transmitted through the polymer films relative to the signal transmitted through quartz. No spectral analysis, however, was carried out. Such short pulses were used in a study by the same authors[119] to analyze the spaced charge distribution in dielectrics. A similar experimental setup (yet one employing piezoelectric recording) was employed in Ref. 118 to measure the thickness of the metallic films. It is, however, obvious that short acoustic pulses are required to measure thin films (of the order of micrometers) given the high sound speed and thermal wave microscopy is likely to be more suitable here.

Broadband acoustic spectroscopy by OA excitation was implemented in Ref. 57 for gases. Figure 7.25(a) shows the OA-signal spectra at two distances from the heated surface of the quartz sample. It is possible to determine the dependence of the sound absorptivity over a broad frequency range from the ratio of these spectra [Fig. 7.25(b)].

The high amplitude of the OA signals permits measurement of sound speed and absorption in highly absorbing media. The acoustic properties of powders were analyzed under OA excitation in Ref. 135. Figure 7.26 shows the dependence of the sound speed and the half-width of the OA pulse propagating in a mixture of ferric oxide (Fe_3O_4) and carbon black powders (at a mean porosity of 49%) on pressure. A 30 MHz recording frequency band was achieved. A spectral analysis of the wave

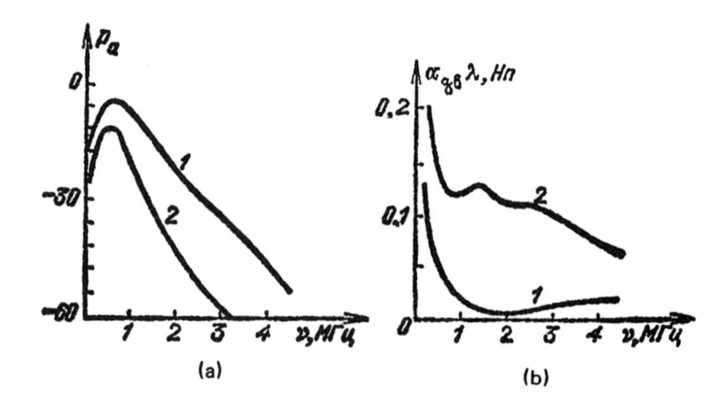

FIG. 7.25. (a) Fourier spectra of OA pulses in a gas at different distances z from the excitation plane: 1, 1.816; 2, 4.356 m; (b) the frequency dependence of the sound absorptivity calculated from these spectra: 1, in pure carbon dioxide; 2, in a CO_2 and H_2O mixture; the partial pressure of water vapor is 20 Torr; total pressure: 1 atm; $T = 22\,°C$.

form of the acoustic pulses would, in principle, make it possible to estimate the dimensions of the powder particles and the dispersion of particle size.

Combining high amplitude and short duration of OA signals, it is possible to effectively utilize such signals to analyze the nonlinear acoustic properties of media. As discussed in Chap. 2, the nonlinear parameter of the media can be determined from the dependence of pulse amplitude or duration on the distance traveled by the wave (or the initial wave amplitude). Figure 7.27 shows sample OA-signal profiles for low and high laser pulse intensities.[136] The change in pulse duration was used to determine the acoustic nonlinearity coefficient of the ethanol in which the OA signal was excited and was propagating.

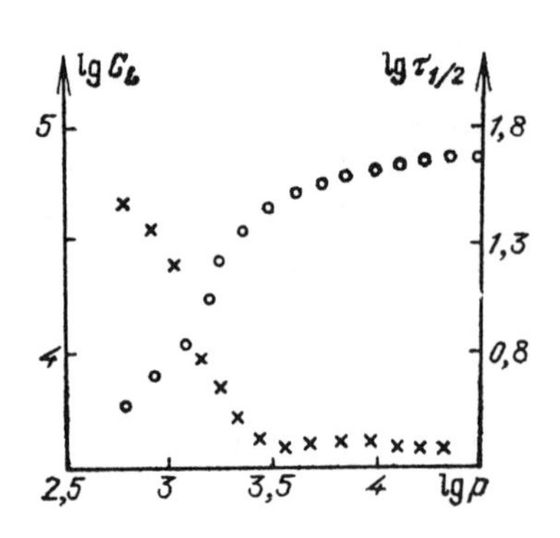

FIG. 7.26. Sound speed c_L (circles) and pulsed half-widths $\tau_{1/2}$ (crosses) in powder vs the hydrostatic pressure measured by the OA method.

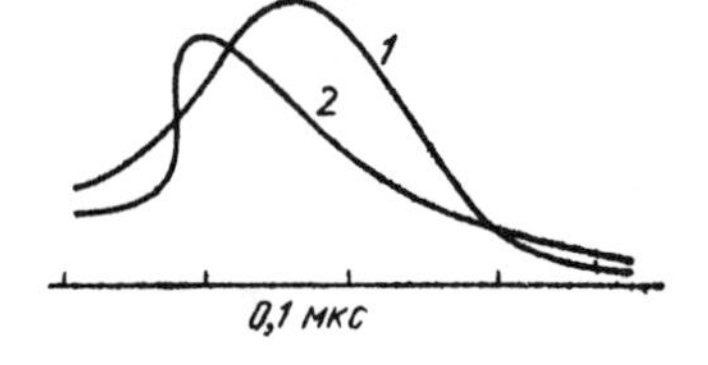

FIG. 7.27. Profiles of OA signals in ethanol excited at low (1) and high (2) CO_2 laser pulse intensities.

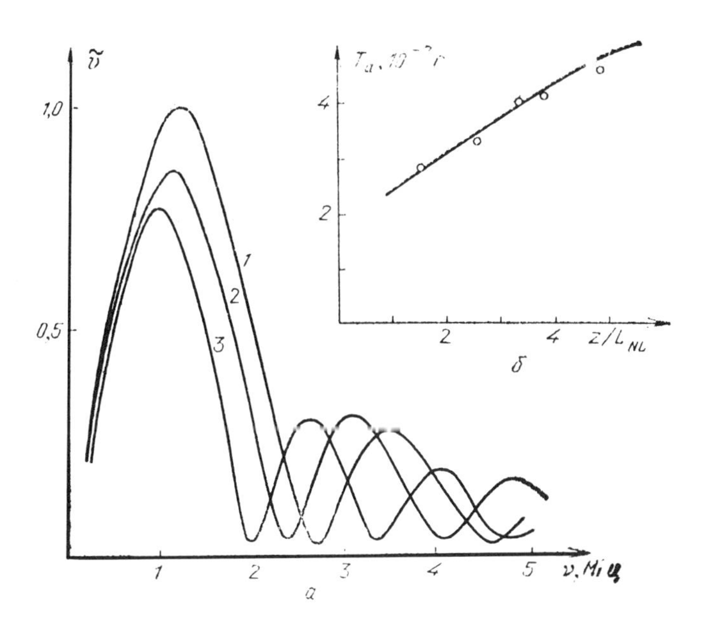

FIG. 7.28. Variation of OA-pulse spectra from pulse propagation in water for z/L_{NL} values of 1.7 (1), 2.6 (2), 4.6 (3). The inset shows the dependence of OA-pulse duration T_a on the distance traveled by the wave recovered from these spectra.

In order to use this scheme it must be verified that the amplitude of the excited OA signal is proportional to the laser pulse energy while its spectrum remains unchanged. In the general case this is not so. Therefore, it is more advisable to develop an OA generator as a source of standard high-amplitude acoustic pulses. It then becomes possible to optimize the generator medium as well as the laser parameters. This idea was implemented in Ref. 125. Transformer oil absorbing *TE* CO_2 laser radiation was employed as the acoustic pulse generator. The acoustic pulse spectrum matched the theoretical pulse spectrum (see Chap. 2) across the entire range of fluences (up to 3 J/cm^2). This revealed the linearity of such an OA generator, which permitted reliable measurement with varying radiation energy.

Figure 7.28(a) shows the transformation of the OA-pulse spectrum with increasing transited distance. Its wave form accurately corresponds to spectrum of *N*-wave. The dependence of OA-signal duration on distance recovered from the minima of this spectrum are accurately reproduced by the theoretical curve [Fig. 7.28(b)].

Nonlinear acoustic waves are characterized by extended, near straight profile sections interconnected by discontinuities (Chap. 2). Hence it is most convenient to measure the nonlinear parameter of these media by means of relation (2.40)

$$\left(\frac{\partial \tau}{\partial p}\right) = \left(\frac{\partial \tau}{\partial p}\right)_0 - \frac{\varepsilon z}{\rho_0 c_0^3}.$$

Figure 7.29 shows the dependence of the slope of the rectilinear section of the OA-signal profile on the distance traveled by the wave in water.[125] The nonlinearity

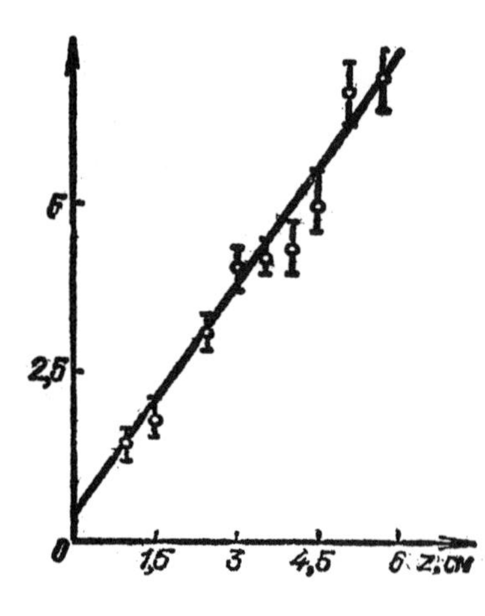

FIG. 7.29. Slope of the rectilinear segment of the OA-signal profile vs the distance traveled by the wave in water.

coefficient ε is determined from the slope of this line. One additional advantage of this method is that the effect of the limited nature of the receive frequency band and dissipation is minimized. The high amplitude and short duration make it possible to use OA pulses to investigate acoustic nonlinear properties of solids.[137]

OA pulse generators can therefore be a convenient tool for broadband acoustic spectroscopy. The short duration of the OA signals makes possible effective utilization of such signals in flaw detection of nontransparent materials. A complete noncontact OA flaw detection method employing optical excitation and detection of sound is also possible. Such a flaw detection technique was implemented with volume[138] and surface[114] waves. This method used a traditional ideology for flaw detection, while optical excitation and recording made possible improvements in its resolution.

In summary, we can state that the OA methods make possible noncontact defect diagnostics; it is advisable to utilize thermal or Rayleigh waves for diagnostics of surface defects and volume waves for diagnostics of bulk defects.

7.6. Optoacoustic Excitation of Ultrashort Pulses

One fundamental feature of OA sound sources is their broadband performance. Thus, the characteristic frequency of the band with the highest sound excitation efficiency $\omega = \alpha c_L$ lies in the range $\omega = 2\pi(10^{10}\text{–}10^{11})\ \text{s}^{-1}$ at high light absorptivities $\alpha \approx 10^5\text{–}10^6\ \text{cm}^{-1}$. Hence, by using ultrashort light pulses for sound excitation, it is possible to excite 10–100 ps acoustic signals.[133] Optical excitation of short acoustic pulses was first reported in Ref. 14. Sound was excited by a mode-locked laser in a metallic film deposited on a crystal bar. These experiments, however, did not record the pulse wave form, but rather measured the amplitudes of the acoustic harmonics at even multiple frequencies of the repetition rate of the light pulses. Such a scheme was subsequently used[118,119] to excite subnanosecond acoustic pulses and record such pulses by piezoelectric and capacitive transducers (Fig. 7.24). Using the anomalously high light absorptivity of hydroxyl-containing liquids (water, ethanol, glycerin) at the wavelength of the YAG-Er laser, subnanosecond acoustic pulses in a liquid were recorded in Ref. 117.

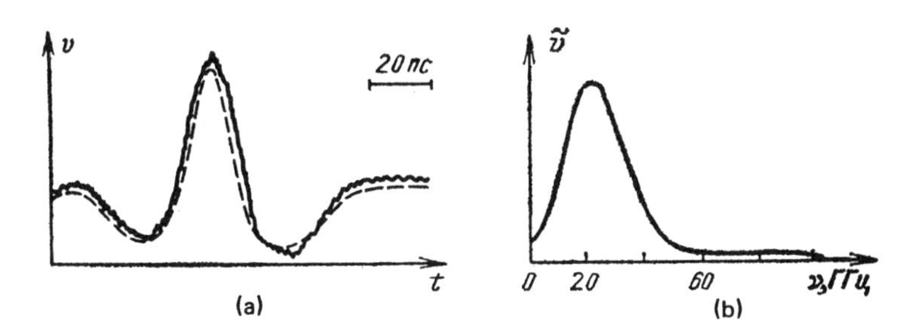

FIG. 7.30. OA-signal profile (a) [dashed line: theory] and its spectrum (b) in a-As_2Te_3.

The receiving frequency band in all experiments listed above was limited to a level of the order of 1 GHz by the recording system. Moreover, hypersonic waves at room temperature cannot propagate significant distances. Two experimental versions are therefore possible: excitation of hypersound in thin films (thickness of the order of a few micrometers) at room temperature or in bulk samples at cryogenic temperatures. It is also obvious that in order to record acoustic pulses shorter than 1 ns it is advisable to use acoustooptic effects: sound modulation of the reflectance, absorptivity, refractive index, etc., and to employ ultrashort laser probe pulses as well. In this case a purely optical sound excitation and recording setup is implemented.

Under thermoelastic excitation of sound, the acoustic wave sources are bulk sources and hence the duration of the deformation pulses is not only limited by the laser pulse duration τ_L but also by the sound transit time through the heating zone (l/c_L). The characteristic dimensions of the heating zone l are determined either by the depth of penetration of light $l \sim \alpha^{-1}$ or by the heat conduction length $l \sim (\chi/\tau_L)^{1/2}$. Phonon heat conduction occurs at velocities below the sound velocity and hence will not broaden the OA pulses. At the same time, heat transfer by electrons in metals and by electron-hole plasma in semiconductors may substantially increase the heating depth, particularly at low temperatures.[140] At room temperature carrier diffusion is largely inhibited by strong scattering by thermal lattice vibrations. It is necessary to simultaneously reduce laser pulse duration and light absorption length to achieve thermoelastic generation of ultrashort deformation pulses at room temperature.

The case of strong absorption and a weak effect of thermodiffusion can be achieved in amorphous semiconductors. Thus, $c_L = 2 \times 10^5$ cm/s, $\chi \sim 10^{-2}$ cm^2/s and $\alpha^{-1} \sim 3 \times 10^{-6}$ cm for a-As_2Te_3 (Ref. 141) for radiation with $\hbar\omega_L = 2$ eV, which yields for the parameter $m_\chi = 0.017$ and $\omega_a = 2\pi \times 10^{10}$ s^{-1}.

Figure 7.30 shows the OA-signal profile observed by modulating the reflection of an ultrashort laser probe pulse in the stroboscopic regime. Since the depth of penetration of the probe radiation is comparable to the length of the deformation wave, the OA-signal profile is smoothed out by the pulsed transient response of the recording system. In this case the temporal resolution was 15–20 ps. The high temporal resolution requires utilizing probe radiation that will at least penetrate 10^{-6} cm into the medium. Therefore, in recording ultrashort light pulses, the temporal resolution is not limited (at present) by the duration of the light probe pulse, but rather by spatial length of the detector medium. It seems that the resolution could be improved by reducing the thickness of the detector film, although this would result in a corresponding reduction in the sensitivity of the recording system (the recording system sensitivity is not rather high). Hence the greatest difficulty experimentally is expanding the recording frequency band.

The thermoelastic sound excitation mechanism in its "pure form" does not produce acoustic pulses shorter than 10–20 ps (even with short laser pulses) due to the fundamental effect of heat diffusion. In this respect the concentration-deformation mechanism may have certain advantages in connection with differences in the dynamics of the plasma and phonon subsystems of the semiconductor. Concentration sound sources can be not only rapidly developed (by reducing τ_L) but can also be rapidly switched off. This is possible due to the strong dependence of the nonlinear recombination time τ_R of the EH pairs on carrier concentration n ($\tau_R \sim n^{-1}$ in the case of bimolecular recombination, $\tau_R \sim n^{-2}$ in the case of Auger recombination). By increasing optical radiation intensity I_0 it is possible to switch off the concentration sound sources for times $\tau_R \lesssim \tau_L$. It then becomes possible to excite acoustic pulses of duration $\tau_a \sim \tau_L$, even if $l/c_L \gg \tau_L$.[142] Such a shortening of OA-pulse duration was observed experimentally in the nanosecond time range.[143] Overall estimates based on the theoretical models developed to date reveal that optical excitation of acoustic pulses of duration $\tau_a \sim 1$–100 ps, spatial length 10^{-2}–1 μm at a pressure amplitude 1–100 kbar is physically attainable.

As discussed above, experiments employing ultrashort acoustic pulses at room temperature can only be carried out on film samples. These pulses can travel macroscopic distances only at crystals at cryogenic temperatures. However, in this case the effect of thermal diffusion increases. Hence, it is necessary to utilize thin strongly absorbing films deposited on the dielectric sample as the generator to limit its effect.

Assume the film and crystal are in thermal contact at the initial time for $T_0 \sim 1$ K. Then a $\tau_L \sim 20$ ps light pulse of intensity $I_0 \sim 100$ MW/cm^2 impacts the film. Then the characteristic electron energy rises to $T_e \sim 100$ K over a time of the order of $10^{-2}\,\tau_L$; in this case the electron-phonon collision time drops to $\tau_{e\text{-ph}} \sim 0.1$ ps.[144] The transit to an equilibrium phonon distribution and expansion of the metal occur with virtually no lag ($\tau_{\text{ph-}e} \lesssim \tau_{e\text{-ph}}$). Since electrons will travel a distance of the order of $v_F \tau_L \sim 10$ μm (v_F is the Fermi velocity) far exceeding film thickness over the laser irradiation time, the film is heated and expands uniformly in volume. As the film expands, a coherent acoustic pulse with a duration of the laser irradiation time τ_L is transmitted to the crystal.

It is significant that the nonequilibrium phonons cannot escape the metallic film over the light pulse time: $\tau_d = d/c_L > \tau_L$. The crystal is not heated and the ultrashort acoustic pulse propagates along cold dielectric crystal. Its amplitude can be determined from standard equations of optoacoustics.[145] Under these conditions the Mach number of the acoustic wave for an aluminum film will be of the order of 5×10^{-4}. We have assumed that the film is held between two crystal samples and hence a unipolar compressional pulse is excited.

Pulse propagation in the crystal can be described[139] by the Korteweg–de Vries equation, since we can neglect absorption of the wave at cryogenic temperatures:[146,147]

$$\frac{\partial v}{\partial z} - \frac{\varepsilon v}{c_L^2}\frac{\partial v}{\partial \tau} - \frac{a^2}{24c_L^3}\frac{\partial^3 v}{\partial \tau^3} = 0.$$

Here v is the vibrational velocity, t is time in the accompanying coordinate system, ε is the nonlinear phonon-phonon interaction parameter, and a is the lattice constant of the crystal. Longitudinal acoustic solitons may be formed from the propagation of the acoustic pulse under these conditions.[139] Their formation condition is satisfaction of the inequality

$$v_a/c_L > a^2/2(c_L\tau_L)^2.$$

Six to nine solitons may form under the conditions outlined above. Soliton width is determined by their amplitude, and for the largest the spatial length may be of the order of

$$\frac{a}{(2v_a/c_L)^{1/2}} \sim 30a.$$

Consequently, its total duration will be approximately 0.1 ps. We note that such pulses can only be formed at a rather high initial signal amplitude.

The OA effect therefore makes it possible to obtain coherent acoustic pulses with the largest possible bandwidth[147]; such pulses yield new information on the interaction of the electron and phonon subsystems in the material.

REFERENCES

1. R. M. White, J. Appl. Phys. **34**, 2123 (1963).
2. R. M. White, J. Appl. Phys. **34**, 3559 (1963).
3. G. A. Askar'yan, A. M. Prokhorov *et al.*, Zh. Eksp. Teor. Fiz. **44**, 2180 (1963) [Sov. Phys. JETP **17**, 1463 (1963)].
4. E. F. Carome, N. A. Clark, and C. E. Moeller, Appl. Phys. Lett. **4**, 95 (1964).
5. G. H. Conners and R. A. Thompson, Appl. Phys. **37**, 3434 (1966).
6. R. Bullough and J. Gilman, J. Appl. Phys. **37**, 3434 (1966).
7. L. S. Gournay, J. Acoust. Soc. Am. **40**, 1322 (1966).
8. J. S. Bushnell and D. J. McCloskey, J. Appl. Phys. **39**, 5541 (1968).
9. C. Yamada, T. Aoki, and M. Katayama, J. Appl. Phys. **40**, 5404 (1969).
10. Hu Chia-Lun, J. Acoust. Soc. Am. **46**, 728 (1969).
11. W. B. Gauster and D. H. Habing, Phys. Rev. Lett. **18**, 1058 (1967).
12. A. Korpel, R. Adlerand, and B. Alpiner, Appl. Phys. Lett. **5**, 86 (1964).
13. D. E. Caddes, C. F. Quate, and C. D. W. Wilkinson, Appl. Phys. Lett. **8**, 309 (1966).
14. M. J. Brienza and A. J. DeMaria, Appl. Phys. Lett. **11**, 44 (1967).
15. R. Adler, IEEE Spectrum **4**, 42 (1967).
16. G. A. Savvinykh, *Neoptika* [*Nonoptics*] (Nauka, Novosibirsk, 1968), p. 415.
17. D. C. Auth, Appl. Phys. Lett. **16**, 521 (1970).
18. H. Eichler and H. Stahl, J. Appl. Phys. **44**, 3429 (1973).
19. R. E. Lee and R. M. White, Appl. Phys. Lett. **12**, 12 (1968).
20. F. V. Bunkin and V. M. Komissarov, Akust. Zh. **19**, 305 (1973) [Sov. Phys. Acoust. **19**, 203 (1973)].
21. A. Rosencwaig, *Photoacoustics and Photoacoustic Spectroscopy* (Wiley, New York, 1980).
22. J. Phys. C **6** (1983).
23. V. P. Zharov and V. S. Letokhov, *Lazernaya optiko-akusticheskaya spektroskopiya* [*Laser Optoacoustic Spectroscopy*] (Nauka, Moscow, 1984).

24. *Optiko-akusticheskiy metod v lazernoy spektrosopii molekulyarnykh gazov [Optoacoustic Method in Laser Spectroscopy of Molecular Gases]* edited by Yu. S. Makushkin (Nauka, Novosibirsk, 1984).
25. IEEE Trans. Ultrason. Ferroelectr. Freq. Control, **UFFC 33** (1986).
26. B. G. Ageev, Yu. H. Ponomarev, and B. A. Tikhomirov, *Nelineynaya optiko-akusticheskaya spektroskopiya molekulyarnykh gazov [Nonlinear Optoacoustic Spectroscopy of Molecular Gases]* (Nauka, Novosibirsk, 1987).
27. A. C. Tam, Rev. Mod. Phys. **58**, 381 (1986).
28. *Fotoakustika i rodstvennye metody: Bibliograficheskiy ukazatel'. Vyp. I. Fotoakustika v kodensirovannykh sredakh [Photoacoustics and Associated Methods: Bibliographic Reference. Vol. 1. Photoacoustics in Condensed Matter]*, 1973–1984, compiler Ya. Ya. Gedrovits (Riga, 1987).
29. C. K. N. Patel and A. C. Tam, Rev. Mod. Phys. **53**, 517 (1981).
30. L. M. Lyamshev and L. V. Sedov, Akust. Zh. **27**, 5 (1981) [Sov. Phys. Acoust. **27**, 4 (1981)].
31. L. M. Lyamshev and K. A. Naugol'nykh, Akust. Zh. **27**, 641 (1981) [Sov. Phys. Acoust. **27**, 641 (1981)].
32. A. A. Karabutov, Usp. Fiz. Nauk **147**, 615 (1985) [Sov. Phys. Usp. **28**, 1042 (1985)].
33. M. W. Sigrist, J. Appl. Phys. **60**, R83 (1986).
34. H. Coufal, Appl. Phys. Lett. **44**, 59 (1984).
35. R. C. C. Liebe, R. S. Moore, and J. K. Whinnery, Appl. Phys. Lett. **5**, 141 (1964).
36. R. L. Swofford, M. E. Long, and A. C. Albrecht, J. Chem. Phys. **65**, 179 (1976).
37. A. C. Boccara, D. Fournier *et al.*, Opt. Lett. **5**, 377 (1980).
38. A. C. Boccara, D. Fournier, and J. Badoz, Appl. Phys. Lett. **36**, 130 (1980).
39. J. C. Murphy and L. C. Aamodt, J. Appl. Phys. **51**, 4580 (1980).
40. S. R. J. Brueck, A. Kildal, and L. J. Belanger, Opt. Commun. **34**, 199 (1980).
41. S. O. Kanstad and P. E. Nordal, Appl. Surf. Sci. **6**, 372 (1980).
42. W. B. Jackson, N. M. Amer *et al.*, Appl. Opt. **20**, 1333 (1918).
43. J. P. Monchalin, J. L. Parpal, and J. M. Gagne, Appl. Phys. Lett. **39**, 391 (1981).
44. J. C. Murphy and L. C. Aamodt, Appl. Phys. Lett. **39**, 519 (1981).
45. P. Charpentier, F. Lepoutre, and L. Bertrand, J. Appl. Phys. **53**, 608 (1982).
46. D. Fournier, A. C. Boccara, and J. Badoz, Appl. Opt. **21**, 74 (1982).
47. I. S. Amer, E. A. Ash *et al.*, Electron. Lett. **17**, 337 (1981).
48. L. C. M. Miranda, Appl. Opt. **22**, 2882 (1983).
49. G. Busse, Infrared Phys. **20**, 419 (1980).
50. G. Busse and K. F. Renk, Appl. Phys. Lett. **42**, 366 (1983).
51. A. Rosencwaig, J. Opsal *et al.*, J. Appl. Phys. **59**, 1392 (1986).
52. C. C. Williams, Appl Phys. Lett. **44**, 1115 (1984).
53. Hutchins D. A. Nondestructive Testing Comm. **1**, 37 (1925).
54. Lucos R., Biguard P. J. Phys. Radium. **3**, 464 (1982).
55. G. P. Davidson and D. C. Emmony, J. Phys. E **13**, 93 (1980).
56. W. Zapka and A. C. Tam, Appl. Phys. Lett. **40**, 310 (1982).
57. A. C. Tam and W. P. Leung, Phys. Rev. Lett. **53**, 560 (1984).
58. B. Sullivan and A. C. Tam, J. Acoust. Soc. Am. **75**, 437 (1984).
59. M. W. Sigrist and F. K. Kneubuhl, J. Acoust. Soc. Am. **64**, 1652 (1978).
60. V. M. Gordienko, A. B. Reshilov, and V. I. Shmal'gaugen, Vest. Moscow State Univ. Phys. Astron. Ser. **19**, 59 (1976).
61. D. C. Emmony, Infrared Phys. **25**, 133 (1985).
62. A. A. Bondarenko, Yu. B. Drobot, and S. V. Kruglov, Defektoskopiya **6**, 85 (1976).
63. C. B. Scruby, R. J. Dewhurst *et al.*, J. Appl. Phys. **51**, 6210 (1980).
64. Yu. O. Barmenkov, V. V. Zosimov et al., Dokl. Akad. Nauk SSSR **290**, 1096 (1986) [Sov. Phys. Dokl. **31** (1986)].
65. D. A. Hutchins and J. D. Macphail, J. Phys. E **18**, 69 (1985).
66. G. M. Sessler, R. Gerhard-Multhaupt *et al.*, J. Appl. Phys. **58**, 119 (1985).
67. A. C. Tam, Appl. Phys. Lett. **45**, 510 (1984).
68. K. L. Vodop'yanov, L. A. Kulevskiy *et al.*, Zh. Eksp. Teor. Fiz. **91**, 114 (1986) [Sov. Phys. JETP **64**, 67 (1986)].
69. C. Thomsen, J. Strait *et al.*, Phys. Rev. Lett. **53**, 989 (1984).
70. J. Opsal and A. Rosencvaig, J. Appl. Phys. **53**. 4240 (1982).

71. A. Rosencwaig, Solid State Techn. **25**, 91 (1982).
72. A. Rosencwaig, Science **218**, 223 (1982).
73. A. N. Morozov and V. Yu. Raevskiy, Zarubezh. Elektr. Tekhn. **2**, 46 (1982).
74. V. O. Rebone, E. P. Isaev *et al.*, Obz. Elektron. tekh. Ser. 8. (2) (1988).
75. A. Rosencwaig, J. Opsal, and D. L. Willenborg, Appl. Phys. Lett. **43**, 166 (1983).
76. C. Karner, A. Mandel, and F. Trager, Appl. Phys. A. **38**, 19 (1985).
77. A. Rosencwaig, J. Appl. Phys. **51**, 2210 (1980).
78. A. Rosencwaig, Electron. Lett. **16**, 928 (1980).
79. R. Tilgner and J. Baumann, J. Nondestr. Eval. **3**, 111 (1982).
80. G. C. Wetsel and F. A. McDonald, Appl. Phys. Lett. **41**, 926 (1982).
81. W. C. Mundy, R. S. Hughes, and C. Carniglia, Appl. Phys. Lett. **43**, 985 (1983).
82. P. Cielo, J. Appl. Phys. **56**, 230 (1984).
83. Y. H. Wong, R. L. Thomas, and J. J. Pouch, Appl. Phys. Lett. **35**, 368 (1979).
84. R. L. Thomas, J. J. Pouch *et al.*, J. Appl. Phys. **51**, 1152 (1980).
85. R. G. Busse and A. Rosencwaig, Appl. Phys. Lett. **36**, 815 (1980).
86. T. Sawada, H. Shimizn, and S. Oda, Jpn. J. Appl. Phys. **20**, L25 (1981).
87. P. Helander, I. Lundstrom, and D. McQueen, J. Appl. Phys. **52**, 1146 (1981).
88. A. C. Tam, J. Opt. Soc. Am. **70**, 581 (1981).
89. R. T. Swimm, Appl. Phys. Lett. **42**, 955 (1983).
90. R. E. Lee and R. M. White, Appl. Phys. Lett. **12**, 12 (1968).
91. A. M. Ledbetter and J. C. Moulder, J. Acoust. Soc. Am. **65**, 840 (1979).
92. F. A. Ash, E. Dieulessaint, and H. Rakouth, Electron. Lett. **16**, 470 (1980).
93. G. Veith and M. Kovatch, Appl. Phys. Lett. **40**, 30 (1982).
94. G. Veith, Appl. Phys. Lett. **41**, 1045 (1982).
95. A. M. Aindow, R. J. Dewhurst, and S. B. Palmer, Opt. Commun. **42**, 116 (1982).
96. Yu. V. Pogorel'skiy, Fiz. Tverd. Tela **24**, 2361 (1982) [Sov. Phys. Solid State **24**, 1340 (1982)].
97. V. V. Krylov and V. I. Pavlov, Akust. Zh. **28**, 836 (1982) [Sov. Phys. Acoust. **28**, 493 (1982)].
98. S. R. J. Brueck, T. F. Deutsch, and D. E. Oates, Appl. Phys. Lett. **43**, 157 (1983).
99. D. Royer and E. Dieulesaint, J. Phys. (Paris) **44**, 79 (1983).
100. S. J. Huard and D. Chardon, J. Phys. (Paris) **44**, 91 (1983).
101. E. P. Velikhov, E. V. Dan'shchikov *et al.*, Pis'ma Zh. Eksp. Teor. Fiz. **38**, 483 (1983) [JETP Lett. **38**, 584 (1983)].
102. A. M. Dykhne and B. P. Rysev, Poverkhn. Fiz. Khim. Mekh. **6**, 17 (1983).
103. A. N. Khodinskiy, L. S. Korochkin, and S. A. Mikhnov, Zh. Prikl. Spektrosk. **38**, 745 (1983).
104. M. Schmidt and K. Dransfeld, Appl. Phys. **A28**, 211 (1982).
105. L. M. Lyamshev and B. I. Chelnokov, Pis'ma Zh. Tekh. Fiz. **8**, 1361 (1982) [Sov. Tech. Phys. Lett. **8**, 585 (1982)].
106. D. Royer and E. Dieulesaint, J. Appl. Phys. **56**, 2507 (1984).
107. S. P. Semin, Akust. Zh. **32**, 225 (1986) [Sov. Phys. Acoust. **32**, 132 (1986)].
108. S. P. Semin, Zh. Tekh. Fiz. **56**, 2224 (1986) [Sov. Phys. Tech. Phys. **31**, 1330 (1986)].
109. S. M. Avanesyan, V. E. Gusev, and N. J. Zheludev, Appl. Phys. A **40**, 3207 (1986).
110. S. M. Avanesyan, S. A. Akhmanov, *et al.* Preprint No. 20, Moscow State University (1986).
111. A. N. Bondarenko, B. Character. Maslov *et al.*, Prib. Tekh. Eksp. **6**, 211 (1975).
112. R. O. Claust and J. H. Cantrell, Jr., Acoust. Lett. **5**, 1 (1981).
113. R. L. Jungerman, J. E. Bower *et al.*, Appl. Phys. Lett. **40**, 313 (1982).
114. R. L. Jungerman, B. T. Khuri-Yakub, and G. S. Kino, Appl. Phys. Lett. **44**, 392 (1984).
115. J. I. Burov, K. P. Branzalov, and D. V. Ivanov, Appl. Phys. Lett. **46**, 141 (1985).
116. L. M. Dorozhkin, M. A. Kulakov *et al.*, Akust. Zh. **31**, 680 (1985) [Sov. Phys. Acoust. **31**, 412 (1985)].
117. F. V. Bunkin, Al. A. Kolomenskiy *et al.*, Akust. Zh. **32**, 21 (1986) [Sov. Phys. Acoust. **32**, 12 (1986)].
118. A. C. Tam, Appl. Phys. Lett. **45**, 510 (1984).
119. R. Garhargd-Multhaupt, G. M. Sessler *et al.*, J. Appl. Phys. **55**, 2969 (1984).
120. A. A. Bondarenko, V. K. Bologdin, and A. I. Kondrat'ev, Akust. Zh. **26**, 828 (1980) [Sov. Phys. Acoust. **26**, 467 (1980)].

121. K. Truell, S. Cheek, and A. Elbaum, *Ul'trazvukovy metody v fizike tverdogo tela* [*Ultrasonic Methods in Solid State Physics*] (Mir, Moscow, 1972).
122. B. Sullivan and A. C. Tam, J. Acoust. Soc. Am. **75**, 437 (1984).
123. C.-Y. Kuo, M. M. F. Vieira, and C.-K. N. Patel, J. Appl. Phys. **55**, 3333 (1984).
124. J. R. M. Viertl, J. Appl. Phys. **51**, 805 (1980).
125. A. A. Karabutov, N. N. Omel'chuk *et al.*, Vestn. Moscow State Univ. Phys. Astron. Ser. **26**, 62 (1985).
126. T. A. Dunina, S. V. Egerev, and K. A. Naugol'nykh, Pis'ma Zh. Tekh. Fiz., **9**, 410 (1983) [Sov. Tech. Phys. Lett. **9**, 176 (1983)].
127. V. V. Filippov and A. N. Khodinskiy, Dokl. Akad. Nauk USSR **30**, 410 (1986).
128. E. B. Cherepetskaya, "Thermooptic excitation and interaction of random waves in dispersionless media," Candidate's Dissertation (MSU, Moscow, 1980).
129. D. A. Hutchins, R. J. Dewhurst *et al.*, Appl. Phys. Lett. **38**, 677 (1981).
130. C. B. Scruby, R. J. Dewhurst *et al.*, J. Appl. Phys. **51**, 6210 (1980).
131. D. A. Hutchins, R. J. Dewhurst, and S. B. Palmer, J. Acoust. Soc. Am. **70**, 1362 (1981).
132. F. V. Bunkin, K. L. Vodop'yanov *et al.*, Izv. Akad. Nauk. SSSR Ser. Fiz. **49**, 558 (1985).
133. S. M. Avanesyan and V. E. Gusev. Izv. Akad. Nauk. SSSR Ser. Fiz. **51**, 248 (1987) [Bulletin of the Academy of Sciences of the USSR. Physical Series, **S1**, 37 (1987)].
134. G. M. Sessler and R. Gerhard-Mulhaupt, J. Appl. Phys. **58**, 119 (1985).
135. W. Imaino and A. C. Tam, Appl. Opt. **22**, 1875 (1983).
136. M. W. Sigrist and V. G. Mikhalevich, Proc. Int Conf. Lasers **1982**, 80.
137. A. A. Karabutov, V. T. Platonenko *et al.*, Vest. Moscow State Univ. Phys. Astron. Ser. **25**, 89 (1984).
138. D. A. Hutchins, F. Nadeau, and P. Cielo, Can. J. Phys. **64**, 1334 (1986).
139. V. E. Gusev and A. A. Karabuitov, *Tez. dokl. IX Vsesoyuzn. akust. konf., sektsiya I* [*Conference Proceedings of the Ninth All-Union Acoustical Conference, Section I*] (Nauka, Moscow, 1983), p. 20.
140. P. Gutfeld, *Fizicheskaya akustika* [*Physical Acoustics*], edited by W. Mason (Mir, Moscow, 1973), p. 5.
141. S. Thomsen, H. T. Grahn *et al.*, Phys. Rev. B **36**, 4129 (1986).
142. S. M. Avanesyan and V. E. Gusev, Kvantovaya Elektron. **13**, 1241 (1986) [Sov. J. Quantum Electron. **16**, 812 (1986)].
143. S. Avanesyan, V. Gusev, and N. Zheludev, Appl. Phys. A **40**, 163 (1986).
144. E. M. Lifshitz and L. P. Pitaevskiy, *Fizicheskaya kinetika. Teoreticheskaya fizika* [*Physical Kinetics. Theoretical Physics*] (Nauka, Moscow, 1979), Vol. 10, p. 528.
145. L. V. Burmistrova, A. A. Karabutov *et al.*, Akust. Zh. **24**, 685 (1978) [Sov. Phys. Acoust. **24**, 369 (1978)].
146. G. Tiuker and V. Rampton, *Giperzvuk v fizike tverdogo tela* [*Hypersound in Solid State Physics*] (Mir, Moscow, 1975), p. 454.
147. S. A. Akhmanov and V. E. Gusev, Usp. Fiz. Nauk **162**(3), 3 (1992) [Sov. Phys. Usp. **35**, March (1992)].

Index